Progress in Inflammation Research

Series Editor

Prof. Dr. Michael J. Parnham
PLIVA
Research Institute
Prilaz baruna Filipovica 25
10000 Zagreb
Croatia

Published titles:
T Cells in Arthritis, P. Miossec, W. van den Berg, G. Firestein (Editors), 1998
Chemokines and Skin, E. Kownatzki, J. Norgauer (Editors), 1998
Medicinal Fatty Acids, J. Kremer (Editor), 1998
Inducible Enzymes in the Inflammatory Response, D.A. Willoughby, A. Tomlinson (Editors), 1999
Cytokines in Severe Sepsis and Septic Shock, H. Redl, G. Schlag (Editors), 1999
Fatty Acids and Inflammatory Skin Diseases, J.-M. Schröder (Editor), 1999
Immunomodulatory Agents from Plants, H. Wagner (Editor), 1999
Cytokines and Pain, L. Watkins, S. Maier (Editors), 1999
In Vivo *Models of Inflammation*, D. Morgan, L. Marshall (Editors), 1999
Pain and Neurogenic Inflammation, S.D. Brain, P. Moore (Editors), 1999
Anti-Inflammatory Drugs in Asthma, A.P. Sampson, M.K. Church (Editors), 1999
Novel Inhibitors of Leukotrienes, G. Folco, B. Samuelsson, R.C. Murphy (Editors), 1999
Vascular Adhesion Molecules and Inflammation, J.D. Pearson (Editor), 1999
Metalloproteinases as Targets for Anti-Inflammatory Drugs, K.M.K. Bottomley, D. Bradshaw, J.S. Nixon (Editors), 1999
Free Radicals and Inflammation, P.G. Winyard, D.R. Blake, C.H. Evans (Editors), 1999
Gene Therapy in Inflammatory Diseases, C.H. Evans, P. Robbins (Editors), 2000
New Cytokines as Potential Drugs, S. K. Narula, R. Coffmann (Editors), 2000
High Throughput Screening for Novel Anti-inflammatories, M. Kahn (Editor), 2000
Immunology and Drug Therapy of Atopic Skin Diseases, C.A.F. Bruijnzeel-Komen, E.F. Knol (Editors), 2000
Inflammatory Processes. Molecular Mechanisms and Therapeutic Opportunities, L.G. Letts, D.W. Morgan (Editors), 2000

Forthcoming titles:
Cellular Mechanisms in Airway Inflammation, C. Page, K. Banner, D. Spina (Editors), 2000
Inflammatory and Infectious Basis of Atherosclerosis, J.L. Mehta (Editor), 2001

Novel Cytokine Inhibitors

Gerry A. Higgs
Brian Henderson

Editors

Springer Basel AG

Editors

Dr. Gerry A. Higgs
Celltech Chiroscience
216 Bath Road
Slough
Berkshire SL1 4EN
UK

Prof. Brian Henderson
Cellular Microbiology Research Group
Eastman Dental Institute
University College London
256 Gray's Inn Road
London WC1X 8LD
UK

Deutsche Bibliothek Cataloging-in-Publication Data
Novel cytokine inhibitors / ed. by Gerry A. Higgs ; Brian Henderson - Basel ; Boston ; Berlin : Birkhäuser, 2000
    (Progress in inflammation research)

The publisher and editor can give no guarantee for the information on drug dosage and administration contained in this publication. The respective user must check its accuracy by consulting other sources of reference in each individual case.

ISBN 978-3-0348-9572-9   ISBN 978-3-0348-8450-1 (eBook)
DOI 10.1007/978-3-0348-8450-1

Originally published by Birkhäuser Verlag, Basel, Switzerland in 2000

Printed on acid-free paper produced from chlorine-free pulp. TCF ∞
Cover design: Markus Etterich, Basel
Cover illustration: The cover picture shows the ribbon structure of a monoclonal antibody binding to two molecules of tumour necrosis factor (TNF). The antibody, CDP571 (HUMICADE™), is effective in reducing the signs and symptoms of rheumatoid arthritis and inflammatory bowel disease (see chapters 6 and 7) and has been genetically engineered to prevent immunogenicity in patients. This picture is reproduced by kind permission of Dee Athwal and Celltech, UK.

9 8 7 6 5 4 3 2 1

Contents

List of contributors

Rodger A. Allen, Celltech Chiroscience, 216 Bath Road, Slough, Berkshire SL1 4EN, UK; e-mail: rallen@celltech.co.uk

Christian Bogdan, Institute for Clinical Microbiology, Immunology, and Hygiene, University of Erlangen-Nuremberg, Wasserturmstrasse 3, D-91054 Erlangen, Germany; e-mail: christian.bogdan@mikrobio.med.uni-erlangen.de

Stanley T. Crooke, Isis Pharmaceuticals, Inc., 2292 Faraday Avenue, Carlsbad, CA 92008, USA; e-mail: scrooke@isisph.com

Peter I. Croucher, Division of Biochemical and Musculoskeletal Medicine, University of Sheffield Medical School, Beech Hill Road, Sheffield S10 2RX, UK

Anthony L. DeFranco, G.W. Hooper Foundation, University of California, San Francisco, CA 94143-0552, USA; e-mail: defranco@socrates.ucsf.edu

Roly Foulkes, Celltech Chiroscience, 216 Bath Road, Slough, Berkshire SL1 4EN, UK; e-mail: rfoulkes@celltech.co.uk

Philip G. Hargreaves, Division of Biochemical and Musculoskeletal Medicine, University of Sheffield Medical School, Beech Hill Road, Sheffield S10 2RX, UK

Brian Henderson, Cellular Microbiology Research Group, Eastman Dental Institute, University College London, 256 Gray's Inn Road, London WC1X 8LD, UK; e-mail: b.henderson@eastman.ucl.ac.uk

Gerry A. Higgs, Research Division, Celltech Chiroscience, 216 Bath Road, Slough, Berkshire SL1 4EN, UK

Ingunn Holen, Division of Biochemical and Musculoskeletal Medicine, University of Sheffield Medical School, Beech Hill Road, Sheffield S10 2RX, UK

John J. Letterio, Lab of Cell Regulation and Carcinogenesis, Building 41, Room C629, National Cancer Institute, National Institutes of Health, Bethesda, MD 20892-5055, USA; e-mail: letterij@pop.nci.nih.gov

Simon Lumb, Celltech Chiroscience, 216 Bath Road, Slough, Berkshire SL1 4EN, UK; e-mail: slumb@celltech.co.uk

Mary Lee MacKichan, Chiron Corp., 4560 Horton St., Emeryville, CA 94608, USA

Ravinder N. Maini, The Kennedy Institute of Rheumatology, 1 Aspenlea Road, London W6 8LH, UK; e-mail: r.maini@ic.ac.uk

Anthony Meager, Division of Immunobiology, National Institute for Biological Standards and Control, Blanche Lane, South Mimms, Potters Bar, Hertfortshire EN6 3QG, UK; e-mail: ameager@nibsc.ac.uk

Raymond J. Owens, Celltech Chiroscience, 216 Bath Road, Slough, Berkshire SL1 4EN, UK; e-mail: rowens@celltech.co.uk

Stephen E. Rapecki, Celltech Chiroscience, 216 Bath Road, Slough, Berkshire SL1 4EN, UK; e-mail: srapecki@celltech.co.uk

Michael F. Smith, Jr., Division of Gastroenterology and Hepatology, University of Virginia School of Medicine, Charlottesville, VA 22908, USA; e-mail: mfs3k@virginia.edu

Amanda Suitters, Celltech Chiroscience, 216 Bath Road, Slough, Berkshire SL1 4EN, UK; e-mail: asuitter@celltech.co.uk

Yoram Vodovotz, Department of Surgery, University of Pittsburgh, 200 Lothrop St., Room E1558, Pittsburgh, PA 15213, USA; e-mail: vodovotzy@msx.upmc.edu

Targets for modulating cytokine responses in inflammatory and infectious diseases

Brian Henderson[1] and Gerry A. Higgs[2]

[1]Cellular Microbiology Research Group, Eastman Dental Institute, University College London, 256 Gray's Inn Road, London WC1X 8LD, UK; [2]Research Division, Celltech Chiroscience, 216 Bath Road, Slough, Berkshire, SL1 4EN, UK

Introduction

Cytokines were discovered in the 1950s by workers studying fever and by those interested in cellular anti-viral defences and in the 1960s by immunologists interested in monocyte-lymphocyte interactions. Further discoveries and rediscoveries were made in the 1970s and 1980s and the term cytokine (first introduced by Cohen in the 1970s) now describes a very large number of proteins and glycoproteins. These proteins are part of the cell-to-cell signalling requirements of the multicellular organism in order to maintain homeostasis. The importance of cytokines is increasingly being delineated by the development of animals in which one or more selected cytokine genes have been disrupted (knocked out) by homologous recombination. Many of these knockout animals show increased sensitivity to infectious microorganisms. Others, such as those where interleukin-2 (IL-2) or interleukin-10 (IL-10) genes have been knocked out, reveal severe local inflammatory responses which appear to be due to aberrant responses to the animal's commensal microflora.

The major impetus to the study of cytokines and the reason that the world's pharmaceutical industry is ploughing billions of dollars into research of cytokines is that these cell-regulatory proteins have been identified as being key mediators of the pathology of human and animal diseases. Much of our understanding of cytokines has come from the study of prototypic diseases such as rheumatoid arthritis, septic shock and atherosclerosis. In these diseases, pro-inflammatory cytokines such as interleukin-1 (IL-1), tumour necrosis factor α (TNFα), interleukin-6 (IL-6), chemokines and growth factors such as transforming growth factor β (TGFβ), fibroblast growth factor (FGF) and platelet-derived growth factor (PDGF), are implicated as being important in disease pathology. It is these, and many other, cytokines that represent the therapeutic targets for the treatment of human and animal diseases. In this book the various means that are being utilised, or could be utilised, to treat infectious and idiopathic diseases through the medium of cytokine manipulation will be discussed. This brief chapter is designed to give an overview of the biology of cytokines and the potential means by which their synthesis or activity can be therapeutically manipulated. The reader is referred to earlier reviews on this topic [1–5].

Cytokines: a brief overview

In a world awash with communication (radio, television, telephone, E-mail), it should not be difficult to convince the reader of the importance of correct communication, and the problems arising when our communication systems go wrong. It is now recognised that all cells, including bacteria, have cell-to-cell communication systems. In mammals three major cell-to-cell communication systems are recognised: neuronal, endocrine and cytokine. It is becoming recognised that these three systems mesh to produce a homeostatic control system. However, the details of the ways these systems mesh still await elucidation. It is probably still too early to make any sweeping statements about the overall physiological role of cytokines. In spite of this, it is probably safe to say that, in contrast to neuronal and endocrine signalling, cytokines are generally important in local cell-to-cell signalling. Of course, this statement, like many general statements about cytokines, will have numerous counter-examples.

In general terms, cytokines are proteins or glycoproteins produced by one cell and able to activate (or inhibit) that cell, another similar cell or distinct cells by virtue of binding to specific high-affinity receptors on the target cell. Thus cytokines can indulge in three types of interaction with cells (Fig. 1). If the producing cell responds to the cytokine, this is known as autocrine signalling. If the cells responding are in the neighbourhood of the producing cell, then we have paracrine signalling (*para*: nearby). Cytokines such as IL-6 may also have a hormone-like role, passing into the blood stream and acting on remote tissues. Interaction of a cytokine with its specific receptor can induce a variety of cellular changes that are shown diagrammatically in Figure 2. This includes activation of the cell with the switching-on of particular patterns of gene transcription. This can result in a variety of cellular changes, including the target cell entering the cell cycle and proliferating, or proliferating and differentiating. Alternatively, cells responding to cytokines, such as members of the TNF family, may undergo apoptosis. The most recently discovered group of cytokines are the chemokines which are important in leukocyte trafficking.

Cytokine nomenclature

The first requirement to understand the "zoology" of cytokines is to learn their names and the "families" to which they belong. This brings the first problem in

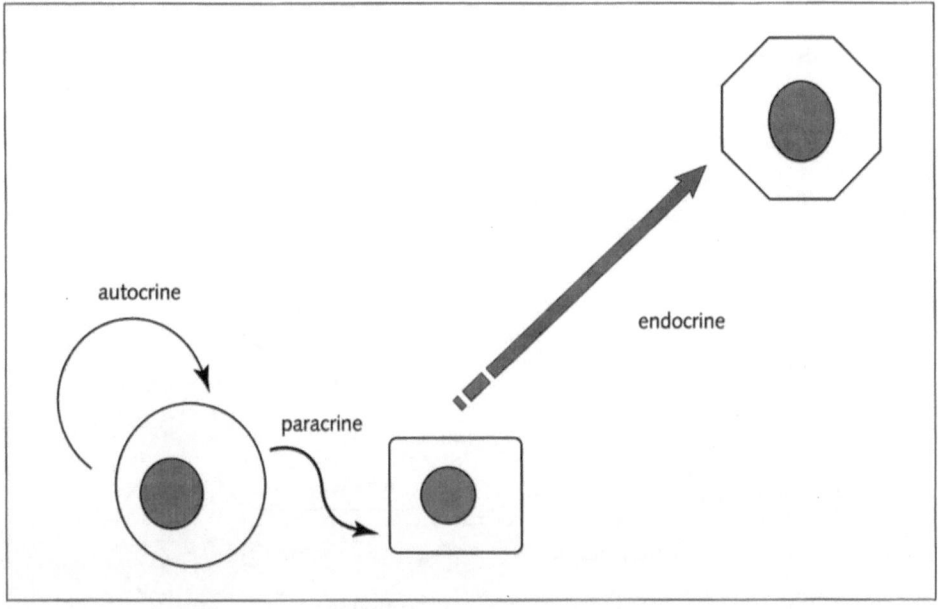

Figure 1
Autocrine, paracrine and endocrine signalling actions of cytokines

understanding these proteins. Generally cytokine nomenclature reflects the first activity described for the molecule. Often this is not the main activity. Thus inter-leukins (*inter*: between and *leuco*: white blood cell) were originally believed to be proteins produced by one leukocyte which interacted with another leukocyte. How-ever, it is now clear that many cells, in addition to leukocytes, can produce various interleukins. Tumour necrosis factor can induce necrosis of certain tumour cell lines but the major actions of this cytokine lie elsewhere.

Be that as it may, it is now conventional to divide cytokines into six groupings or families (Tab. 1). These are the interleukins, cytotoxic cytokines (e.g., TNFα), interferons, colony-stimulating factors, growth factors and chemokines. Many of the interleukins (e.g., IL-3, 4, 5, 6, 7, 9, 13) are also classified as belonging to the haemopoietin receptor family. There is significant overlap between these groupings. For example, IL-8 is a chemokine and IL-3 is a colony-stimulating factor (CSF). A more satisfactory subdivision is on the basis of the receptors to which the cytokines bind. It has to be remembered that cytokines have no inherent biological activity and can only stimulate cells by binding to a specific cell surface receptor. Each recep-tor will have associated with it a particular set of interacting proteins which trans-duce the intracellular signal once the receptor binds to the cytokine.

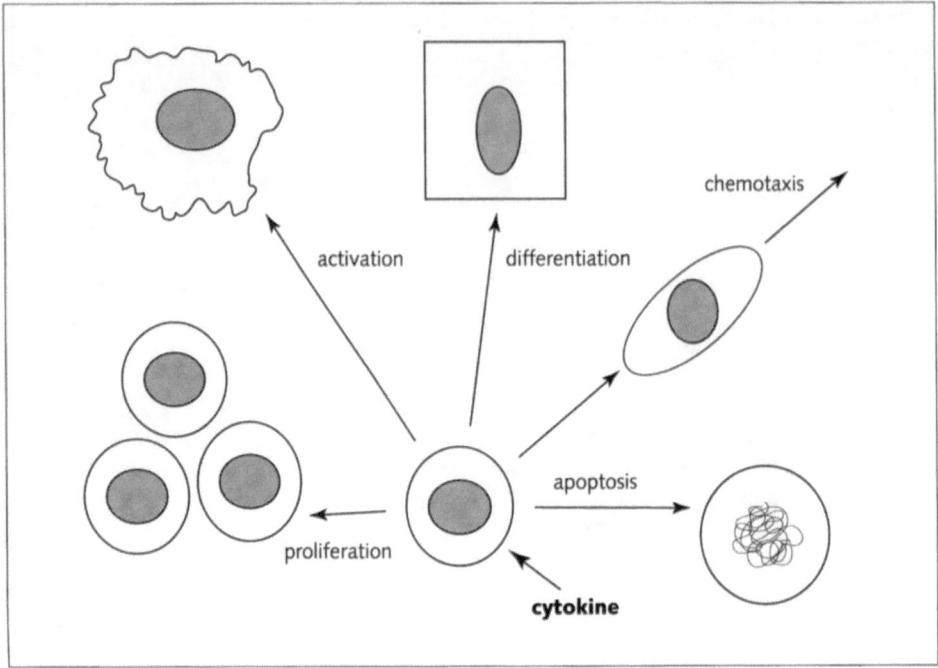

Figure 2
Biological functions of cytokines

Biological actions of cytokines

Cytokines have major physiological roles in the homeostatic control of the multi-cellular organism. One key role is in our defence mechanisms against infectious agents. The generation of the very large numbers of circulating leukocytes (mono-cytes, dendritic cells, lymphocytes, natural killer cells, etc.) is dependent on the interplay between interleukins, CSFs, TNF family members, growth factors and chemokines. Once formed, the surveillance actions of these leukocytes and their defensive measures against invading viruses, bacteria, protozoa, etc., are dependent on the generation of the proper network of cytokines. The inflammatory response is a complex process involving leukocytes, fixed-tissue cells (e.g., fibroblasts, vascular endothelial cells) and soluble mediators such as the complement system. The interactions between these various cells are controlled largely by the production and release of a variety of cytokines. Two key pro-inflammatory cytokines are IL-1 and TNFα. As will be discussed in the chapter by Croucher et al. in this volume, both of these cytokines require to be cleaved by specific proteases to produce the

Table 1 - Cytokine groupings

Group	Example
Interleukins	IL-1, IL-2, IL-18, etc.
Tumour necrosis factor family	TNFα, lymphotoxin, CD27L, etc.
Interferons	α-, β-, γ-interferons
Colony stimulating factors	IL-3, M-CSF, G-CGF, GM-CSF, etc.
Growth factors	TGFα, TGFβ, PDGF, etc.
Chemokines	IL-8, RANTES, MIP-1α, Eotaxin, etc.

active cytokine. In the case of IL-1, the protease is caspsase-1 (also known as pro-IL-1 converting enzyme or ICE) and the protease cleaving the pro-form of TNFα are members of the cell surface-associated proteases known as the ADAMs (a disintegrin and a metalloproteinase). The TNFα-converting enzyme (TACE) is ADAM 17. The induction, processing and release of IL-1 and TNFα can be seen as the initiating event in inflammation. These cytokines are recognised by many cell types, including fixed-tissue macrophages and vascular endothelial cells, and begin the inflammatory process. A key action of these cytokines is their ability to induce the synthesis of other cytokines. Among the most important are the chemokines (such as IL-8) and the cytokine-like hormone, IL-6. Production of these cytokines ensures that the correct leukocytes are attracted into the site of inflammation and that the proper whole-body signalling is induced in order to promote the complex physiological signalling system known as the acute phase response. In the normal protective inflammatory response, the network of cytokines required to induce inflammation, kill the invading parasite, initiate wound healing and resolve the inflammation, is switched on in the correct order and with the correct kinetics. However, if either the order or the kinetics of cytokine production are aberrant, then pathology can ensue. Perhaps the simplest example of this (although we are still unable to therapeutically manage it) is septic shock. The release of bacterial factors (e.g., lipopolysaccharide (LPS)) into the body results in the catastrophic production of pro-inflammatory cytokines, IL-1 and TNFα, without the counter-regulatory production of inhibitory cytokines like IL-10. It is the over-production of IL-1 and TNFα that produces the cardiovascular changes and end-organ failure characteristic of septic shock. IL-1 and TNFα are also the major therapeutic targets in patients with rheumatoid arthritis and are also perceived to be major targets in many other infectious and idiopathic diseases. In the next section the methods which are being developed to therapeutically modulate cytokines will be described.

Therapeutic modulation of cytokines

There are a number of potential points for therapeutic intervention in the pathways leading to cytokine induction, expression, activation, receptor binding and intracellular signalling (Tab. 2). One of the commonest means of inducing cytokine genes has been by the challenge of cells with LPS derived from bacterial cell walls. Ligation of LPS by CD14 (a high-affinity LPS-receptor) is a potent mechanism for inducing cytokine genes such as TNF and IL-1 (see the chapter by MacKichan and DeFranco, this volume). This is the mechanism which triggers septic shock and attempts have been made to neutralise bacterial toxins in order to prevent septic shock and its cardiovascular consequences.

Inflammatory cytokines are not only produced in response to infections, however. In allergic and autoimmune diseases, binding at the antigen receptors (FcRs) or the T cell receptor (TCR) activates signalling pathways which result in cytokine gene induction. Signalling along these pathways is controlled by specific protein kinases and transcription factors which, in turn, can be regulated by intracellular messengers such as cyclic adenosine monophosphate (cAMP). The inhibition of the intracellular enzymes which regulate cytokine synthesis offers a number of targets to suppress cytokines (see the chapter by Allen and Rapecki, this volume)). Similarly, cytokine production can be inhibited by designing specific antisense oligonucleotides to prevent the cytosolic expression of particular cytokines from their mRNA (see the chapter by Crooke, this volume).

IL-1 and TNF are produced as inactive pro-forms which are sometimes expressed on the cell surface or may be secreted as soluble proteins. The production of active cytokines requires the activity of specific proteases such as ICE and TACE, which clearly are targets for new drugs which will down-regulate cytokine activity (see the chapter by Croucher et al., this volume).

The classical pharmacological approach to blocking the actions of a key transmitter or mediator has been to develop specific receptor antagonists. With the large polypeptide cytokines, however, this has proved difficult. No small-molecule cytokine-receptor antagonists have emerged and this may be due to the difficulties in blocking high-affinity interactions which are multi-point in nature. In contrast, the neutralisation of cytokines using biological approaches has achieved marked successes. Highly specific, high-affinity monoclonal antibodies to a number of cytokines, including IL-1, TNF and IL-5 have been generated and shown to have efficacy in various pre-clinical disease models (see the chapter by Suitters and Foulkes, this volume). A TNF-blocking antibody produced marked symptomatic improvement in patients with rheumatoid arthritis or inflammatory bowel disease and this represents one of the first anti-cytokine therapies to be marketed (see the chapter by Maini, this volume). A similar approach to cytokine neutralisation is the recombinant generation of fusion proteins which incorporate high-affinity cytokine receptors and single-chain immunoglobulins in the same molecule. The

Table 2 - Therapeutic targets to control cytokines

---

1 Inhibition of cytokine gene induction: e.g., by elevation of cAMP

2 Inhibition of cytokine gene transcription: e.g., by blocking transcription factors

3 Inhibition of cytokine expression: e.g., by antisense

4 Inhibition of cytokine processing: e.g., TACE and ICE inhibitors

5 Neutralisation of cytokines: e.g., by antibodies or receptor-fusion proteins

6 Cytokine receptor antagonists: e.g., IL-1ra, blocking antibodies

7 Inhibitors of cytokine signalling: e.g., inhibitors of p38 kinase

8 Inhibitors of cytokine receptor expression

9 Naturally-occurring regulator cytokines: e.g., IL-10

10 Microbial anti-cytokine factors

TNF-receptor/Ig fusion protein is an example of the clinical validity of this approach and it is likely that a number of other therapeutic molecules using this concept will emerge (see the chapter by Meager, this volume). A third biological approach to the therapeutic neutralisation of cytokines has utilised the endogenous interleukin-1 receptor antagonist (IL-1ra) (see the chapter by Smith, this volume).

If cytokine induction, expression, synthesis, activation and receptor binding has occurred, the one remaining point for possible therapeutic intervention remains the signalling from the cytokine receptor itself. Cytokines induce the expression of a range of inflammatory molecules, including leukocyte adhesion molecules and metalloproteinases. The signalling pathways which transduce the cytokine signals are regulated by mitogen-activated protein kinases (MAPKs) and one in particular, p38 MAPK, has proved pivotal in TNF signalling. Novel inhibitors of p38 are showing promising therapeutic activity in models of arthritis and autoimmune disease (see the chapter by Owens and Lumb, this volume).

Finally, some cytokines themselves have potential therapeutic properties. For example, TGFβ and IL-10 suppress some inflammatory processes and regulate immunity and the host's response to infection (see the chapter by Bogdan et al., this volume). Also, some cytokines derived from microorganisms have anti-inflammatory properties, which have presumably evolved to help establish them in the host's tissues (see the chapter by Henderson, this volume; [6]). It is possible that these naturally regulating cytokines could be harnessed to form the basis of novel therapies.

References

1   Henderson B, Blake S (1992) Therapeutic potential of cytokine manipulation. *Trends Pharmacol Sci* 13: 145–152

2   Higgs GA (1993) The therapeutic potential of monoclonal antibodies in inflammatory diseases. *Curr Opin Invest Drugs* 2(1): 21–35

3   Henderson B, Poole S (1994) Modulation of cytokine function: therapeutic applications. *Adv Pharmacol* 25: 53–115

4   Henderson B (1996) Overview of cytokine modulation. In: B Henderson, MW Bodmer (eds): *Therapeutic modulation of cytokines*, CRC Press, Boca Raton, 81–90

5   Higgs GA (1997) Novel approaches to the inhibition of cytokine responses in asthma. *J Pharm Pharmacol* 49 (3): 25–31

6   Henderson B, Poole S, Wilson M (1998) *Bacteria-cytokine interactions in health and disease*. Portland Press, London

Cell signaling and cytokine induction by lipopolysaccharide

Mary Lee MacKichan[1] and Anthony L. DeFranco[2]

[1]Chiron Corp., 4560 Horton St., Emeryville, CA 94608, USA; [2]G.W. Hooper Foundation, University of California, San Francisco, CA 94143-0552, USA

Overview

Multicellular organisms have evolved molecular systems to recognize the structural hallmarks of pathogens and an arsenal of responses to contain and ultimately eliminate invaders. Lipopolysaccharide (LPS) derived from the cell walls of Gram-negative bacteria is an important initiator of such "innate" immune responses. This recognition normally leads to a robust local immune response to invading bacteria which is protective; however, if LPS is present in the blood, the response becomes systemic and can lead to endotoxic shock and death. The ancient threat of sepsis remains an important clinical problem today and motivates much of the interest in the molecular biology of LPS responses. Here we review recent literature relevant to LPS-induced signaling in myeloid cells, with emphasis on the contribution of these signals to cytokine expression.

Recognition of LPS induces macrophages to express many different types of proinflammatory molecules, including promoters of cell migration, adhesion, phagocytosis, and bactericidal activity, but a critical element in the genesis of septic shock is the induced expression of the cytokines tumor necrosis factor alpha (TNFα) and interleukin-1 (IL-1). The signaling reactions leading to this induced cytokine expression offer potential targets for therapeutic intervention to prevent the associated damage to tissues, and ultimately the organism, that can result during a systemic immune response to a Gram-negative bacterial infection.

Although multiple LPS-binding proteins exist in humans, signaling in response to LPS is principally initiated by the formation of a complex between LPS and LPS-binding protein (LBP), a constituent of serum, and subsequent transfer of LPS to CD14, a high-affinity LPS-receptor expressed on myeloid cells. CD14 is a glycophosphatidylinositol (GPI)-anchored receptor and is believed to require an as yet unidentified coreceptor to transduce a signal to the cytoplasm. The binding of LPS to CD14 causes an increase in intracellular tyrosine phosphorylation, and this is required for most ensuing signaling and changes in macrophage gene expression and function. The tyrosine kinase responsible for signal initiation is not known, and the principal candidates for this role, the Src and Syk family tyrosine kinases, have recently been shown to be dispensable for LPS responsiveness. The initial targets of tyrosine phosphorylation also remain to be thoroughly characterized, but further downstream, LPS-induced signaling leads to the activation of mitogen-activated protein kinases (MAPKs) by phosphorylation on conserved tyrosine and threonine residues in their activation loops. At least three MAPK families are activated by LPS: the ERK MAPKs and the stress-activated MAPKs, JNK and p38. MAPKs in turn phosphorylate members of several transcription factor families on serine and threonine residues, including bZIP and Ets proteins, and activate other protein kinases as well, notably the MAPK-associated protein (MAPKAP) kinases. A separate signaling pathway initiated by LPS leads to the phosphorylation and degradation of inhibitors of Rel/NF-κB transcription factors (IκBs), allowing translocation of Rel dimers to the nucleus, where they activate transcription *via* κB sites on DNA. These pathways, and other signals, cooperate to initiate transcription and, in some cases, increase translation of cytokines genes. A picture of the signals elicited by LPS recognition is emerging; however, important gaps remain in our knowledge of the molecules involved in initiation of LPS signaling through CD14.

LPS receptors and binding proteins

Several cell types make responses to high (microgram/ml) concentrations of LPS, including endothelial cells, B cells and possibly T cells, while only myeloid cells are responsive to very low (picogram/ml) concentrations of LPS [1]. This increased sensitivity depends on the expression of CD14, a 50–55 kDa GPI-linked glycoprotein (reviewed in [1]), as demonstrated by several lines of evidence (Fig. 1). First, antibodies to CD14 block LPS responses, including cytokine production, *in vitro* [2]and *in vivo* [3], and non-myeloid cells become responsive to low concentrations of LPS upon stable transfection of CD14 [4, 5]. In mouse models of septic shock, overexpression of a human CD14 transgene confers hypersensitivity to the effects of LPS [6]. Conversely, targeted deletion of CD14 has a protective effect, rendering mice resistant to shock induced by Gram-negative bacteria or LPS [7].

In CD14$^{-/-}$ mice, even ten times the LD$_{100}$ of LPS injected intraperitoneally failed to induce symptoms of endotoxemia. In a second murine model of septic shock, in which mice are sensitized to the effects of LPS by D-galactosamine (GalN) treatment, which impairs liver function, CD14$^{-/-}$ mice also survived a normally lethal dose of LPS, although the protective effect of CD14 deletion was not absolute, as CD14$^{-/-}$ mice succumbed to endotoxic shock when the LPS dose was increased tenfold [7]. Monocytes isolated from mice lacking CD14 did not secrete detectable levels of TNFα or IL-6 *in vitro* in response to LPS concentrations up to 100 ng/ml; at higher concentrations of LPS, expression of these cytokines was detectable, but substantially reduced relative to controls [7]. These results argue convincingly for the

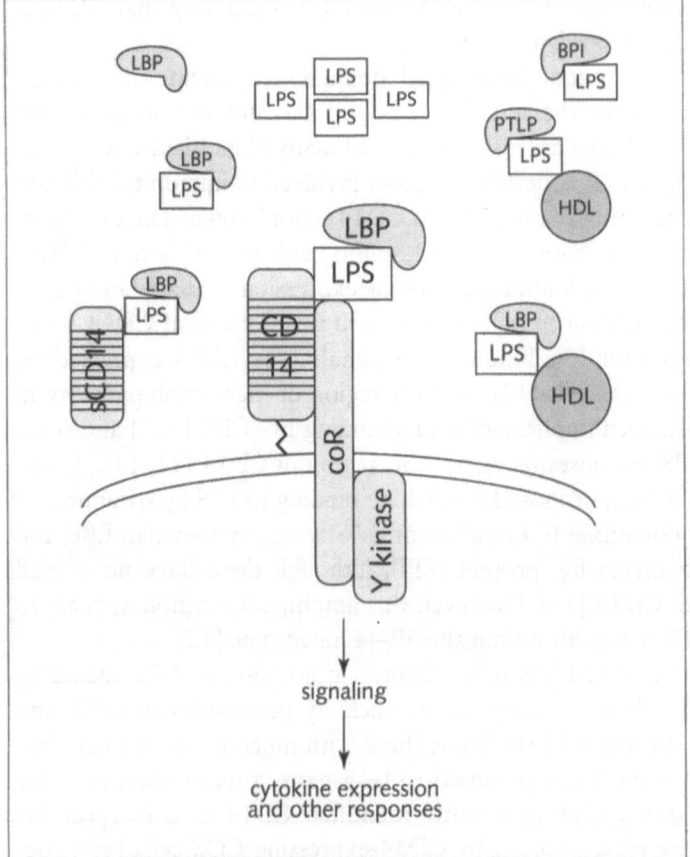

Figure 1
Lipopolysaccharide (LPS) receptor and binding proteins. LPS-binding protein (LBP) binds LPS from aggregates in serum and catalytically transfers LPS to CD14, a high-affinity GPI-linked receptor on myeloid cells and the principal mediator of LPS-induced signaling and cytokine expression. LBP can also transfer LPS to soluble CD14 (sCD14), which can activate other cell types such as endothelia. Other lipid transfer proteins, bactericidal/permeability-increasing protein (BPI) and phospholipid transfer protein (PLTP) can bind and neutralize LPS. Both LBP and PLTP transfer LPS to high-density lipoprotein particles (HDL), which may represent a mechanism for clearance of LPS.

importance of CD14 in LPS-induced cytokine expression and Gram-negative sepsis, although CD14-independent signaling pathways also exist and can mediate responses to high concentrations of LPS. Antibody blocking studies also suggested a signal-independent role for CD14 in clearance of bacteria [8], but CD14$^{-/-}$ mice actually show less bacterial dissemination than wild-type mice [7], suggesting that CD14 is not required for this function *in vivo*.

CD14 is composed of a unique N-terminal domain, ten leucine-rich repeats (LRRs), and a GPI anchor in the membrane-bound but not the soluble form (sCD14) of the receptor [9]. LRRs are also a feature of many plant disease resistance (R) gene products and of two *Drosophila* receptors involved in immunity, Toll and 18-wheeler [10], but the function of the LRRs in CD14 is not known. The LPS-binding domain is contained in the N-terminal 152 amino acids (aa) of human CD14, and this region is sufficient for LPS-induced cytokine expression mediated by sCD14 binding [11]. Alanine substitution of blocks of 5 aa in this portion of CD14, identified aa 39-44 as essential for LPS binding and signaling by CD14 expressed on Chinese hamster ovary (CHO) cells [12]. A short region of high amphipathicity in CD14, aa 57-64, has also been implicated in LPS binding by sCD14 [13] and some antibodies that block LPS responses recognize this region of CD14 [12, 14]. A similar amphipathic loop has been proposed to mediate binding to LPS by structures as diverse as the antibiotic polymixin B, *Limulus* anti-LPS factor, mammalian LBP, and bactericidal/permeability-increasing protein (BPI), although these have no overall sequence homology with CD14 [15]. However, this amphipathic region appears to be less important for LPS recognition than the 39-44 aa epitope [12].

It should be noted that CD14 has other ligands in addition to LPS, including Gram-positive bacterial cell wall components, such as peptidoglycan (PG) and lipoteichoic acid (LTA), although CD14 binds these with much lower affinity than LPS [16, 17]. Thus CD14 has been proposed to be a pattern recognition receptor [18]. In addition, expression cloning recently identified CD14 as a receptor for apoptotic cells, which are phagocytosed by CD14-expressing COS cells [19], suggesting a previously unsuspected function for CD14 in engulfment of dying cells by macrophages. However, binding of apoptotic cells to macrophages did not induce cytokine secretion [19], indicating that this interaction differs from the CD14-mediated recognition of LPS.

Because CD14 lacks a cytoplasmic domain, it is not thought to initiate signaling in response to LPS on its own, but rather through a hypothesized coreceptor. Such a scheme is used by some cytokine receptors, which have ligand-binding subunits that do not signal but do cooperate with signal-transducing coreceptors [20]. Similarly, several GPI-linked proteins have recently been found to serve a ligand-binding function and signal through one or more transmembrane proteins; these include ciliary neurotrophic factor receptor alpha [21] and two different GPI-linked receptor/ligand complexes that signal through a common receptor tyrosine kinase, Ret [22, 23]. However, in contrast to many GPI-linked molecules [24], CD14 retains its ability to mediate signaling in response to LPS when the GPI linkage is replaced by a heterologous transmembrane domain, despite the resulting change in membrane localization from Triton-X100 insoluble membrane "rafts" to the Triton-X100 soluble membrane compartment [25, 26].

A soluble form of CD14 (sCD14) found in serum can promote the response of endothelial cells to LPS, although these cells lack membrane-tethered CD14 [1]. A specific receptor for sCD14/LPS is thought to confer this response and could be identical to the hypothesized signal-transducing coreceptor utilized by GPI-linked CD14 on macrophages. Although the latter idea has not been proven, it is supported by the finding that LPS-dependent ^{125}I-sCD14 binding is reduced in CHO cells expressing membrane-bound CD14 [27]. Cross-linking of sCD14/LPS to labeled cells has been used in attempts to capture the sCD14 receptor, resulting in the identification of an 80-kDa protein in one case [28], and a 216 kDa structure in another [27]. A third potential CD14 coreceptor of 69-kDa was identified on the basis of its recognition by an antibody with the properties of a CD14 agonist [29]. However, no information is available regarding the molecular identity of any of these candidate sCD14 receptors or CD14-signaling subunits.

Several receptors with broad ligand specificity bind LPS with low affinity. These include β2 integrins, which have been proposed to serve a signal-transducing function for several GPI-linked receptors [30], and members of the scavenger receptor family (reviewed in [31]). The β2 integrin complement receptors, CR3 (CD11b/CD18) and CR4 (CD11c/CD18), when expressed in CHO cells, confer binding to Gram-negative bacteria and activation of NF-κB and TNFα synthesis in response to high concentrations of LPS [32–34]. Interestingly, resonance energy transfer microscopy experiments reveal that LPS induces association of CR3 with CD14 [35], suggesting CR3 could be a signal-transducing coreceptor for CD14. However, the cytoplasmic domains of CR3 are not required to confer LPS-responsiveness in CHO cells [33], and myeloid cells from patients lacking the gene encoding CD18 respond normally to LPS [36]. Thus these integrins are unlikely to be the principal transducers of CD14-mediated signals.

Scavenger receptors (SR) of the A class, SR-AI, -AII, and Marco, generated by alternative splicing of the same gene, have been shown to bind LPS, as well as other polyanionic ligands [31]. Although LPS partial structures can compete with another scavenger receptor ligand for binding to SR-A on transfected CHO cells, there is no effect of a competing scavenger receptor ligand on LPS-induced TNFα secretion by RAW 264.7 murine macrophages [37]. This result suggests that rather than participating in signal transduction and cytokine expression, SR-As might contribute to clearance of LPS and bacteria. In support of this idea, mice lacking the SR-A gene are more susceptible to endotoxic shock [38]or lethal infection with *Listeria monocytogenes* [39] than wild-type controls, and produce greater amounts of proinflammatory cytokines upon LPS treatment [38]. However, clearance of bacteria from plasma is normal in an SR-A$^{-/-}$ mouse line [39], implying that SR-A is not solely responsible for this function.

Binding of LPS by CD14 is facilitated by LBP in serum, which catalytically loads LPS onto CD14 (reviewed in [40]). Data obtained in LBP$^{-/-}$ mice generated by two different groups support the notion that LBP is necessary for LPS-binding by CD14 at low concentrations and is likely to be essential for CD14-dependent LPS-induced signaling. *In vitro*, sera from wild-type but not LBP$^{-/-}$ mice could support the transfer of fluorescently labeled LPS to CD14-expressing CHO cells [41], and cells in whole blood from LBP$^{-/-}$ mice are an order of magnitude less responsive to LPS than cells in whole blood from heterozygotes, as measured by TNFα production [42]. *In vivo* results with LBP$^{-/-}$ mice depended on the model system of endotoxic shock used. LBP$^{-/-}$ mice sensitized with GalN produced little TNF in response to LPS and were protected from endotoxic shock [41]. However, a sublethal dose of LPS elicited TNF expression in a second strain of LBP$^{-/-}$ mice challenged directly with endotoxin, a result in apparent contradiction to results obtained *in vitro* in whole blood from these mice [43]. LBP is clearly required for normal resistance to Gram-negative bacterial infection. Mice lacking LBP were unable to eliminate an experimentally administered injection of live *Salmonella typhimurium* and, despite the absence of endotoxaemia, died within 5 days with high bacteria counts in liver and other tissues, while control animals survived [41]. Interestingly, LBP is not required for clearance of LPS from the circulation [41].

LBP is one member of a family of lipid transfer proteins [44] that includes two other proteins that bind LPS, namely BPI and phospholipid transfer protein (PLTP). Unlike LBP, BPI is not a normal constituent of serum, but rather is secreted by polymorphonuclear leukocytes (reviewed in [45]). Also in contrast to LBP, which speeds the binding of LPS to CD14, BPI acts as an antagonist for LPS binding to CD14 and initiation of signaling [45]. BPI has a higher affinity for LPS than LBP, suggesting it could dampen CD14-dependent signaling in response to LPS. A protective effect of recombinant BPI or its N-terminal domain has been demonstrated in animal studies and most recently in Phase II clinical trials in humans [45]. PLTP has also been shown to bind LPS and mediate its transfer to high-density lipoprotein (HDL) particles but not to CD14, and PLTP can block LPS responses of neutrophils *in vitro* [46]. PLTP likely also acts as a competitor of LBP for binding to LPS, interfering with CD14-dependent signaling and cytokine expression.

LBP can transfer LPS from micelles or sCD14 to HDL particles *in vitro* [43, 47] as well as to CD14, suggesting a potential role for HDL in LPS responses or in clearance of LPS. Reconstituted HDL particles (rHDL) inhibit TNFα secretion in response to LPS in human whole blood *in vitro* [48] and in humans *in vivo* [49]. These and other results (e.g. [50, 51]) suggest that high levels of circulating HDL or LDL (low-density lipoprotein) particles may reduce the chance of systemic LPS responses by neutralizing LPS and/or reducing macrophage responsiveness. Despite the ability of LBP to transfer LPS to HDL *in vitro*, LBP$^{-/-}$ mice have normal serum cholesterol and triglycerides and the pattern of serum lipoproteins in LBP$^{-/-}$ mice is no different from controls even when animals are fed a high-fat diet [41]. Thus LBP appears to have no unique role in cellular lipid metabolism, while serum lipoprotein particles indirectly influence responses to LPS.

Although the transfer of LPS to CD14 by LBP appears to be the central initial event leading to LPS signaling and inflammatory responses, there is evidence for one or more CD14-independent pathways. As mentioned above, studies in CD14$^{-/-}$ mice have demonstrated that CD14 is dispensable for induced expression of TNFα and IL-1 by macrophages and monocytes in response to high (≥ 100 ng/ml) LPS concentrations [7, 52]. Even in CD14$^{-/-}$ mice injected with low doses of LPS, acute phase proteins are expressed at wild-type levels [53]. The CD14-independent pathway involved in the responses of CD14-deficient myeloid cells may not be identical to that used by CD14-negative cells in wild-type animals that respond to CD14/LPS complexes, as the two pathways differ in their serum and LPS-concentration dependence [7]. The elucidation of CD14-independent pathways may be facilitated by the availability of CD14$^{-/-}$ mice, and cloning of the putative CD14 coreceptor might also shed light on this area.

LPS-induced tyrosine kinase activity

One of the earliest responses in macrophages treated with LPS is an increase in protein tyrosine phosphorylation [54] (reviewed in [55]). This response is mediated by CD14 [56] and is required for many downstream events. The wide range of LPS responses sensitive to tyrosine kinase inhibitors suggests a tyrosine kinase might be responsible for all LPS signal initiation. Studies with pharmacological inhibitors demonstrate a requirement for tyrosine kinase activity in the LPS-stimulated release of arachidonic acid metabolites [54], activation of MAPKs [57–59], production of proinflammatory cytokines [58, 60–62] and nitric oxide [58], increased macrophage tumoricidal activity *in vitro* [63], and LPS-induced lethality in a mouse model [58]. It should be noted that in some studies responses to LPS were not abrogated by blockers of tyrosine kinase activity. For example, in CHO cells transfected with CD14, tyrosine kinase inhibitors reduced but did not block activation of NF-κB by LPS [64]. However, in similar studies, κB-binding activity in LPS-treated human monocytes was reduced to basal levels in the presence of the tyrosine kinase inhibitor herbimycin A [65]. On balance, the evidence suggests tyrosine kinase activity is required for LPS-CD14-mediated signal initiation.

The critical target of these tyrosine kinase inhibitors is not known, but could be identical to the hypothesized CD14 coreceptor or an associated nonreceptor tyrosine kinase. Initially, *src* family tyrosine kinases were proposed to serve the latter function in LPS-signal initiation, as the kinases Hck, Lyn and Fgr could be coimmunoprecipitated with CD14 or were activated by LPS treatment [61, 62, 66]. However, macrophages from mice lacking all three of these *src* kinases respond normally to LPS, indicating they are not needed for LPS-signal transduction [67]. Similarly, Syk tyrosine kinase, which plays a central role in B-cell antigen and Fc receptor signaling, becomes tyrosine phosphorylated in response to LPS stimulation [68], but macrophages derived from the foetal liver of *syk⁻ᐟ⁻* and wild-type embryos make similar responses to LPS [69].

Targets of the initial tyrosine kinase phosphorylation induced by LPS include the adapter protein Shc [68], SHIP, a lipid phosphatase [68], and Vav, a putative guanine nucleotide (GTP) exchange factor for Rho family GTPases [61]. Studies to determine whether any of these is required for downstream responses and changes in gene expression in LPS-stimulated cells have not been published, although we observed that macrophages from *vav⁻ᐟ⁻* mice exhibited roughly normal LPS signaling and cytokine expression (M.T. Crowley, A.L. DeFranco, unpublished observations). It is now clear that there are at least three Vav-like proteins, however, so the involvement of these proteins in LPS signaling remains possible.

Role of MAPKs in LPS responses

Among the signaling events activated by LPS are the three mitogen-activated protein kinase (MAPK) pathways. The MAPKs are conserved from yeast to man and are activated by many signals, including growth and "stress" stimuli (reviewed in [70]). The three MAPK families, ERK, JNK and p38, each have multiple members, encoded by separate genes from which additional isoforms are generated by alternate splicing. MAPKs are activated by dual-specificity (tyrosine/threonine) kinases known as MEKs (MAPK/ERK kinases), which in turn are activated by serine/threonine phosphorylation by upstream MEKKs (MEK kinases) (Fig. 2).

ERK

Regulation of the ERK MAPK cascade is relatively well understood and has been described in many systems. ERK1 and ERK2 are activated by MEK1 and MEK2, which are downstream of Raf, or in some cases other MEKKs [70]. Both ERK1 and ERK2 are rapidly activated in myeloid cells upon LPS exposure (within 10–15 min) [57, 59, 71]. This activation is sensitive to tyrosine kinase inhibitors [57] and, at low LPS concentrations, to blocking antibodies to CD14 [56, 72]. LPS treatment has also been shown to increase the activity of the kinases MEK1 and -2 in RAW 264.7 and BAC-1.2F5 macrophage lines [73, 74].

The MEK1/2 kinase Raf-1 is phosphorylated and activated in LPS-treated BAC-1.2F5 macrophages [74, 75], and is likely to be responsible for activation of the MEK-ERK pathway in these cells, although similar Raf-1 activation by LPS was not detected by another group using a second macrophage line [76]. In mammalian cells, activation of the Raf-MEK-ERK kinase cascade commonly requires the small GTPase Ras, although other less well-defined signals are also needed to fully activate Raf-1 (reviewed in [77]) and Ras-independent activation of Raf has also been

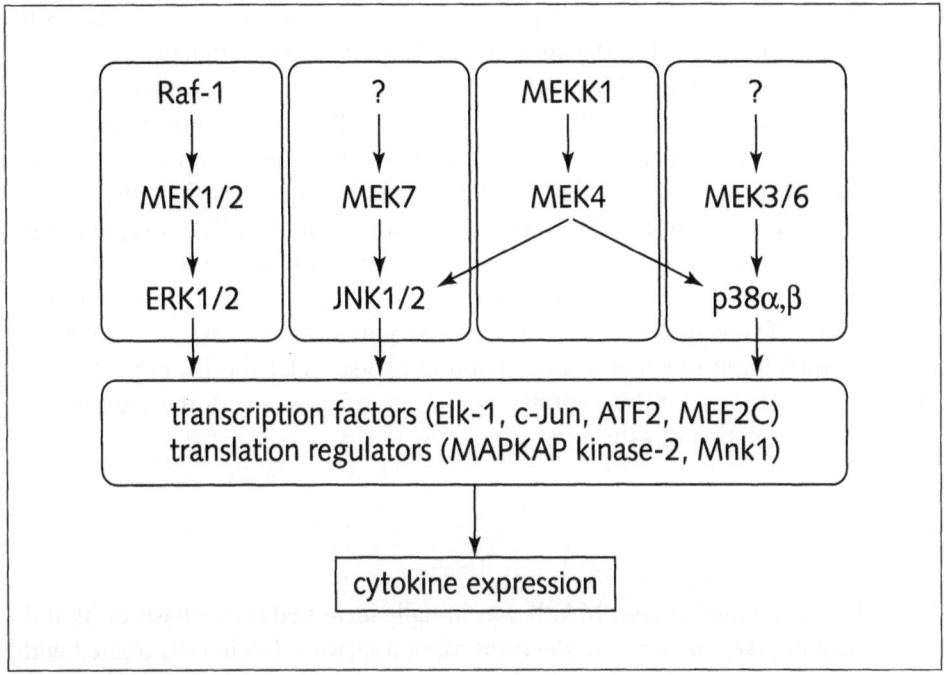

Figure 2

Mitogen-activated protein kinase (MAPK) cascades activated by LPS. Each cascade consists of one (or possibly more) MEKK (e.g., Raf), which phosphorylates specific downstream MEKs (e.g., MEK1/2) on serine residues, activating the MEKs to phosphorylate MAPKs (e.g., ERK1/2) on threonine and tyrosine residues. The activated MAPKs in turn phosphorylate and activate multiple transcription factor substrates and additional serine/threonine kinases, and these contribute to the LPS-induced increases in transcription and translation of cytokine and other proinflammatory genes.

reported [78, 79]. Direct assessment of Ras status in LPS-stimulated monocytes showed an increase in active, GTP-bound Ras relative to controls [61], suggesting activation of the ERK MAPK pathway by LPS may be Ras-dependent. In support of this idea, dominant-negative forms of Ras in RAW 264.7 cells blocked LPS-induced activation of the TNF promoter [80]. Although no increase in Ras GTP binding was detected in LPS-stimulated BAF-1.2F5 macrophages [75], Ras may already be quite active in transformed cell lines, and the Ras activation observed in monocytes treated with LPS [74] may more accurately reflect the physiological situation.

The contribution of the Raf-MEK-ERK pathway to LPS-induced responses has been assessed by expression of an inducible Raf-estrogen receptor fusion protein (Raf-ER) in RAW 264.7 macrophages [81], and by pharmacological inhibition of

MEK1/2 activity with PD98059 [82] or expression of dominant-negative Raf [80]. ERK activation by Raf-ER in the absence of other LPS-induced signals caused only a small (5%) increase in TNFα mRNA and protein synthesis, perhaps in part because of the inability of Raf-ER to activate NF-κB [81]. In the converse situation, when ERK activation was blocked by expression of dominant-negative Raf, activation of the TNF promoter was inhibited [80]. Similarly, when PD98059 was used to block ERK activation during LPS stimulation, expression of TNFα, IL-1, and IL-6 was reduced in a dose-dependent manner, although cytokine expression was not completely abrogated [82] (J. Hambleton, A.J. Finn, A.L. DeFranco, unpublished observations). These findings demonstrate a requirement for ERK1/2 activity to achieve normal levels of cytokine expression in LPS-stimulated cells, but ERK activation is not sufficient for this expression, a result in keeping with the combinatorial nature of regulation of gene expression.

JNK

JNK (c-Jun N-terminal kinase) MAPK was initially identified on the basis of its ability to phosphorylate and activate the transcription factor c-Jun in cells treated with ultraviolet light (reviewed in [83, 84]). Three genes encoding JNKs have been cloned from mammalian cells, *jnk1*, -2 and -3, and with alternatively spliced forms at least 10 different isoforms have been described [85]. These JNKs are activated by a wide variety of stress stimuli. Exposure to LPS rapidly activates JNK in bone-marrow-derived macrophages and macrophage cell lines in a CD14- and tyrosine kinase-dependent manner [86, 87]. At least two JNK isoforms, of 46 and 54 kDa, are activated by LPS in RAW 264.7 macrophages, with the 46-kDa activity predominating [86]. Although both JNK1 and JNK2 isoforms of these molecular weights are produced by alternate splicing [85], immunoblotting suggests JNK1 (also called SAPKγ) isoforms are responsible for most c-Jun N-terminal kinase activity in RAW 264.7 cells [87]. Interestingly, a role for JNK in LPS responses is evolutionarily conserved. A homologue of JNK, DJNK, is similarly activated by LPS in *Drosophila* cell lines [88], suggesting genetic approaches in flies may yield further insights into the role of JNK in innate immunity in mammals [83].

The best-characterized upstream activator of JNKs in mammalian cells is MEK4 (also called MKK4, SEK1 or JNKK1) [89–91]. LPS treatment rapidly increases MEK4 activity [73, 87] and, to a lesser extent, a second JNK kinase activity in RAW 264.7 cells [87]. This second MEK may be a target of the anti-inflammatory action of glucocorticoids, as pretreatment of macrophages with glucocorticoids blocked most LPS-induced JNK activity (70%) without reducing MEK4 activity [73]. Targeted deletion of MEK4 demonstrated that in some cases activation of JNK, e.g., by osmotic shock, is mediated by a MEK4-independent pathway [92]. Interestingly, the glucocorticoid dexamethasone also inhibits activation of JNK by osmotic shock, suggesting this stimulus and LPS activate a common glucocorticoid-sensitive MEK distinct from MEK4.

More recently, MEK7 has been identified as a specific upstream activator of JNK [83]. MEK7 is a homologue of *hemipterous* (*hep*), a *Drosophila* MEK upstream of DJNK, and MEK7 can complement developmental defects caused by certain *hep* alleles [93]. The relative contributions of MEK4 and MEK7 to JNK activation may be cell-type dependent, as TNF treatment of U937 cells increased MEK7 activity more than MEK4 activity [94], yet TNF and IL-1 treatment did not induce JNK or p38 activity in MEK4$^{-/-}$ embryonic fibroblasts [95].

Multiple MEKKs are capable of activating the JNK pathway [70]. MEKK1, in particular, is a strong activator of JNK, although this may also depend on cell type [70]. In RAW 264.7 cell extracts, the major peak of LPS-induced MEK4 kinase activity coeluted with fractions containing an anti-MEKK1 immunoreactive species, suggesting MEKK1 may be the upstream activator of the MEK4-JNK cascade in LPS-treated macrophages [87]. However, other fractions also appear to contain some MEKK activity [87], and many serine/threonine kinases in addition to MEKK1 have been shown to be capable of activating JNK in other systems. These include mixed lineage kinases (MLKs), TGFβ-activated kinase (TAK), and tumor progression locus 2 (Tpl-2), which can all phosphorylate and activate MEK4 (reviewed in [96]). Cdc42 and Rac, members of the Rho family of small GTPases, have also been shown to act upstream of JNK [83, 84], although their specific involvement in LPS signaling has not been demonstrated.

p38

The p38 family of MAPKs was identified on the basis of its LPS-induced tyrosine phosphorylation in a pre-B-cell line transfected with CD14 [97], and independently by virtue of its binding to, and inactivation by, pyridinyl imidazole compounds [98]. These compounds, e.g., SB203580, block LPS-induced expression of the cytokines TNFα, IL-1, IL-6 and IL-10 [82, 98, 99], suggesting p38 activity plays an important role in LPS cytokine induction. In RAW 264.7 macrophages and in CHO cells expressing CD14, LPS treatment increases p38 kinase activity toward recombinant ATF-2, a transcription factor [73, 87, 100]. Multiple p38-related MAPKs (p38α–δ, and their alternatively spliced isoforms) have been described in mammalian cells [70, 101], but nothing is known regarding their individual contributions to LPS-induced p38 activity.

p38 MAPKs are also activated by other stress stimuli not associated with induction of inflammatory responses [84], including osmotic shock, which also activates yeast (HOG1) and *Drosophila* (D-p38) p38 homologues [84], as well as JNK. LPS treatment of Drosophila S2* cells increases p38 activity [102]. Surprisingly, inhibition of the p38 pathway with dominant-negative D-p38α or SB203580 in these cells increases LPS-induced expression of antimicrobial peptide mRNAs [102], suggesting D-p38s are negative regulators of some LPS responses in this system.

MEK3 and MEK6 have been identified as upstream activators of p38α and -γ [70], while only MEK6 can also activate the p38β2 isoform [103]. Results in MEK4$^{-/-}$ mice suggest p38 is also a substrate of MEK4 in some cells [95], but not all [104]. In an LPS-stimulated murine macrophage line, p38 activity increased tenfold within 15 min, while a twofold increase in MEK3 and MEK6 activity were apparent only at 30 and 60 min, respectively [73]. MEK4 was highly active at earlier times (15 min) [73], suggesting that MEK4 may be the principal activator of p38 in LPS-stimulated cells. The MEKKs responsible for activation of p38 kinase cascades are not clearly defined, and several MEKKs may coordinately activate both JNK and p38 [70, 96]. Candidate MEKKs upstream of p38 include MEKK1 [105] MEKK3, TAK1 [106], ASK-1 [107], and PAK1 [108] (see also [96]), but it remains to be determined which, if any, of these mediates p38 activation by LPS.

MAPK targets

Following activation, MAPKs translocate to the nucleus, where they phosphorylate multiple transcription factors and other serine/threonine kinases which mediate their ultimate effects on gene expression and function [84]. MAPKs can alter gene expression at the level of translation [98, 109] as well as transcription, and some enzymes, notably cytosolic phospholipase A2, may be regulated directly by MAPK-mediated phosphorylation [110].

The best-characterized transcription factor substrates of MAPKs in mammals, and in simpler organisms as well, are members of the Ets, bZIP and MADS box families of DNA-binding proteins [111]. Phosphorylation of these proteins by MAPKs often increases their transactivating potential, but can also increase DNA binding or complex formation, or alter protein stability [84, 111]. Members of all of these transcription factor families are thought to be activated by LPS *via* MAPKs, as well as by other mechanisms, and most cytokine and other proinflammatory genes contain their corresponding DNA binding sites. (For a comprehensive summary of LPS-inducible elements in cytokine promoters see [112].) In some cases a requirement for a specific MAPK for LPS-induced gene expression has been demonstrated. However, some transcription factors are substrates for more than one MAPK [84, 111], and this possible redundancy, as well as the coordinate activation of MAPKs by LPS, may mask the contribution of individual MAPKs to LPS-induced expression in some cases.

Ets proteins, which contain a winged-helix-turn-helix DNA-binding motif, can activate gene expression upon phosphorylation by MAPKs [84, 113]. Ets activity usually requires formation of a complex of Ets proteins with other transcription factors, often AP-1 or MADS box proteins, such as serum response factor (SRF), which activates the c-fos promoter in combination with Ets ternary complex factors [114].

Ets proteins are downstream targets of the ERK MAPK cascade in *Drosophila* and *C. elegans*, as well as in mammals [111]. LPS, alone or with interferon gamma (IFNγ) or phorbol 12-myristate 13-acetate (PMA), has been shown to increase the DNA binding and/or transcriptional activity of the Ets proteins Elk-1, SAP, PU.1, and a novel Ets-2 related protein, GLp109 [74, 75, 115, 116]. However, Ets proteins are not phosphorylated exclusively by the ERK MAPKs, as both JNK and p38 also phosphorylate Elk-1 on similar sites [85, 101], and LPS-induced SAP activity in BAC-1.2F5 cells did not require activation of ERK [75].

c-Fos, c-Jun and related AP-1 binding proteins belong to the bZIP family of transcription factors, which are characterized by a basic DNA binding domain and a leucine zipper dimerization domain. As mentioned above, JNK MAPK was initially identified on the basis of its ability to bind c-Jun and phosphorylate its N-terminal transactivation domain, increasing its transactivating potential and stability [84, 117]. LPS treatment has been shown to induce JNK activity and AP-1-dependent transcription of a reporter gene in RAW 264.7 macrophages [86, 87]. The TNFα promoter contains adjacent Ets and AP-1-related sites, and *via* these sites c-Jun and c-Ets proteins cooperatively induce transactivation of a reporter gene driven by the TNF promoter [118]. This suggests that LPS-induced phosphorylation of c-Jun, *via* JNK, and Ets proteins, *via* ERK, contribute to transactivation of the TNFα gene. Other proinflammatory genes may be similarly regulated by these transcription factors. For example, DNA footprinting and gel-retardation experiments showed that LPS stimulation induced binding of c-Jun and Ets proteins to the tissue factor promoter in THP-1 cells [119].

ATF-2, a member of the ATF/CREB bZIP subfamily, is phosphorylated by p38 in LPS-stimulated CD14-transfected CHO cells [100], and is also a substrate for JNK *in vitro* [120]. Phosphorylation of ATF proteins by MAPKs increases their transactivating potential [111]. Results in an ATF-2-deficient mouse demonstrate a requirement for this transcription factor in LPS-induced expression of E-selectin [121]. ATF-2 may also contribute to TNF expression in LPS-stimulated cells [122], although this has not been examined in myeloid cells from ATF-2$^{-/-}$ mice.

Binding sites for CCAAT/enhancer-binding proteins (C/EBPs), another subfamily of bZIP transcription factors, are found in the promoters of many cytokine and acute phase response genes and have been implicated in their LPS-induced transactivation (see [112, 123] and references therein). C/EBPβ was initially cloned on the basis of its ability to activate IL-6 expression [124], and the promoters of IL-1 and TNF are also responsive to C/EBPβ [125, 126] . Ectopic C/EBPβ expression permitted LPS-induced expression of IL-6 mRNA in a lymphoblastoid cell line, and antisense mRNA to C/EBPβ blocked this response in a macrophage line [127]. While targeted deletion of C/EBPβ did not alter LPS-induced production of TNFα or IL-6 in monocytes [128], this result is likely to reflect functional redundancy of C/EBP family members rather than a lack of requirement for C/EBP activity for LPS responses [129].

LPS alters both the expression and activity of C/EBP proteins, and activates cooperating transcription factors [123]. C/EBPβ has been proposed to cooperate with c-Jun to activate TNFα expression in U937, a myelomonocytic cell line, in response to treatment with LPS and PMA [130]. To our knowledge, MAPK-dependent phosphorylation directly regulating C/EBP activity has not been demonstrated in LPS-stimulated cells. However, in other systems, ERK and p38 have been implicated in the phosphorylation and activation of C/EBPβ [131] and CHOP, another C/EBP related factor, respectively [132], suggesting the possiblity that MAPK pathways may similarly affect C/EBP proteins in LPS-stimulated macrophages.

A search for p38 substrates, using a catalytically inactive p38 as bait in the yeast two-hybrid system, identified the MADS box transcription factor MEF2C as a potential p38 target [133]. Although originally identified in muscle cells, MEF2C is expressed in myeloid cells, and its phosphorylation and binding activity increase in LPS-treated RAW 264.7 cells [133]. Results of an in-gel kinase assay suggest p38 is the major MEF2C kinase in these cells. Although MEF2C has not been shown to transactivate cytokine genes, LPS-induced phosphorylation of MEF2C has been shown to increase transactivation of the c-Jun promoter [133], thus contributing indirectly to cytokine expression.

MAPKs also activate the serine/threonine kinase p90^{Rsk} and several MAPKAP kinases, which in turn phosphorylate additional downstream targets, including transcription factors [70, 134]. The MAPKAP-related kinases MAPKAP kinase-2, -3, and -5 [135–138] and MNK1 and -2 [139, 140] are phosphorylated and activated by p38, as well as ERK or JNK in some cases [137, 138]. MAPKAP kinase-2, although initially identified as a substrate for ERK [141], has been shown to be activated and to phosphorylate heat shock proteins in response to LPS in a p38-dependent fashion [82, 142]. MAPKAP kinase-2 may have additional substrates important for LPS responses, such as CREB and ATF-1, which are phosphorylated by MAPKAP kinase-2 in fibroblast growth factor-stimulated cells [143]. MAPKAP kinase-3 (also called 3pK), like the transcription factor MEF2C, was isolated in a yeast two-hybrid screen with a catalytically inactive p38 [136], and subsequently shown to be activated by TNFα or hyperosmotic shock [136]. However, a role for this MAPKAP kinase or the newly cloned MAPKAP kinase-5 in LPS signaling remains to be tested.

In vitro, the MAPKAP-related kinase MNK1 phosphorylates elongation factor E1F-4E [139], which enhances the affinity of E1F-4E for the mRNA 5' cap structure and presumably increases translation (reviewed in [109]). LPS increases phosphorylation of elongation factor E1F-4E, and this depends at least in part on TNFα synthesis [144]. TNFα itself has been shown to activate MNK1 in HeLa cells *via* the p38 MAPK pathway, although the ERK MAPKs can also serve this function [139, 140].

Translation of TNFα and IL-1 mRNAs is enhanced in LPS-treated cells, and this stimulus-dependent regulation is mediated at least in part through specific sequences in the 3' untranslated regions (UTRs) of these mRNAs ([145, 146]). p38 has been reported to contribute to LPS-induced cytokine expression at the level of translation, based on results with pyridinyl imidazole compounds [98, 147]. However, p38 likely affects LPS-induced cytokine transcription as well, as studies with these same p38 inhibitors demonstrate a requirement for p38 for transactivation of cytokine promoters in response to other stimuli [99, 148].

There is evidence that JNK activity may also contribute to LPS-induced increases in cytokine translation. Expression of kinase-dead JNK3 (SAPKβ) in RAW 264.7 cells decreased LPS-induced translation of a reporter gene bearing the 3' UTR of the TNFα gene [73]. In a complementary experiment, expression of wild-type JNK3 overcame dexamethasone-induced inhibition of LPS-activation of a TNF translational reporter [73]. A similar requirement for JNK activity to stabilize IL-2 mRNA in a T-cell line has been posited, based on results with a pyridinyl imidazole inhibitor and expression of dominant-negative MKK and MEKK proteins [149].

IκB degradation and NF-κB/Rel activation

Rel/NF-κB transcription factors bind specifically as homo- or heterodimers to κB motifs and play a major role in immunity (reviewed in [150, 151]). Many components in the pathway leading to κB-dependent transactivation identified in mammals have homologues in flies and even plants (reviewed in [10]). Most proinflammatory genes require Rel/NF-κB proteins for LPS-induced expression [112], and increased nuclear κB-binding activity in human peripheral blood mononuclear cells is correlated with a lethal outcome in sepsis [152]. Rel proteins and their inhibitors are proven targets for agents that block cytokine expression. The anti-inflammatory properties of glucocorticoids and salicylates depend at least in part on their ability to inhibit induction of nuclear Rel activity [153–155]. Similarly, the inhibitory effects of IL-10 and -11 on LPS-induced expression of proinflammatory cytokines have been shown to be mediated by inhibition of nuclear translocation of NF-κB [156, 157].

Rel proteins are characterized by a conserved N-terminal DNA binding and dimerization domain called the Rel homology domain (RHD) [150]. In mammals, five Rel proteins are known; three family members, RelA (p65), c-Rel and RelB, also have unique C-terminal transactivation domains. Two other Rel proteins, NF-κB1 (p105/p50) and NF-κB2 (p100/p52), are initially synthesized as precursors and then proteolytically processed to yield κB-binding subunits that consist of little more than their RHDs.

Targeted deletion of murine genes encoding Rel proteins, singly and in combination, has unequivocally demonstrated their importance in immunity (reviewed in [158, 159]). Although there is some redundancy among Rel proteins, each has a unique function that cannot be provided by other members of the family. Most studies of Rel-deficient mice have focused on the effects on lymphocytes [158]; however, alterations in cytokine production by LPS-stimulated macrophages in several of these strains have also been reported. In peritoneal macrophages from mice lacking both NF-κB1 and NF-κB2, stimulation with LPS and IFNγ elicits almost no IL-6 or GM-CSF production, although TNFα expression is no different from wild-type controls [160]. Similarly, macrophages from mice with a targeted deletion of the transactivation domain of c-Rel secrete less IL-1β and granulocyte macrophage-colony stimulating factor (GM-CSF) in response to LPS/IFNγ treatment, among other defects [161]. In mice lacking any c-Rel protein, TNFα expression in response to LPS treatment is reduced in resident and peritoneal macrophages, although expression of IL-6 and CSFs is enhanced, suggesting Rel proteins can have both positive and negative effects on transcription and this also depends on the macrophage population examined [162]. Thus, Rel family members play important roles in LPS induction of many cytokines.

Rel dimers are tightly regulated by their interaction with the inhibitory IκBs, of which there are three principal forms, α, β and ε. IκBs mask the Rel nuclear localization signals, sequestering Rel dimers in the cytoplasm [150]. IκBα, β and ε proteins and the labile C-termini of p105 and p100 precursors all contain multiple copies of an ankyrin repeat motif, and these repeats are required for the association of IκBs with Rel proteins [150]. IκBα is the most studied of the Rel inhibitors, and its importance in immune homeostasis is underscored by the phenotype of IκBα$^{-/-}$ mice, which suffer from dermatitis accompanied by high levels of TNFα mRNA in the skin [163]. However, LPS treatment of IκBα$^{-/-}$ embryonic fibroblasts can still elicit a considerable increase in nuclear κB-binding activity, suggesting other IκBs also regulate LPS-induced Rel activity [164]. The p105 precursor also binds other Rel subunits, especially p50, retaining them in the cytoplasm [165]. In macrophages from mice lacking the p105 protein, but still expressing the corresponding p50 DNA-binding activity, LPS-stimulated cytokine secretion is dramatically reduced, probably due to the increased concentration of nuclear p50 homodimers, which bind DNA but do not activate transcription [166].

Upon treatment with LPS and other stimuli, including proinflammatory cytokines, IκB proteins are inducibly phosphorylated on two closely spaced N-terminal serines and rapidly degraded *via* the ubiquitin-26S proteasome pathway [167, 168] (Fig. 3). Despite the wide variety of stimuli that activate NF-κB, all appear to

---

Figure 3

The IκB/Rel regulatory pathway. Activation of nuclear κB-binding activity by TNFα and IL-1 involves receptor-associated kinases (RIP and IRAK) and adaptor proteins (TRAF2 and -6). The TRAF molecules interact with NIK, a MEKK-related serine/threonine kinase, which in turn phosphorylates IκB kinases (IKKs) and activates the multi-subunit IKK complex. MEKK-

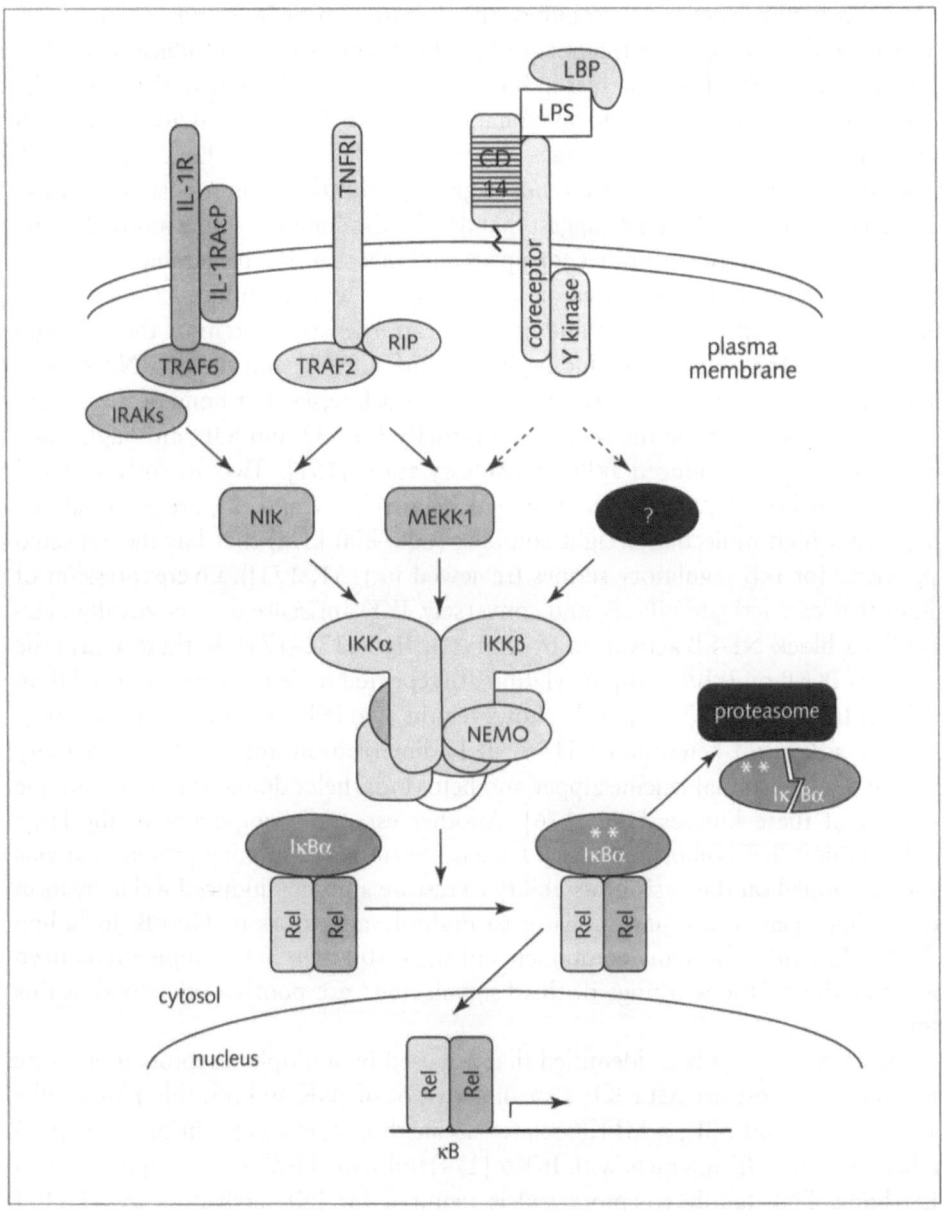

1 is also activated by TNF and IL-1 and can similarly activate the IKKs. The IKK complex has a molecular weight of ~ 700 kDa, and includes IKKα, β, a putative adaptor protein, NEMO, and other subunits. The activated IKKs phosphorylate IκB proteins on conserved N-terminal serine residues, targeting them for degradation in the proteasome, and releasing associated Rel dimers to the nucleus where they activate κB-dependent transcription.

use this common mechanism to induce the degradation of IκBs and release of Rel proteins to the nucleus. LPS treatment of peritoneal macrophages induces phosphorylation of these regulatory serines in IκBα and -β proteins, leading to their total disappearance within 20 min and maximal nuclear κB binding activity within 1 h [169]. In human THP-1 macrophages, IκBα was undetectable within 30 min of LPS treatment, while IκBβ and -ε were not degraded until 60–90 min post-stimulation [170]. These differing kinetics suggest that despite the common serine motif, the various IκBs have differing affinities for upstream kinases or the proteasome.

The recognition that all activators of nuclear Rel activity induce a similar phosphorylation of IκBα serines 32 and 36 led to an intensive search for the kinase(s) responsible. Many kinases, including PKCs, Raf-1, PKA, and the ds-RNA dependent kinase PKR, were proposed as possible IκBα kinases, but none of these were ultimately found to have the required specificity for S32 and S36, although some may contribute to induced IκBα phosphorylation [151]. Two recently isolated homologous kinases, termed IKKα and IKKβ (or IKK1 and -2), are essential elements in a high molecular weight complex (600–800 kDa) that has the expected specificity for IκB regulatory serines (reviewed in [151, 171]). Overexpression of either IKK can activate NF-κB, and conversely, IKK antisense or catalytically inactive IKKs block NF-κB activation by TNFα or IL-1 [172–174]. In these assays the effects of IKKβ on IκBα phosphorylation are reported to be more pronounced than those of IKKα (e.g., [175]). In cells, however, the two IKKs are thought to act coordinately as an α/β heterodimer [176]. IKK dimerization and full kinase activity require the C-terminal leucine zipper and helix-loop-helix domains that are unique features of these kinases [173, 176]. Another essential component of the large multi-subunit IKK complex is NEMO, a leucine zipper-containing protein that was recently cloned on the basis of its ability to restore stimulus-induced Rel activation to cell lines genetically unresponsive to multiple activators of NF-κB, including LPS [177]. Evidently, many components of the ~ 700 kDa IKK complex remain to be isolated, and the workings of this "signalsome" are poorly understood at this time.

Two kinases have been identified that are used by multiple receptors to activate IKK activity. These are MEKK1, also an activator of JNK and possibly p38, as discussed above, and NIK, a MEKK-related serine/threonine kinase. In the yeast two-hybrid system, NIK interacts with IKKα [174]and with TRAF2, an adaptor protein that binds TNF family receptors and is required for JNK activation by TNFR I [178]. NIK is also activated by receptors for IL-1 *via* TRAF6. At present, it is not known whether LPS also requires NIK to activate the IKK complex.

The second upstream activator of IKKs identified in TNF-treated cells is MEKK1, which is also activated by LPS [87]. Dominant-negative MEKK1 constructs gave mixed results in κB-dependent reporter gene assays [179–181]. However, MEKK1 more recently has been found to associate with IKK activity through multiple chromatographic purification steps [175] and can phosphorylate and activate IKKβ *in vitro* and *in vivo* [180, 182, 183]. In addition, the Tax protein of human T-cell leukemia virus I has recently been shown to bind MEKK1, and MEKK1 activity is required for Tax-induced activation of NF-κB [183]. An appealing hypothesis is that MEKK1 has a similar role in activation of both stress-activated MAPKs and NF-κB in response to LPS as well as TNFα. However, there is as yet no evidence for this hypothesis, and it is equally plausible that LPS uses unique enzymes to activate the common IKK complex.

Other signaling pathways activated by LPS

We have described the principal known signals contributing to LPS-induced cytokine expression in macrophages: tyrosine kinase activity, the MAPKs and their targets, and Rel/NF-κB transcription factors. Other signaling pathways have been shown to be activated by LPS, but their relationship to cytokine expression is not well established, and will be described in less detail. These signals include membrane lipids, such as those generated by the action of phosphatidylinositol 3-kinase (PI3K), and ceramide.

LPS has been shown to activate PI3K in monocytes [184] and macrophages [200]. Studies with pharmacological inhibitors of PI3K implicate this enzyme in the activation of p70^{S6kinase}, regulation of nitric oxide (NO) production [200], and CD14-dependent phagocytosis [185]. A small stimulatory effect of the PI3K inhibitor wortmannin on TNFα synthesis in LPS-treated peritoneal macrophages has been reported [186]; however, similar studies in RAW 264.7 macrophages revealed no such effect on cytokine production [200].

Ceramide, which is generated by cleavage of membrane sphingomyelin by specific phospholipases (reviewed in [187]), has also been linked to LPS signaling. LPS was reported to activate a ceramide-activated protein kinase [188], since identified as KSR1 (kinase suppressor of Ras) [189], and a LPS-hyporesponsive mouse strain, C3H/HeJ, was reported to be unresponsive to ceramide analogues as well, as assessed by cytokine mRNA expression [190]. However, in a transformed macrophage line derived from a C3H/HeJ mouse, we found that responses to ceramide analogues were not impaired [201]. Furthermore, ceramide analogues did not elicit TNFα expression in either mutant or wild-type macrophage lines [201], suggesting this effect is not common to all macrophages. Our results indicate LPS does not act merely as a ceramide mimic, as has been suggested [191], although it may activate ceramide-dependent signaling [201].

LPS has also been shown to increase other second messengers, including intracellular calcium [192], and activate cellular enzymes, like phospholipases [193, 194], and G proteins [195]. Although these pathways have important functions in macrophage biology, a clear link to cytokine expression has not been made.

Conclusion

CD14 on macrophages has emerged as an important receptor for LPS in mammals. The proteins that transmit CD14-mediated signals to the cytoplasm and initiate signaling through tyrosine phosphorylation are no doubt of equal importance in sepsis, and their identification would open new avenues of investigation and possible therapeutic intervention. Despite our incomplete knowledge of the nature of the high-affinity LPS receptor, downstream signaling molecules that regulate cytokine expression have been identified. Some of these, notably p38 and NF-κB, are targets of anti-inflammatory drugs already in use. As further substrates of the MAPKs are identified, and the regulators of IκB activity are defined, new candidates for drug targets will likely emerge, perhaps allowing more specific inhibition of LPS-dependent activation of these pathways.

The genetics of the *Drosophila* immune system has been an important source of insight into mechanisms of innate immune responses. For example, the Toll receptor, which activates Rel-related proteins in *Drosophila* and has homology to the cytoplasmic domain of the IL-1 receptor, has recently been shown to have at least five human homologues [196, 197]. Similarly, a kinase required for IL-1 receptor signaling, IRAK, was also isolated on the basis of its homology to Pelle, a *Drosophila* kinase downstream of Toll [198], as well as by purification [99]. Although genetic studies in these simpler organisms cannot substitute for studies in mammals, they may continue to lend insight into fundamental schemes in innate immunity.

Genetic and biochemical data have outlined an evolutionarily conserved cassette of signaling responses to inflammatory stimuli, involving the stress-activated kinases and Rel proteins. These coordinately activate transcription and increase translation of cytokines and a wide range of proinflammatory molecules. A variety of receptors activate these signaling cascades in a similar fashion. How specificity is conferred on these responses remains obscure and, along with identification of the full CD14 receptor complex, are important remaining issues in the field of LPS-induced signaling and endotoxic shock.

Note added in proof: Since submission of this review, it has been discovered that members of the Toll family of receptors, especially Toll-like receptor 4, are likely to be the transmembrane signaling receptors for LPS hypothesized in the text and shown as the "coreceptor" in Figure 1 [202].

References

1   Ulevitch RJ, Tobias PS (1995) Receptor-dependent mechanisms of cell stimulation by bacterial endotoxin. *Ann Rev Immunol* 13: 437–457

2   Wright SD, Ramos RA, Tobias PS, Ulevitch RJ, Mathison JC (1990) CD14, a receptor for complexes of LPS and LPS binding protein. *Science* 249: 1431–1433

3   Leturcq DJ, Moriarty AM, Talbott G, Winn RK, Martin TR, Ulevitch RJ (1996) Antibodies against CD14 protect primates from endotoxin-induced shock. *J Clin Invest* 98: 1533–1538

4   Lee J-D, Kato K, Tobias PS, Kirkland TN, Ulevitch RJ (1992) Transfection of CD14 into 70Z/3 cells dramatically enhances the sensitivity to complexes of lipopolysaccharide (LPS) and LPS binding protein. *J Exp Med* 175: 1697–1705

5   Golenbock DT, Liu Y, Millham FH, Freeman MW, Zoeller RA (1993) Surface expression of human CD14 in Chinese hamster ovary fibroblasts imparts macrophage-like responsiveness to bacterial endotoxin. *J Biol Chem* 268: 22055–22059

6   Ferrero E, Jiao D, Tsuberi BZ, Tesio L, Rong GW, Haziot A, Goyert SA (1993) Transgenic mice expressing human CD14 are hypersensitive to lipopolysaccharide. *Proc Natl Acad Sci USA* 90: 2380–2384

7   Haziot A, Ferrero E, Kontgen F, Hijiya N, Yamamoto S, Silver J, Stewart CL, Goyert SM (1996) Resistance to endotoxin shock and reduced dissemination of gram-negative bacteria in CD14-deficient mice. *Immunity* 4: 407–414

8   Gegner JA, Ulevitch RJ, Tobias PS (1995) Lipopolysaccharide (LPS) signal transduction and clearance: Dual roles for LPS binding protein and membrane CD14. *J Biol Chem* 270: 5320–5325

9   Ferrero E, Hsieh C-L, Francke U, Goyert SM (1990) CD14 is a member of the family of leucine-rich proteins and is encoded by a gene syntenic with multiple receptor genes. *J Immunol* 145: 331–336

10   Medzhitov R, Janeway CA Jr (1998) An ancient system of host defense. *Curr Opin Immunol* 10: 12–15

11   Juan TS, Kelley MJ, Johnson DA, Busse LA, Hailman E, Wright SD, Lichenstein HS (1995) Soluble CD14 truncated at amino acid 152 binds lipopolysaccharide (LPS) and enables cellular response to LPS. *J Biol Chem* 270: 1382–1387

12   Stetler F, Bernheiden M, Menzel R, Jack RS, Witt S, Fan X, Pfister M, Schutt C (1997) Mutation of amino acids 39-44 of human CD14 abrogates binding of lipopolysaccharide and *Escherichia coli*. *Eur J Biochem* 24: 100–109

13   Juan TS-C, Hailman E, Kelley MJ, Busse LA, Davy E, Empig CJ, Narhi LO, Wright SD, Lichenstein HS (1995) Identification of a lipopolysaccharide binding domain in CD14 between amino acids 57 and 64. *J Biol Chem* 270: 5219–5224

14   Viriyakosol S, Kirkland TN (1995) A region of human CD14 required for lipopolysaccharide binding. *J Biol Chem* 270: 361–368

15   Hoess A, Watson S, Siber GR, Liddington R (1993) Identification of the LPS binding domain of an endotoxin neutralising protein, Limulus anti-LPS factor. *EMBO J* 12: 3351–3356

16   Weidemann B, Brade H, Rietschel ET, Dziarski R, Bazil V, Kusumoto S, Flad HD, Ulmer AJ (1994) Soluble peptidoglycan-induced monokine production can be blocked by anti-

CD14 monoclonal antibodies and by lipid A partial structures. *Infect Immun* 62: 4709–4715

17 Cleveland MG, Gorham JD, Murphy TL, Tuomanen E, Murphy KM (1996) Lipoteichoic acid preparations of gram-positive bacteria induce interleukin-12 through a CD14-dependent pathway. *Infect Immun* 64: 1906–1912

18 Pugin J, Heumann D, Tomasz A, Kravchenko VV, Akamatsu Y, Nishijima M, Glauser MP, Tobias PS, Ulevitch RJ (1994) CD14 is a pattern recognition receptor. *Immunity* 1: 509–516

19 Devitt A, Moffatt OD, Raykundalia C, Capra JD, Simmons DL, Gregory CD (1998) Human CD14 mediates recognition and phagocytosis of apoptotic cells. *Nature* 392: 505–509

20 Stahl N, Yancopoulos GD (1993) The alphas, betas and kinases of cytokine receptor complexes. *Cell* 74: 587–590

21 Treanor JJS, Goodman L, de Sauvage F, Stone DM, Poulsen KT, Beck CD, Gray C, Armanini MP, Pollock RA, Hefti F et al (1996) Characterization of a multicomponent receptor for GDNF. *Nature* 382: 80–83

22 Buj-Bello A, Adu J, Pinon LGP, Horton A, Thompson J, Rosenthal A, Chinchetru M, Buchman VL, Davies AM (1997) Neurturin responsiveness requires a GPI-linked receptor and the Ret receptor tyrosine kinase. *Nature* 387: 721–724

23 Klein RD, Sherman D, Ho WH, Stone D, Bennett GL, Moffat B, Vandlen R, Simmons L, Gu Q, Hongo JA et al (1997) A GPI-linked protein that interacts with Ret to form a candidate neurturin receptor. *Nature* 387: 717–721

24 Brown D (1993) The tyrosine kinase connection: how GPI-anchored proteins activate T cells. *Curr Opin Immunol* 5: 349–354

25 Lee J-D, Kravchenko V, Kirkland TN, Han J, Mackman N, Moriarty A, Leturcq D, Tobias PS, Ulevitch RJ (1993) Glycosyl-phosphatidylinositol-anchored or integral membrane forms of CD14 mediate identical cellular responses to endotoxin. *Proc Natl Acad Sci USA* 90: 9930–9934

26 Pugin J, Kravchenko VV, Lee J-D, Kline L, Ulevitch RJ, Tobias PS (1998) Cell activation mediated by glycosylphosphatidylinositol-anchored or transmembrane forms of CD14. *Infect Immun* 66: 1174–1180

27 Vita N, Lefort S, Sozzani P, Reeb R, Richards S, Borysiewicz LK, Ferrarar P, Labeta MO (1997) Detection and biochemical characteristics of the receptor for complexes of soluble CD14 and bacterial lipopolysaccharide. *J Immunol* 158: 3457–3462

28 Schletter J, Brade H, Brade L, Kruger C, Loppnow H, Kusumoto S, Rietschel ET, Flad H-D, Ulmer AJ (1995) Binding of lipopolysaccharide (LPS) to an 80-kilodalton membrane protein of human cells is mediated by soluble CD14 and LPS-binding protein. *Infect Immun* 63: 2576–2580

29 Fukuse S, Maeda T, Webb DR, Devens BH (1995) A 69-kDa membrane protein associated with lipopolysaccharide (LPS)-induced signal transduction in the human monocytic cell line THP-1. *Cell Immunol* 164: 248–254

30 Petty HR, Todd RF III (1996) Integrins as promiscuous signal transduction devices. *Immunol Today* 17: 209–212

31 Pearson AM (1996) Scavenger receptors in innate immunity. *Curr Opin Immunol* 8: 20–28

32 Ingalls RR, Golenbock DT (1995) CD11c/CD18, a transmembrane signaling receptor for lipopolysaccharide. *J Exp Med* 181: 1473–1479

33 Ingalls RR, Arnaout MA, Golenbock DT (1997) Outside-in signaling by lipopolysaccharide through a tailless integrin. *J Immunol* 159: 433–438

34 Medvedev AE, Flo T, Ingalls RR, Golenbock DT, Teti G, Vogel SN, Espevik T (1998) Involvement of CD14 and complement receptors CR3 and CR4 in nuclear factor-κB activation and TNF production induced by lipopolysaccharide and Group B streptococcal cell walls. *J Immunol* 160: 4535–4542

35 Zarewych DM, Lindzelskii AL, Todd RF III, Petty HR (1996) LPS induces CD14 association with complement receptor type 3, which is reversed by neutrophil adhesion. *J Immunol* 156: 430–433

36 Wright SD, Detmers PA, Aida Y, Adamowski R, Anderson DC, Chad Z, Kabbash LG, Pabst MJ (1990) CD18-deficient cells respond to lipopolysaccharide *in vitro*. *J Immunol* 144: 2566–2571

37 Hampton RY, Golenbock DT, Penman M, Krieger M, Raetz CRH (1991) Recognition and plasma clearance of endotoxin by scavenger receptors. *Nature* 352: 342–344

38 Haworth R, Platt N, Keshav S, Hughes D, Darley E, Suzuki H, Kurihara Y, Kodama T, Gordon S (1997) The macrophage scavenger receptor type A is expressed by activated macrophages and protects the host against lethal endotoxic shock. *J Exp Med* 186: 1431–1439

39 Suzuki H, Kurihara Y, Takeya M, Kamada N, Kataoka M, Jishage K, Ueda O, Sakaguchi H, Higashi T, Suzuki T (1997) A role for macrophage scavenger receptors in atherosclerosis and susceptibility to infection. *Nature* 386: 292–296

40 Ulevitch RJ, Tobias PS (1994) Recognition of endotoxin by cells leading to transmembrane signaling. *Curr Opin Immunol* 6: 125–130

41 Jack RS, Fan W, Bernheiden M, Rune G, Ehlers M, Weber A, Kirsch G, Mentel R, Furll B, Freudenberg M et al (1997) Lipopolysaccharide-binding protein is required to combat a murine Gram-negative bacterial infection. *Nature* 389: 742–745

42 Wurfel MM, Monks BG, Ingalls RR, Dedrick RL, Delude R, Zhou D, Lamping N, Schumann RR, Thieringer R, Fenton MJ et al (1997) Targeted deletion of the lipopolysaccharide (LPS)-binding protein gene leads to profound supression of LPS responses *ex vivo*, whereas *in vivo* responses remain intact. *J Exp Med* 186: 2051–2056

43 Wurfel MM, Kunitake ST, Lichenstein H, Kane JP, Wright SD (1994) Lipopolysaccharide (LPS)-binding protein is carried on lipoproteins and acts as a cofactor in the neutralization of LPS. *J Exp Med* 180: 1025–1035

44 Tall A (1995) Plasma lipid transfer proteins. *Ann Rev Biochem* 64: 235–257

45 Elsbach P, Weiss J (1998) Role of the bactericidal/permeability-increasing protein in host defence. *Curr Opin Immunol* 10: 45–49

46 Hailman E, Albers JJ, Wolfbauer G, Tu A-Y, Wright SD (1996) Neutralization and transfer of lipopolysaccharide by phospholipid transfer protein. *J Biol Chem* 271: 12172–12178

47 Wurfel MM, Hailman E, Wright SD (1995) Soluble CD14 acts as a shuttle in the neutralization of lipopolysaccharide (LPS) by LPS-binding protein and reconstituted high density lipoprotein. *J Exp Med* 181: 1743–1754

48 Parker TS, Levine DM, Chang JC, Laxer J, Coffin CC, Rubin AL (1995) Reconstituted high-density lipoprotein neutralizes gram-negative bacterial lipopolysaccharides in human whole blood. *Infect Immun* 63: 253–258

49 Pajkrt D, Doran JE, Koster F, Lerch PG, Arnet B, van der Poll T, ten Cate JW, van Deventer SJ (1996) Antiinflammatory effects of reconstituted high-density lipoprotein during human endotoxemia. *J Exp Med* 184: 1601–1608

50 Ohlsson BG, Englund MCO, Karlsson A-LK, Knutsen E, Erixon C, Skribeck H, Liu Y, Bondjers G, Wieklund O (1996) Oxidized low density lipoprotein inhibits lipopolysaccharide-induced binding of nuclear factor-κB to DNA and the subsequent expression of tumor necrosis factor-α and interleukin-1β in macrophages. *J Clin Invest* 98: 78–89

51 Neatea MG, Demacker PNM, Kullberg BJ, Boerman OC, Verschueren K, Stalenhoef AFH, van der Meer JWM (1996) Low-density lipoprotein receptor-deficient mice are protected against lethal endotoxemia and severe gram-negative infections. *J Clin Invest* 97: 1366–1372

52 Perera P-Y, Vogel SN, Detore GR, Haziot A, Goyert SM (1997) CD14-dependent and CD14-independent signaling pathways in murine macrophages from normal and CD14 knockout mice stimulated with lipopolysaccharide or taxol. *J Immunol* 158: 4422–4429

53 Haziot A, Lin XY, Zhang F, Goyert SM (1998) The induction of acute phase proteins by lipopolysaccharide uses a novel pathway that is CD14-independent. *J Immunol* 160: 2570–2572

54 Weinstein SL, Gold MR, DeFranco AL (1991) Bacterial lipopolysaccharide stimulates protein tyrosine phosphorylation in macrophages. *Proc Natl Acad Sci USA* 88: 4148–4152

55 DeFranco AL, Crowley MT, Finn A, Hambleton J, Weinstein SL (1997) The role of tyrosine kinases and MAP kinases in LPS-induced signaling. In: J Levin (ed): Endotoxin and sepsis: Molecular mechanisms of pathogenesis, host resistance, and therapy. Wiley-Liss, Inc, New York, 119–136

56 Weinstein SL, June CH, DeFranco AL (1993) Lipopolysaccharide-induced protein tyrosine phosphorylation in human macrophages is mediated by CD14. *J Immunol* 151: 3829–3838

57 Weinstein SL, Sanghera JS, Lemke K, DeFranco AL, Pelech SL (1992) Bacterial lipopolysaccharide induces tyrosine phosphorylation and activation of mitogen-activated protein kinases in macrophages. *J Biol Chem* 267: 14955–14962

58 Novogrodsky A, Vanichkin A, Patya M, Gazit A, Osherov N, Levitzki A (1994) Prevention of lipopolysaccharide-induced lethal toxicity by tyrosine kinase inhibitors. *Science* 264: 1319–1322

59   Ding A, Sanchez E, Nathan CF (1993) Taxol shares the ability of bacterial lipopolysaccharide to induce tyrosine phosphorylation of microtubule-associated protein kinase. *J Immunol* 151: 5596–602

60   Shapira L, Takashiba S, Champagne C, Amar S, Van Dyke TE (1994) Involvement of protein kinase C and protein tyrosine kinase in lipopolysaccharide-induced TNF-α and IL-1β production by human monocytes. *J Immunol* 153: 1818–1824

61   Geng Y, Gulbins E, Altman A, Lotz M (1994) Monocyte deactivation by interleukin 10 *via* inhibition of tyrosine kinase activity and the ras signaling pathway. *Proc Natl Acad Sci USA* 91: 8602–8606

62   Beaty CD, Franklin TL, Uehara Y, Wilson CB (1994) Lipopolysaccharide-induced cytokine production in human monocytes: Role of tyrosine phosphorylation in transmembrane signal transduction. *Eur J Immunol* 24: 1278–1284

63   Dong Z, O'Brian CA, Fidler IJ (1993) Activation of tumoricidal properties in macrophages by lipopolysaccharide requires protein-tyrosine kinase activity. *J Leukoc Biol* 53: 53–60

64   Delude RL, Fenton MJ, Savedra R Jr, Perera PY, Vogel SN, Thieringer R Golenbock, DT (1994) CD14-mediated translocation of nuclear factor-κB induced by lipopolysaccharide does not require tyrosine kinase activity. *J Biol Chem* 269: 22253–22260

65   Geng Y, Zhang B, Lotz M (1993) Protein tyrosine kinase activation is required for lipopolysaccharide induction of cytokines in human blood monocytes. *J Immunol* 151: 6692–6700

66   Stefanova I, Corcoran ML, Horak EM, Wahl LM, Bolen JB, Horak ID (1993) Lipopolysaccharide induces activation of CD14-associated protein tyrosine kinase p53/56lyn. *J Biol Chem* 268: 20725–20728

67   Meng F, Lowell CA (1997) Lipopolysaccharide (LPS)-induced macrophage activation and signal transduction in the absence of src-family kinases hck, fgr, lyn. *J Exp Med* 185: 1661–1670

68   Crowley MT, Harmer SL, DeFranco AL (1996) Activation-induced association of a 145-kDa tyrosine-phosphorylated protein with shc and syk in B lymphocytes and macrophages. *J Biol Chem* 271: 1145–1152

69   Crowley MT, Costello PS, Fitzer-Attas CJ, Turner M, Meng F, Lowell C, Tybulewicz VLJ, DeFranco AL (1997) A critical role for syk in signal transduction and phagocytosis mediated by Fc-γ receptors on macrophages. *J Exp Med* 186: 1027–1039

70   Robinson MJ, Cobb MH (1997) Mitogen-activated protein kinase pathways. *Curr Opin Cell Biol* 9: 180–186

71   Dong Z, Qi X, Fidler IJ (1993) Tyrosine phosphorylation of mitogen-activated protein kinases is necessary for activation of murine macrophages by natural and synthetic bacterial products. *J Exp Med* 177: 1071–1077

72   Liu MK, Herrera-Velit P, Brownsey RW, Reiner NE (1994) CD14-dependent activation of protein kinase C and mitogen-activated protein kinases (p42 and p44) in human monocytes treated with bacterial lipopolysaccharide. *J Immunol* 153: 2642–2652

73   Swantek JL, Cobb MH, Geppert TD (1997) Jun N-terminal kinase/stress-activated pro-

tein kinase (JNK/SAPK) is required for lipopolysaccharide stimulation of tumor necrosis factor α (TNFα) translation: glucocorticoids inhibit TNFα translation by blocking JNK/SAPK. *Mol Cell Biol* 17: 6274–6282

74   Reimann T, Buscher D, Hipskind RA, Krautwald S, Lohmann-Matthes ML, Baccarini M (1994) Lipopolysaccharide induces activation of the Raf-1/MAP kinase pathway. A putative role for Raf-1 in the induction of the IL-1β and the TNF-α genes. *J Immunol* 153: 5740–5749

75   Buscher D, Hipskind RA, Krautwald S, Reimann T, Baccarini M (1995) Ras-dependent and -independent pathways target the mitogen-activated protein kinase network in macrophages. *Mol Cell Biol* 15: 466–475

76   Guthridge CJ, Eidlen D, Arend WP, Gurierrez-Hartmann A, Smith MF Jr (1997) Lipopolysaccharide and Raf-1 kinase regulate secretory interleukin-1 receptor antagonist gene expression by mutally antagonistic mechanisms. *Mol Cell Biol* 17: 1118–1128

77   Morrison DK, Cutler RE Jr (1997) The complexity of Raf-1 regulation. *Curr Opin Cell Biol* 9: 174–179

78 . Stancato LF, Sakatsume M, David M, Dent P, Dong F, Petricoin E, Krolewski JJ, Silvennoinen O, Saharinen P, Pierce J et al (1997) Beta interferon and oncostatin M activate Raf-1 and mitogen-activated protein kinase through a JAK1-dependent pathway. *Mol Cell Biol* 17: 3833–3840

79   Ueda Y, Hirai S-I, Osada S-I, Suzuki A, Mizuno K, Ohno S (1996) Protein kinase C delta activates the MEK-ERK pathway in a manner independent of Ras and dependent on Raf. *J Biol Chem* 271: 23512–23519

80   Geppert TD, Whitehurst CE, Thompson P, Beutler B (1994) Lipopolysaccharide signals activation of tumor necrosis factor biosynthesis through the ras/raf-1/MEK/ MAPK pathway. *Mol Med* 1: 93–103

81   Hambleton J, McMahon M, DeFranco AL (1995) Activation of Raf-1 and mitogen-activated protein kinase in murine macrophages partially mimics lipopolysaccharide-induced signaling events. *J Exp Med* 182: 147–154

82   Foey AD, Parry SL, Williams LM, Feldmann M, Foxwell BMJ, Brennan FM (1998) Regulation of monocyte IL-10 synthesis by endogenous IL-1 and TNF-α: Role of the p38 and p42/44 mitogen-activated protein kinases. *J Immunol* 160: 920–928

83   Ip YT, Davis RJ (1998) Signal transduction by the c-Jun N-terminal kinase (JNK) – from inflammation to development. *Curr Opin Cell Biol* 10: 205–219

84   Su B, Karin M (1996) Mitogen-activated protein kinase cascades and regulation of gene expression. *Curr Opin Immunol* 8: 402–411

85   Gupta S, Barrett T, Whitmarsh AJ, Cavanagh J, Sluss HK, Berijard B, Davis RJ (1996) Selective interaction of JNK protein kinase isoforms with transcription factors. *EMBO J* 15: 2760–2770

86   Hambleton J, Weinstein SL, Lem L, DeFranco AL (1996) Activation of c-Jun N-terminal kinase in bacterial lipopolysaccharide-stimulated macrophages. *Proc Natl Acad Sci USA* 93: 2774–2778

87   Sanghera JS, Weinstein SL, Aluwalia M, Girn J, Pelech SL (1996) Activation of multiple

proline-directed kinases by bacterial lipopolysaccharide in murine macrophages. *J Immunol* 156: 4457–4465

88   Sluss HK, Han Z, Barrett T, Davis RJ, Ip YT (1996) A JNK signal transduction pathway that mediates morphogenesis and an immune response in Drosophila. *Genes Dev* 10: 2745–2758

89   Derijard B, Raingeaud J, Barrett T, Wu I-H, Han J, Ulevitch RJ, Davis RJ (1995) Independent human MAP kinase signal transduction pathways defined by MEK and MKK isoforms. *Science* 267: 682–685

90   Sanchez I, Hughes RT, Mayer BJ, Yee K, Woodgett JR, Avruch J, Kyriakis JM, Zon LI (1994) Role of SAPK/ERK kinase-1 in the stress-activated pathway regulating transcription factor c-Jun. *Nature* 372: 794–798

91   Lin A, Minden A, Martinetto H, Claret F-X, Lange-Carter C, Mercurio F, Johnson GL, Karin M (1995) Identification of a dual specificity kinase that activates the Jun kinases and p38-Mpk2. *Science* 268: 286–290

92   Nishina H, Fischer KD, Radvanyl L, Shahinian A, Hakem R, Rubie EA, Bernstein A, Mak TW, Woodgett JR, Penninger JM (1997) Stress-signalling kinase Sek1 protects thymocytes from apoptosis mediated by CD95 and CD3. *Nature* 385: 350–353

93   Holland PM, Suzanne M, Campbell JS, Noselli S, Cooper JA (1997) MKK7 is a stress-activated mitogen-activated protein kinase functionally related to hemipterous. *J Biol Chem* 272: 24994–24998

94   Moriguchi T, Toyoshima F, Masuyama N, Hanafusa H, Gotoh Y, Nishida E (1997) A novel SAPK/JNK kinase, MKK7, stimulated by TNFα and cellular stresses. *EMBO J* 16: 7045–7053

95   Ganiatsas S, Kwee L, Fujiwara Y, Perkins A, Ikeda T, Labow MA, Zon LI (1998) SEK1 deficiency reveals mitogen-activated protein kinase cascade crossregulation and leads to abnormal hepatogenesis. *Proc Natl Acad Sci USA* 95: 6881–6886

96   Fanger GR, Gerwins P, Widmann C, Jarpe MB, Johnson GL (1997) MEKKs, GCKs, MLKs, PAKs, TAKs, and Tpls: upstream regulators of the c-Jun amino-terminal kinases? *Curr Opin Genet Dev* 7: 67–74

97   Han J, Lee JD, Bibbs L, Ulevitch RJ (1994) A MAP kinase targeted by endotoxin and hyperosmolarity in mammalian cells. *Science* 265: 808–811

98   Lee JC, Laydon JT, McDonnell PC, Gallagher TF, Kumar S, Green D, McNulty D, Blumenthal MJ, Heys JR, Landvatter SW et al (1994) A protein kinase involved in the regulation of inflammatory cytokine biosynthesis. *Nature* 372: 739–746

99   Beyaert R, Cuenda A, Berghe WV, Plaisance S, Lee JC, Haegeman G, Cohen P, Fiers W (1996) The p38/RK mitogen-activated protein kinase pathway regulates interleukin-6 synthesis in response to tumor necrosis factor. *EMBO J* 15: 1914–1923

100 Raingeaud J, Gupta S, Rogers JS, Dickens M, Han J, Ulevitch RJ, Davis RJ (1995) Proinflammatory cytokines and environmental stress cause p38 mitogen-activated protein kinase activation by dual phosphorylation on tyrosine and theonine. *J Biol Chem* 270: 7420–7426

101 Jiang Y, Gram H, Zhao M, New L, Gu J, Feng L, Di Padova F, Ulevitch RJ, Han J (1997)

Characterization of the structure and function of the fourth member of p38 mitogen-activated protein kinases, p38δ. *J Biol Chem* 272: 30122–30128

102 Han ZS, Enslen H, Hu X, Meng X, Wu I-H, Barrett T, Davis RJ, Ip YT (1998) A conserved p38 mitogen-activated protein kinase pathway regulates *Drosophila* immunity gene expression. *Mol Cell Biol* 18: 3527–3539

103 Enslen H, Raingeaud J, Davis RJ (1998) Selective activation of p38 mitogen-activated protein (MAP) kinase isoforms by the MAP kinases MKK3 and MKK6. *J Biol Chem* 273: 1741–1748

104 Yang D, Tournier C, Wysk M, Lu HT, Xu J, Davis RJ, Flavell RA (1997) Targeted disruption of the MKK4 gene causes embryonic death, inhibition of c-Jun NH2-terminal kinase activation, and defects in AP-1 transcriptional activity. *Proc Natl Acad Sci USA* 94: 3004–3009

105 Guan Z, Buckman SY, Pentland AP, Templeton DJ, Morrison AR (1998) Induction of cyclooxygenase-2 by activated MEKK1->SEK1/MKK4->p38 mitogen-activated protein kinase pathway. *J Biol Chem* 273: 12901–12908

106 Shirakabe K, Yamaguchi K, Shibuya H, Irie K, Matsuda S, Moriguchi T, Gotoh Y, Matsumoto K, Nishida E (1997) TAK1 mediates the ceramide signaling to stress-activated protein kinase/c-Jun N-terminal kinase. *J Biol Chem* 272: 8141–8144

107 Ichijo H, Nishida E, Irie K, ten Dijke P, Saitoh M, Moriguchi T, Takagi M, Matsumoto K, Miyazono K, Gotoh Y (1997) Induction of apoptosis by ASK1, a mammalian MAPKK that activates SAPK/JNK and p38 signaling pathways. *Science* 275: 90–94

108 Zhang S, Han J, Sells MA, Chernoff J, Knaus UG, Ulevitch RJ, Bokoch GM (1995) Rho family GTPases regulate p38 mitogen-activated protein kinase through the downstream mediator Pak1. *J Biol Chem* 270: 23934–23926

109 Sonenberg N, Gingras A-C (1998) The mRNA 5' cap-binding protein eIF4E and control of cell growth. *Curr Opin Cell Biol* 10: 268–275

110 Lin L-L, Wartmann M, Lin AY, Knopf JL, Seth A, Davis RJ (1993) cPLA2 is phosphorylated and activated by MAP kinase. *Cell* 72: 269–278

111 Treisman R (1996) Regulation of transcription by MAP kinase cascades. *Curr Opin Cell Biol* 8: 205–215

112 Sweet MJ, Hume DA (1996) Endotoxin signal transduction in macrophages. *J Leukocyte Biol* 60: 8–26

113 Wasylyk B, Hagman J, Gutierrez-Hartmann A (1998) Ets transcription factors: nuclear effectors of the Ras-MAP-kinase signaling pathway. *Trends Biochem Sci* 23: 213–216

114 Treisman R (1994) Ternary complex factors: Growth factor regulated transcriptional activators. *Curr Opin Genet Dev* 4: 96–101

115 Ma X, Neurath M, Gri G, Trinchieri G (1997) Identification and characterization of a novel Ets-2-related nuclear complex implicated in the activation of the human interleukin-12 p40 gene promoter. *J Biol Chem* 272: 10389–10395

116 Shackelford R, Adams DO, Johnson SP (1995) IFN-γ and lipopolysaccharide induce DNA binding of transcription factor PU.1 in murine tissue macrophages. *J Immunol* 154: 1374–1382

117 Karin M, Hunter T (1995) Transcriptional control by protein phosphorylation: Signal transmission from the cell surface to the nucleus. *Curr Biol* 5: 747–757

118 Kramer B, Wiegmann K, Kronke M (1995) Regulation of the human TNF promoter by the transcription factor Ets. *J Biol Chem* 270: 6577–6583

119 Groupp ER, Donovan-Peluso M (1996) Lipopolysaccharide induction of THP-1 cells activates binding of c-Jun, Ets, and Egr-1 to the tissue factor promoter. *J Biol Chem* 271: 12423–12430

120 Gupta S, Campbell D, Derijard B, Davis RJ (1995) Transcription factor ATF2 regulation by the JNK signal transduction pathway. *Science* 267: 389–93

121 Reimold AM, Grusby MJ, Kosaras B, Fries JWU, Mori R, Maniwa S, Clauss IM, Collins T, Sidman RL, Glimcher MJ, Glimcher LH (1996) Chondrodysplasia and neurological abnormalities in ATF-2-deficient mice. *Nature* 379: 262–265

122 Newell CL, Deisseroth AB, Lopez-Berestein G (1994) Interaction of nuclear proteins with an AP-1/CRE-like promoter sequence in the human TNF-α gene. *J Leukocyte Biol* 56: 27–35

123 Wedel A, Ziegler-Heitbrock HW (1995) The C/EBP family of transcription factors. *Immunobiol* 193: 171–185

124 Akira S, Isshiki H, Sugita T, Tanabe O, Kinoshita S, Nishio Y, Nakajima T, Hirano T, Kishimoto T (1990) A nuclear factor for IL-6 expression (NF-IL6) is a member of a C/EBP family. *EMBO J* 9: 1897–1906

125 Natsuka S, Akira S, Nishio Y, Hashimoto S, Sugita T, Isshiki H, Kishimoto T (1992) Macrophage differentiation-specific expression of NF-IL6, a transcription factor for interleukin 6. *Blood* 79: 460–466

126 Zhang Y, Rom WN (1993) Regulation of the interleukin-1β (IL-1β) gene by mycobacterial components and lipopolysaccharide is mediated by two nuclear factor-IL6 motifs. *Mol Cell Biol* 13: 3831–7

127 Bretz JD, Williams SC, Baer M, Johnson PF, Schwartz RC (1994) C/EBP related protein 2 confers lipopolysaccharide-inducible expression of interleukin 6 and monocyte chemoattractant protein 1 to a lymphoblastic cell line. *Proc Natl Acad Sci USA* 91: 7306

128 Tanaka T, Akira S, Yoshida K, Umemoto M, Yoneda Y, Shirafuji N, Fujiwara H, Suematsu S, Yoshida N, Kishimoto T (1995) Targeted disruption of the NF-IL6 gene discloses its essential role in bacteria killing and tumor cytotoxicity by macrophages. *Cell* 80: 353–61

129 Hu H-M, Baer M, Williams SC, Johnson PF, Schwartz RC (1998) Redundancy of C/EBPα, -β, and -δ in supporting the lipopolysaccharide-induced transcription of IL-6 and monocyte chemoattractant protein-1. *J Immunol* 160: 2334–2342

130 Zagariya A, Mungre S, Lovis R, Birrer M, Neww S, Thimmapaya B, Pope R (1998) Tumor necrosis factor α gene regulation: enhancement of C/EBPβ-induced activation by c-Jun. *Mol Cell Biol* 18: 2815–2824

131 Nakajima T, Kinoshita S, Sasagawa T, Sasaki K, Naruto M, Kishimoto T, Akira S (1993) Phosphorylation at threonine-235 by a ras-dependent mitogen-activated protein kinase

cascade is essential for transcription factor NF-IL6. *Proc Natl Acad Sci USA* 90: 2207–2211

132 Wang X, Ron D (1996) Stress-induced phosphorylation and activation of the transcription factor CHOP (GADD153) by p38 MAP kinase. *Science* 272: 1347–1349

133 Han J, Jiang Y, Li Z, Kravchenko VV, Ulevitch RJ (1997) Activation of the transcription factor MEF2C by the MAP kinase p38 in inflammation. *Nature* 386: 296–299

134 Cano E, Mahadevan LC (1995) Parallel signal processing among mammalian MAPKs. *Trends Biochem Sci* 20: 117–122

135 Rouse J, Cohen P, Trigon S, Morange M, Alonso-Llamazares A, Zamanillo D, Hunt T, Nebreda AR (1994) A novel kinase cascade triggered by stress and heat shock that stimulates MAPKAP kinase-2 and phosphorylation of the small heat shock proteins. *Cell* 78: 1027–1037

136 McLaughlin MM, Kumar S, McDonnell PC, van Horn S, Lee JC, Livi GP, Young PR (1996) Identification of a mitogen-activated protein (MAP) kinase-activated protein kinase-3, a novel substrate of CSBP p38 MAP kinase. *J Biol Chem* 271: 8488–8492

137 Ludwig S, Engel K, Hoffmeyer A, Sithanandam G, Neufeld B, Palm D, Gaestel M, Rapp UR (1996) 3pK, a novel mitogen-activated protein (MAP) kinase-activated protein kinase is targeted by three MAP kinase pathways. *Mol Cell Biol* 16: 6687–6697

138 Ni H, Wang XS, Diener K, Yao Z (1998) MAPKAPK5, a novel mitogen-activated protein kinase (MAPK)-activated protein kinase, is a substrate of the extracellular-regulated kinase (ERK) and p38 kinase. *Biochem Biophys Res Comm* 243: 492–496

139 Waskiewicz AJ, Flynn A, Proud CG, Cooper JA (1997) Mitogen-activated protein kinases activate the serine/threonine kinases Mnk1 and Mnk2. *EMBO J* 16: 1909–1920

140 Fukunaga R, Hunter T (1997) MNK1, a new MAP kinase-activated protein kinase, isolated by a novel expression screening method for identifying protein kinase substrates. *EMBO J* 16: 1921–1933

141 Stokoe D, Campbell DG, Nakielny S, Hidaka H, Leevers SJ, Marshall C, Cohen P (1992) MAPKAP kinase-2; a novel protein kinase activated by mitogen-activated protein kinase. *EMBO J* 11: 3985–3994

142 Cuenda A, Rouse J, Doza YN, Meier R, Cohen P, Gallagher TF, Young PR, Lee JC (1995) SB 203580 is a specific inhibitor of a MAP kinase homologue which is stimulated by cellular stresses and interleukin-1. *FEBS Lett* 364: 229–233

143 Tan Y, Rouse J, Cariati S, Cohen P, Comb MJ (1996) FGF and stress regulate CREB and ATF-1 *via* a pathway involving p38 MAP kinase and MAPKAP kinase-2. *EMBO J* 15: 4629–4642

144 Haas DW, Shepherd VL, Hagedorn CH (1992) Lipopolysaccharide stimulates phosphorylation of eukaryotic initiation factor-4F in macrophages and tumor necrosis factor participates in this event. *Second Messengers Phosphoproteins* 14: 163–171

145 Han J, Brown T, Beutler B (1990) Endotoxin-responsive sequences control cachectin/tumor necrosis factor biosynthesis at the translational level. *J Exp Med* 171: 465–475

146 Kern JA, Warnock LJ, McCafferty JD (1997) The 3' untranslated region of IL1β regulates protein production. *J Immunol* 158: 1187–1193

147 Pritchett W, Hand A, Sheilds J, Dunnington D (1995) Mechanism of action of bicyclic imidazoles defines a translational regulatory pathway for tumor necrosis factor α. *J Inflamm* 45: 97–105

148 Rincon M, Enslen H, Raingeaud J, Recht M, Zapton T, Su MS-S, Penix LA, Davis RJ, Flavell RA (1998) Interferon-γ expression by Th1 effector T cells mediated by the p38 MAP kinase signaling pathway. *EMBO J* 17: 2817–2829

149 Chen C-Y, Del Gatto-Konczak F, Wu Z, Karin M (1998) Stabilization of interleukin-2 mRNA by the c-Jun NH2-terminal kinase pathway. *Science* 280: 1945–1949

150 Ghosh S, May MJ, Kopp EB (1998) NF-κB and rel proteins: Evolutionarily conserved mediators of immune responses. *Ann Rev Immunol* 16: 225–260

151 May MJ, Ghosh S (1998) Signal transduction through NF-κB. *Immunol Today* 19: 80–88

152 Böhrer H, Qiu F, Zimmermann T, Zhang Y, Jllmer T, Männel D, Böttiger BW, Stern DM, Waldherr R, Saeger HD et al (1997) Role of NFκB in the mortality of sepsis. *J Clin Invest* 100: 972–985

153 Auphan N, DiDonato JA, Rosette C, Helmberg A, Karin M (1995) Immunosuppression by glucocorticoids: Inhibition of NF-κB activity through induction of IκB synthesis. *Science* 270: 286–289

154 Scheinman RI, Cogswell PC, Lofquist AK, Baldwin AS Jr (1995) Role of transcriptional activation of IκBα in mediation of immunosuppression by glucocorticoids. *Science* 270: 283–286

155 Kopp E, Ghosh S (1994) Inhibition of NF-κB by sodium salicylate and aspirin. *Science* 265: 956–959

156 Wang P, Wu P, Siegel MI, Egan RW, Billah MM (1995) Interleukin (IL)-10 inhibits nuclear factor κB (NFκB) activation in human monocytes. *J Biol Chem* 270: 9558–9563

157 Trepicchio WL, Want L, Bozza M, Dorner AJ (1997) IL-11 regulates macrophage effector function through the inhibition of nuclear factor-κB. *J Immunol* 159: 5661–5670

158 Gerondakis S, Grumont R, Rourke I, Grossman M (1998) The regulation and roles of Rel/NF-κB transcription factors during lymphocyte activation. *Curr Opin Immunol* 10: 355–359

159 Sha WC (1998) Regulation of immune responses by NF-κB/Rel transcription factors. *J Exp Med* 187: 143–146

160 Iotsova V, Caamano J, Loy J, Yang Y, Lewin A, Bravo R (1997) Osteopetrosis in mice lacking NF-κB1 and NF-κB2. *Nature Med* 3: 1285–1289

161 Carrasco D, Cheng J, Lewin A, Warr G, Yang H, Rizzon C, Rosas F, Snappe C, Bravo R (1998) Multiple hemopoietic defects and lymphoid hyperplasia in mice lacking the transcriptional activation domain of the c-rel protein. *J Exp Med* 187: 973–984

162 Grigoriadis G, Zhan Y, Grumont RJ, Metcalf D, Handman E, Cheers C, Gerondakis S (1996) The Rel subunit of NF-κB-like transcription factors is a positive and negative reg-

ulator of macrophage gene expression: distinct roles for Rel in different macrophage populations. *EMBO J* 15: 7099–7107

163 Klement JF, Rice NR, Car BD, Abbondanzo SJ, Powers GD, Bhatt PH, Chen CH, Rosen CA, Stewart CL (1996) IκBα deficiency results in a sustained NF-κB response and severe widespread dermatitis in mice. *Mol Cell Biol* 16: 2341–2349

164 Beg AA, Sha WC, Bronson RT, Baltimore D (1995) Constitutive NF-κB activation, enhanced granulopoiesis, and neonatal lethality in IκBα-deficient mice. *Genes Dev* 9: 2736–2746

165 Rice NR, MacKichan ML, Israel A (1992) The precursor of NF-κB p50 has IκB-like functions. *Cell* 71: 243–253

166 Ishikawa H, Claudio E, Dambach D, Raventos-Suarez C, Ryan C, Bravo R (1998) Chronic inflammation and susceptibility to bacterial infections in mice lacking the polypeptide (p)105 precursor (NF-κB1) but expressing p50. *J Exp Med* 187: 985–996

167 Cordle SR, Donald R, Read MA, Hawiger J (1993) Lipopolysaccharide induces phosphorylation of MAD3 and activation of c-rel and related NF-κB proteins in human monocytic THP-1 cells. *J Biol Chem* 268: 11803–11810

168 Whiteside ST, Ernst MK, LeBail O, Laurent-Winter C, Rice N, Israel A (1995) N- and C-terminal sequences control degradation of MAD3/IkBa in response to inducers of NF-κB activity. *Mol Cell Biol* 15: 5339–5345

169 Velasco M, Diaz-Guerra MJM, Martin-Sanz P, Alvarez A, Bosca L (1997) Rapid up-regulation of IκBβ and abrogation of NF-κB activity in peritoneal macrophages stimulated with lipopolysaccharide. *J Biol Chem* 272: 23025–23030

170 Whiteside ST, Epinat J-C, Rice NR, Israel A (1997) I kappa B epsilon, a novel member of the IκB family, controls RelA and cRel NF-κB activity. *EMBO J* 16: 1413–1426

171 Baeuerle PA (1998) Pro-inflammatory signaling: Last pieces in the NF-κB puzzle? *Curr Biol* 8: R19–R22

172 DiDonato JA, Hayakawa M, Rothwarf DM, Zandi E, Karin M (1997) A cytokine-responsive IκB kinase that activates the transcription factor NF-κB. *Nature* 388: 548–54

173 Woronicz JD, Gao X, Cao Z, Rothe M, Goeddel DV (1997) IκB kinase-β: NF-κB activation and complex formation with IκB kinase-α and NIK. *Science* 278: 866–9

174 Régnier CH, Song HY, Gao X, Goeddel DV, Cao Z, Rothe M (1997) Identification and characterization of an IκB kinase. *Cell* 90: 373–383

175 Mercurio F, Zhu H, Murray BW, Shevchenko A, Bennett BL, Li J, Young DB, Barbosa M, Mann M, Manning A et al (1997) IKK-1 and IKK-2: cytokine-activated IκB kinases essential for NF-κB activation [see comments]. *Science* 278: 860–6

176 Zandi E, Rothwarf DM, Delhase M, Hayakawa M, Karin M (1997) The IκB kinase complex (IKK) contains two kinase subunits, IKKα and IKKβ, necessary for IκB phosphorylation and NF-κB activation. *Cell* 91: 243–252

177 Yamaoka S, Courtois G, Bessia C, Whiteside ST, Weil R, Agou F, Kirk HE, Kay RJ, Israel A (1998) Complementation cloning of NEMO, a component of the IκB kinase complex esential for NF-κB activation. *Cell* 93: 1231–1240

178 Malinin NL, Boldin MP, Kovalenko AV, Wallach D (1997) MAP3K-related kinase involved in NF-κB induction by TNF, CD95 and IL-1. *Nature* 385: 540–544

179 Hirano M, Osada S-I, Aoki T, Hirai S-I, Hosaka M, Inoue J-i, Ohono S (1996) MEK kinase is involved in tumor necrosis factor α-induced NF-κB activation and degradation by IκBα. *J Biol Chem* 271: 13234–13238

180 Lee FS, Hagler J, Chen ZJ, Maniatis T (1997) Activation of the IκBα kinase complex by MEKK1, a kinase of the JNK pathway. *Cell* 88: 213–222

181 Liu Z-G, Hsu H, Goeddel DV, Karin M (1996) Dissection of TNF receptor 1 effector functions: JNK activation is not linked to apoptosis while NF-κB activation prevents cell death. *Cell* 87: 565–576

182 Nakano H, Shindo M, Sakon S, Nishinaka S, Mihara M, Yagita H, Okumura K (1998) Differential regulation of IκB kinase α and β by two upstream kinases, NF-κB-inducing kinase and mitogen-activated protein kinase/ERK kinase kinase-1. *Proc Natl Acad Sci USA* 95: 3537–3542

183 Yin M-J, Christerson LB, Yamamoto Y, Kwak Y-T, Xu S, Mercurio F, Barbosa M, Cobb MH, Gaynor RB (1998) HTLV-1 Tax protein binds to MEKK1 to stimulate IκB kinase activity and NF-κB activation. *Cell* 93: 875–884

184 Herrera-Velit P, Reiner NE (1996) Bacterial lipopolysaccharide induces the association and coordinate activation of p53/56lyn and phosphatidylinositol-3-kinase in human monocytes. *J Immunol* 156: 1157–1165

185 Schiff DE, Kline L, Soldau K, Lee JD, Pugin J, Tobias PS, Ulevitch RJ (1997) Phagocytosis of gram-negative bacteria by a unique CD14-dependent mechanism. *J Leukocyte Biol* 62: 786–94

186 Park YC, Lee CH, Kang HS, Chung HT, Kim HD (1997) Wortmannin, a specific inhibitor of phosphatidylinositol-3-kinase, enhances LPS-induced NO production from murine peritoneal macrophages. *Biochem Biophys Res Comm* 240: 692–696

187 Hannun YA, Obeid LM (1997) Ceramide and the eukaryotic stress response. *Biochem Soc Trans* 25: 1171–1175

188 Joseph CK, Wright SD, Bornmann WG, Randolph JT, Kumar ER, Bittman R, Liu J, Kolesnick RN (1994) Bacterial lipopolysaccharide has structural similarity to ceramide and stimulates ceramide-activated protein kinase in myeloid cells. *J Biol Chem* 269: 17606–17610

189 Zhang Y, Yao B, Delikat S, Bayoumy S, Lin X-H, Basu S, McGinley M, Chan-Hui P-Y, Lichenstein H, Kolesnick R (1997) Kinase suppressor of ras is ceramide-activated protein kinase. *Cell* 89: 63–72

190 Barber SA, Perera P-Y, Vogel SN (1995) Defective ceramide response in C3H/HeJ (Lpsd) macrophages. *J Immunol* 155: 2303–2305

191 Wright SD, Kolesnick RN (1995) Does endotoxin stimulate cells by mimicking ceramide? *Immunol Today* 16: 297–302

192 Letari O, Nicosia S, Chiavaroli C, Vacher P, Schlegel W (1991) Activation by bacterial lipopolysaccharide causes changes in the cytosolic free calcium concentration in single peritoneal macrophages. *J Immunol* 147: 980–983

193 Hempel WM, DeFranco AL (1991) Expression of phospholipase C isozymes by murine B lymphocytes. *J Immunol* 146: 3713–3720

194 Natarajan V, Iwamoto GK (1994) Lipopolysaccharide-mediated signal transduction through phospholipase D activation in monocytic cell lines. *Biochim Biophys Acta* 1213: 14–20

195 Daniel-Issakani S, Siegel AM, Strulovici B (1989) Lipopolysaccharide response is linked to the GTP binding protein, Gi2, in the promonocytic cell line U937. *J Biol Chem* 264: 20240–20247

196 Rock FL, Hardiman G, Timans JC, Kastelein RA, Bazan JF (1998) A family of human receptors structurally related to Drosophila Toll. *Proc Natl Acad Sci USA* 95: 588–593

197 Medzhitov R, Preston-Hurlburt P, Janeway CA Jr (1997) A human homologue of the Drosophila Toll protein signals activation of adaptive immunity. *Nature* 388: 394–397

198 Trofimova M, Sprenkle AB, Green M, Sturgill TW, Goebl MG, Harrington MA (1996) Developmental and tissue-specific expression of mouse pelle-like protein kinase. *J Biol Chem* 271: 17609–17612

199 Cao Z, Henzel WJ, Gao X (1996) IRAK: a kinase associated with the interleukin-1 receptor. *Science* 271: 1128–1131

200 Weinstein SL, Finn AJ, Dave S, Meng F, Lowell CA, Sanghera JS, DeFranco AL (2000) Phosphatidylinositol 3-kinase and mTOR mediate lipopolysaccharide-stimulated nitric oxide production in macrophages via interferon-β. *J Leuk Biol* 67: 405–414

201 MacKichan ML, DeFranco AL (1999) Role of ceramide in lipopolysaccharide (LPS)-induced signaling. LPS increases ceramide rather than acting as a structural homolog. *J Biol Chem* 274: 1767–1775

202 Anderson KV (2000) Toll-signaling pathways in the innate immune response. *Curr Opin Immunol* 12: 13–19

Regulation of cytokine production by inhibitors of cell signalling

Rodger A. Allen and Stephen E. Rapecki

Celltech Chiroscience, 216 Bath Road, Slough, Berkshire SL1 4EN, UK

Introduction

Cytokines are soluble messengers that play a pivotal role in regulating immune responses. They operate within a network where their effects are pleiotropic and their function is often redundant. Cytokines typically affect adjacent cells or even the cell of origin and act as paracrine, juxtacrine, autocrine or intracrine mediators. As with any biological system this regulation is in a delicate balance. An organism responds to disturbances in its physiological homeostasis by mounting an acute phase response, which is characterised by dramatic changes in the concentration of some plasma proteins termed acute phase proteins. IL-6 has been identified *in vitro* and *in vivo* as the major hepatocyte stimulating factor. Other pro-inflammatory cytokines, such as IL-1, IL-12 and TNFα, are produced to orchestrate a cellular response to trauma or invasion of the body by pathogenic organisms. The cytokines mediate a wide range of symptoms associated with trauma and infection such as fever, anorexia, tissue wasting, and immunomodulation. Inappropriate or over-production of these same cytokines is associated with mortality and pathology in a wide range of diseases, such as malaria, sepsis, rheumatoid arthritis, inflammatory bowel disease, cancer and AIDS. As well as pro-inflammatory cytokines, there are anti-inflammatory cytokines, including IL-1ra, IL-4, IL-10, IL-11, IL-13 and TGFβ, whose function is to ameliorate the potentially harmful effects of pro-inflammatory mediators, thus restricting the magnitude and duration of the inflammatory response.

The immune system's ability to distinguish between self and non-self relies on the regulation of cellular responses to antigens presented by major histocompatibility complex (MHC) molecules to T cells. These responses include those of T helper cells, which can be divided into two populations. Functionally distinct T helper (Th)-cell subsets, known as Th1 and Th2 cells, are characterised by the patterns of cytokines they produce. IL-12 and IFNγ induce Th1 differentiation *in vitro*. Th1 cells secrete primarily IL-2, IFNγ and TNFα and are involved in the cellular immune

Novel Cytokine Inhibitors, edited by Gerry A. Higgs and Brian Henderson response, delayed type hypersensitivity, induction of the antibody isotype IgG2a, and defence against intracellular pathogens, such as viruses. Imbalance in favour of Th1 cytokines leads to inflammatory diseases such as rheumatoid arthritis, inflammatory bowel disease and multiple sclerosis. IL-4 is the primary cytokine that induces Th2 differentiation. Th2 cells secrete primarily IL-4, IL-5, IL-6, IL-10 and IL-13 and are involved in the antibody immune response, isotype switching to IgG1, IgG2b and IgE, allergic reactions, and defence against extracellular pathogens, such as parasites. Th2 cytokines can be regarded as part of a down-regulatory mechanism for inappropriate Th1 responses. Imbalance in favour of Th2 cytokines leads to diseases such as systemic lupus erythematosus, progressive systemic sclerosis or allergic diseases.

This central role for cytokines in regulating the immune response, and their involvement in pathological processes when in imbalance, makes their therapeutic manipulation attractive as targets in immune and inflammatory diseases. Cytokines are produced by a wide range of cells including granulocytes, lymphocytes, monocytes/macrophages, fibroblasts, endothelial and epithelial cells, and there are numerous approaches that can be taken to modulate the production of any particular cytokine. This review will focus primarily on two cytokines, IL-2 and TNFα, produced by two cell types, the T cell and the monocyte. The review will give an overview of the major approaches being taken to inhibit cytokine production or function, again focusing on two classes of inhibitors, PDE4 (phosphodiesterase 4) inhibitors and inhibitors of p56^{Lck}, a member of the Src-kinase family.

T-cell receptor (TCR) signalling

The T cell antigen receptor is a complex multi-subunit structure composed of a ligand-binding heterodimer (αβ or γδ) that recognises peptides bound to major histocompatibility complex molecules [1], and the non-polymorphic CD3ε, CD3γ, CD3δ and TCRζ chains [2]. These non-polymorphic chains are required for receptor assembly, cell surface expression and signalling [3]. Located in the cytoplasmic domains of these CD3 and TCR chains are immunoreceptor tyrosine-based activation motifs (ITAMs) [4, 5], three of which are in each of the TCRζ chains, and one in each of the CD3 chains [6]. Multiple ITAMs may provide the capacity to amplify the signal generated by the TCR, as it has been demonstrated that the signal strength generated by recombinant signalling proteins is dependant upon the number of ITAMs present [7]. The ITAMs may also play a role in determining the specificity of signalling, as it has been reported that they can differentially bind Zap-70, phospholipase Cγ1, phosphatidylinositol 3-kinase and Shc [8–11].

Members of at least four families of tyrosine kinases are involved in the transduction of the signal generated from the TCR. Src-kinase family members, p56^{Lck}

(Lck) and p59^{Fyn} (Fyn), play a role in TCR signalling [5, 12, 13]. The central importance of Lck in TCR signalling was demonstrated by a mutant of the Jurkat cell line, termed JCaM1, which failed to give a calcium flux upon TCR stimulation. Analysis of this mutant, which was also defective in the induction of tyrosine phosphorylation, revealed that the expression of functional Lck tyrosine kinase was lacking [14]. Expression of wt Lck in JCaM1 restored the ability of the cell to respond to TCR stimulation [14]. Lck knockout mice demonstrated a reduced proliferative response in T cells to both CD3 and TCRαβ crosslinking [15], and inhibition of TCR-mediated signal transduction was demonstrated by microinjection of anti-Lck antibodies into T cells [16]. Immunoprecipitation studies demonstrated that Fyn associates with the TCR directly [17], and co-expression studies revealed an interaction with the cytoplasmic domains of CD3ε and γ chains as well as TCRζ and η chains [18]. A function for Fyn in TCR signalling was suggested by a rapid, transient two- to four-fold increase in its kinase activity following TCR stimulation [19]; overexpression of Fyn in T cells augments responses to TCR stimulation, including tyrosine phosphorylation and IL-2 production [20, 21].

The Syk family of protein tyrosine kinases includes Zap-70 and Syk. The essential role of Zap-70 in TCR signalling and T-cell development has been established by studies in human patients with a rare immunodeficiency syndrome resulting from mutations in Zap-70, and in mice deficient in Zap-70 expression [22–26]. In transfection studies with Zap-70, it was shown that its association with TCRζ, and its activation, was dependent on the presence of a Src family kinase, either Fyn or Lck [27]. This was confirmed with the Lck-deficient T-cell line JCaM1 where Zap-70 is neither recruited to the TCR nor tyrosine-phosphorylated [28]. Syk, like Src family kinases, augments TCR-triggered tyrosine phosphorylation of CD3/ζ, and, unlike Zap-70, can stimulate tyrosine phosphorylation of a ζ-bearing chimera in transiently transfected Cos-1 cells [29]. These results suggest that Syk may have a unique role in T cells as a consequence of its ability to efficiently phosphorylate multiple components of the TCR signalling cascade. The signalling defects (including protein tyrosine phosphorylation, calcium flux and IL-2 transcription) in the P116 cell line, a Syk- and Zap-70-negative somatic mutant derived from the Jurkat T-cell line, were fully reversible by re-introduction of catalytically active Syk or Zap-70; however, in contrast to Zap-70, Syk expression triggered a significant degree of cellular activation in the absence of TCR ligation [30]. The two Syk family members expressed in T cells are structurally similar but differ in terms of biochemical activity and regulation and have distinct characteristics in T cell signalling [31].

Csk, a member of another protein tyrosine kinase family, negatively regulates Src family tyrosine kinases and thus T cell signalling [5]. Csk phosphorylates the carboxy-terminal tyrosine in Lck and Fyn, maintaining these proteins in an inactive state. Dephosphorylation is mediated by the phosphatase CD45, allowing Lck and Fyn to be activated by TCR stimulation [5]. Another class of tyrosine kinase, the Tec family, is also involved in TCR signalling [32–34]. The Tec-family kinases preferentially expressed in T cells are Itk and Tec. Itk appears to be downstream of Lck in the T cell signalling cascade, and it is recruited to different signalling complexes following ligation of CD28 or the TCR complex [35]. From double knockouts in mice it was concluded that Itk and Fyn make independent contributions to TCR-induced T cell activation [36], and that Itk functions downstream of, or in parallel to, Zap-70 to facilitate IP_3 production [37]. However, the defect in T cell signalling in Itk-deficient mice is very modest [38]. Tec is activated following TCR/CD3 or CD28 ligation and interacts with CD28 receptors in an activation-dependent manner. Tec can phosphorylate p62(Dok), a CD28 substrate, whereas Itk cannot, giving the two Tec-family kinases different roles in T cell activation [38]. The details of the signalling cascade are still evolving and have been the subject of numerous reviews over the last few years [39–43].

Characterisation of pathways downstream of the protein tyrosine kinases has been dependent on the identification of kinase substrates. PLCγ1 was the first tyrosine kinase substrate, other than the TCR subunits, to be identified in T cells [44]. PLCγ1 is activated by phosphorylation [5] leading to phosphatidylinositol metabolism. Calcium mobilisation results from generation of IP_3 and the calcium channel IP_3 receptor is also regulated by Fyn-mediated tyrosine phosphorylation [45]. Tyrosine phosphorylation of Slp-76 (SH2 domain leukocyte protein), a substrate of Zap-70 [46], mediates its association with Vav [47-50]. TCR stimulation induces tyrosine-phosphorylated proteins, including Slp-76, Cbl and Shc, to associate with the Grb2-Sos complex [51-56] which is involved in Ras regulation [57, 58]. The importance of the Ras/mitogen-activated protein kinase (MAPK) pathway in TCR signalling has been demonstrated by several studies in T cell anergy [59, 60]. These cells have a block in the activation of the Ras/MAPK pathway and fail to produce IL-2. It has been shown recently that the Lck SH3 domain is required for activation of the MAPK pathway [61].

T-cell cytokine expression is largely inducible and initiated by cross-linking of the T-cell antigen receptor and/or other accessory cell-surface molecules followed by activation of the major signalling pathways and transcription factors. These include a calcium-dependent and cyclosporin-sensitive pathway that regulates the nuclear factor of activated T cells (NFAT), a protein kinase C-dependent pathway that regulates transcription factor activator protein-1 (AP-1) and nuclear factor kappa B (NF-κB), and a p21ras-activated MAPK cascade that affects AP-1 and other transcription factors [53].

Lipopolysaccharide (LPS) signalling

The signalling pathways utilised by LPS to produce cytokines will only be covered briefly here because this topic is covered in detail in the previous chapter.

The response of cells to LPS is dependent on the expression of CD14. This is based on anti-CD14 antibodies being capable of blocking cytokine production both *in vitro* [62] and *in vivo* [63], transfection of CD14 conferring LPS-responsiveness on CD14-negative cells [64, 65] and overexpression of a human CD14 transgene conferring hypersensitivity to effects of LPS in mouse models of septic shock [66]. Because CD14, a 50–55 kDa GPI-linked glycoprotein [67], lacks a cytoplasmic domain, it is not thought to initiate signalling in response to LPS on its own, but *via* a hypothesised co-receptor. Binding of LPS to CD14 is facilitated by LPS-binding protein (LBP) [68]; in fact, LBP may be essential for CD14-dependent signalling [69, 70]. On the other hand, bacterial/permeability-increasing factor (BPI), secreted by polymorphonuclear leukocytes [71], has a higher affinity for LPS than LBP and can act as an antagonist of LPS binding to, and signalling via, CD14 [71]. In addition to the CD14-mediated effects, it is believed that high concentrations of LPS can utilise CD14-independent pathways. It has been demonstrated that TNFα and IL-1 can be produced by high concentrations of LPS in monocytes and macrophages from CD14$^{-/-}$ mice [72, 73], and even low doses of LPS elicit wild-type levels of acute phase proteins in these mice [74].

One of the earliest CD14-dependent signalling responses to LPS in macrophages is protein tyrosine phosphorylation [75, 76]. Inhibitor studies demonstrate an involvement of tyrosine kinase activity in the LPS-stimulated release of arachadonic acid metabolites [75], activation of MAPKs [77–79] and production of pro-inflammatory cytokines [78, 80–82]. Src-family tyrosine kinases were thought to play a role in LPS signalling as Hck, Lyn and Fgr were coimmunoprecipitated with CD14 or were activated by LPS stimulation [81–83]. However, macrophages from mice lacking all three of these Src kinases respond normally to LPS, indicating that they are not essential for LPS signalling [84]. Similarly, Syk becomes tyrosine-phosphorylated in response to LPS activation [85], but macrophages derived from the foetal liver of Syk$^{-/-}$ and wild-type mice give similar responses to LPS [86].

One group of kinases that is activated by LPS is the MAPKs. The three MAPK families, Erk, Jnk and p38, each have multiple members encoded by separate genes, and are activated by dual-specificity tyrosine/threonine kinases known as MEKs, which are themselves activated by serine/threonine phosphorylation by upstream MEKKs. LPS rapidly activates both Erk1 and Erk2 [77, 79, 87] and this activation is sensitive to tyrosine kinase inhibitors [77]. Similarly, LPS rapidly activates Jnk in bone marrow-derived macrophages and macrophage cell lines in a CD14- and tyrosine kinase-dependent manner [88, 89]. The p38 family of MAPKs was identified on the basis of its LPS-induced tyrosine phosphorylation in a pre-B-cell line transfected with CD14 [90]. p38 plays an important role in the LPS-induced expression of TNFα, IL-1, IL-6 and IL-10 [91–93].

Following activation, MAPKs are translocated to the nucleus, where they phosphorylate multiple transcription factors and other serine/threonine kinases that mediate their effects on gene expression and function [94].

Transcription factors

The signalling pathways and target genes that become activated upon cell surface stimulation are currently being delineated, but AP-1 is proving to be an important regulator of nuclear gene expression in leukocytes. AP-1 is comprised of multiple protein complexes that form between the protein products of the proto-oncogenes c-fos and c-jun, and their related gene family members. Although they are often among the first genes to be transcribed after stimulation of the cell, fos and jun genes are usually only transiently expressed, and do not require *de novo* protein synthesis. AP-1 function is regulated primarily by phosphorylation [95–97]. Activation of naïve T cells requires two signalling events [98]. One signal is delivered after ligation of the TCR by an antigen presented in association with MHC molecules on an antigen-presenting cell. The second, or co-stimulatory, signal is then required to stimulate T cells to produce IL-2 and other cytokines and to proliferate [99]. Unlike the TCR pathway, the main co-stimulatory signal from CD28 involves the phosphatidylinositol-3-kinase and acid sphingomyelinase pathways [100-102]. These two signalling pathways, of the TCR and CD28, have been found to converge at the level of Jnk activation [103]. Co-stimulation is required for enhanced production of IL-2 and other cytokines, such as TNFα, and augments AP-1 transcriptional activity in T cells both *in vitro* [104, 105] and *in vivo* [106]. The IL-2 gene was one of the first cellular genes shown to have an AP-1 site within its promoter [107]. Subsequently, AP-1 has been shown to play an integral role in the regulation of the IL-2 gene, either alone or in combination with other transcription factors, such as NFAT and the octamer proteins (Oct-1 and Oct-2) [108–110].

The transcription factor NFATp (nuclear factor of activated T cells, pre-existing) is likely to regulate the cyclosporin-sensitive transcription of cytokine genes during the immune response. NFATp is a cytosolic protein that translocates, under the regulation of calcium and calcineurin, to the nucleus following activation. This translocation is inhibited by cyclosporin A (CsA) and FK506. The targets of both drugs are distinct classes of intracellular receptors (immunophilins); CsA binds the cyclophilins and FK506 binds FK-binding proteins [111]. The phosphatase activity of the calcium- and calmodulin-dependent serine/threonine phosphatase calcineurin is inhibited by CsA-cyclophilin and FK506-FKBP12 complexes [112].

NF-κB is a protein transcription factor, first identified by Sen and Baltimore [113], that functions to enhance the transcription of a variety of genes, including cytokines such as TNFα, IL-1 and IL-6 [114, 115]. Activation and regulation of NF-κB is tightly controlled by a group of inhibitory proteins (IκB) that sequester NF-κB in the cytoplasm of immune/inflammatory effector cells. NF-κB activation involves phosphorylation, ubiquitination and proteolysis of IκB, liberating NF-κB to migrate to the nucleus where it binds to specific promotor sites and activates gene transcription [116–118].

Many cellular functions are regulated by the cAMP-dependent pathway. Transcriptional regulation, upon stimulation of this pathway, is mediated by a family of cAMP-responsive nuclear factors [119]. These factors contain the basic domain/leucine zipper motifs and bind as dimers to cAMP-response elements (CRE). The function of CRE-binding proteins (CREBs) is modulated by phosphorylation by several kinases, including the cAMP-dependent PKA [120] and p90[rsk] [121]. Cross-talk between the mitogenic signalling pathways and cAMP-responsive transcription has been established [122], which supports the concept of converging signalling within the PKA and PKC pathways in the cytoplasm [123–125] and in the nucleus [126].

Tyrosine kinase inhibitors

CsA has been a revolution in treating autoimmune diseases, in particular in transplantation. CsA, FK506 and the structurally-related rapamycin all bind to the immunophilins which are peptidyl-prolyl cis-trans isomerases catalysing the interconversion of peptidyl-prolyl imide bonds in peptide and protein substrates [127]. CsA binds cyclophilin, while FK506 and rapamycin bind FK506-binding protein or $FKBP_{12}$ [127, 128]; however, the prolyl isomerase activity of these immunophilins is not essential for their immunosuppressive effects. The CsA-cyclophilin and FK506-$FKBP_{12}$ complexes bind to and inhibit calcineurin, a calcium/calmodulin-dependent protein phosphatase [129–132]. The rapamycin-$FKBP_{12}$ complex does not bind calcineurin [130, 131] but acts at a later point in the signalling cascade by binding the mammalian target of rapamycin (mTOR) [128, 133]. All of these inhibitor-immunophilin complexes are potent inhibitors of T-cell proliferation; CsA and FK506 inhibiting IL-2 transcription [134], and rapamycin inhibiting IL-2 receptor expression [134] and signalling [128, 134]. While these immunosuppressive drugs are widely used, they are associated with side-effects, primarily nephrotoxicity [135, 136], which is driving the search for drugs with similar potencies but without the unwanted side-effects. One area of active research is the development of tyrosine kinase inhibitors.

An important first consideration is whether it is possible to selectively inhibit individual tyrosine kinases with similar catalytic domains [137, 138]. The evidence is that it is possible. The most extensively studied protein tyrosine kinase (PTK) inhibitors include the tyrphostins [139], the isoflavin genistein [140] and herbimycin A [141]. Results from studies with PTK inhibitors confirm previous reports that tyrosine kinases play a crucial role in transduction of TCR signalling [142-144]. Herbimycin A has been shown to markedly inhibit both the resting and stimulated levels of phosphotyrosine-containing proteins, including PLCγ1 and the TCRζ chain, and prevent activation of PLC by anti-TCR monoclonal antibodies [145].

Genistein had a much less pronounced effect than herbimycin A on the appearance of tyrosine phosphoproteins, and was less specific [145]. A reporter construct, NF-AT/CAT permitted study of the effects of PTK inhibitors on a subset of signals involved in T-cell activation which are normally insufficient to activate IL-2 gene expression [146]. This study showed that genistein and tyrphostin inhibited the activation of the reporter whether stimulation was *via* anti-CD3 antibodies, PHA or phorbol ester and calcium ionophore, which bypass many of the early events induced by TCR engagement [146]. Furthermore, genistein specifically enhanced TCR signalling at low concentrations, leading to the conclusion that there is a genistein-sensitive PTK that is involved in down-regulation of TCR signalling [146]. The value of many of these early studies was limited because the specificity of many of these inhibitors were not known, and of course specificity must be viewed in the context of the continual discovery of new tyrosine kinases [147].

More specific kinase inhibitors are currently under active investigation; these include inhibitors of MAPKs. These are a family of kinases that are arranged in three linear pathways (Erk/MAPK, Jnk/SAPK and p38) that are differentially activated by mitogenic stimuli (including growth factors and phorbol esters), cytokines (such as TNFα and IL-1) and stresses (such as heat shock, osmotic shock, ultraviolet light, or LPS). Each pathway has three well-established levels of kinase in the central portion of their signalling cascade, generically termed MAPKKK (or MEKK), MAPKK (or MEK) and MAPK. These are all dual-specificity kinases, the MEKs being tyrosine/threonine kinases, while the MEKKs and MAPKs are serine/threonine kinases. Inhibitors of p38 kinase are being developed as new anti-inflammatory agents for diseases such as rheumatoid arthritis and asthma [148]. They are potent inhibitors of TNFα and IL-1 release from LPS-stimulated monocytes [149] and demonstrate inhibition, albeit weaker, of CD28-dependent IL-2 production and IL-2-dependent T-cell proliferation [150]. Inhibitors of p38 are covered in detail in another chapter of this book.

Another active area of research is the inhibition of Src-family tyrosine kinases, p56[Lck] and p59[Fyn], and Syk-family tyrosine kinase Zap-70. It was felt that specific inhibition of these targets would selectively affect TCR signalling in an adult population, potentially leading to the development of therapeutic agents with utility in transplant rejection, autoimmunity and allergic diseases [147].

WIN 61651, an ATP-competitive p56[Lck] tyrosine kinase inhibitor developed by Sterling Winthrop [151], inhibited both the autophosphorylation of p56[Lck] (IC_{50} 24 μM) and the phosphorylation of an exogenous peptide (IC_{50} 18 μM). It demonstrated specificity for p56[Lck] over serine/threonine kinases, such as PKC and PKA, and over some other kinases, including erbB2, EGF receptor kinase and insulin receptor kinase, but was equipotent for inhibition of P56[Lck] and PDGF receptor kinase. WIN 61651 inhibited the tyrosine phosphorylation of cellular proteins when JB2.7 cells were stimulated with antibodies to CD3 and CD4. The lowest IC_{50} values (47 μM) were observed for proteins with molecular weights of 57 kDa and

63 kDa. The inhibitor demonstrated dose-dependent inhibition of IL-2 production of cells activated by either anti-CD3/phorbol myristate acetate (PMA) (13 µM) or anti-CD3/anti-CD4/anti-CD28 (9.8 µM), in the absence of significant cellular toxicity as measured by trypan blue exclusion. It was also demonstrated that WIN 61651 inhibited antigen-driven T-cell activation in the mixed lymphocyte reaction (MLR) (IC$_{50}$ 6.5 µM), and that exogenous IL-2 was able to fully reconstitute cellular proliferation. However, it was noted that the IC$_{50}$ values in the functional T-cell activation assays were considerably lower than those observed in whole cell phosphorylation assays and even somewhat lower than those observed in cell-free p56^{Lck} assays. A number of possible explanations were put forward, the most likely of which is that WIN 61561 was not completely specific for p56^{Lck}.

A group from Abbott has recently described an isothiazolone compound (A-125800) which is a slightly more potent p56^{Lck} inhibitor than WIN 61651 [152]. It inhibits p56^{Lck} activity with an IC$_{50}$ of 1–7 µM, is equipotent in blocking ZAP-70 activity, but is 50–100 times less potent against the catalytic activities of p38 and Jnk MAPKs. A-125800 blocked activation-dependent TCR tyrosine phosphorylation and intracellular calcium mobilisation in Jurkat T cells (IC$_{50}$ 35 µM) and blocked T-cell proliferation in response to alloantigen (IC$_{50}$ 14 µM) and anti-CD3/CD28-induced IL-2 secretion (IC$_{50}$ 2.2 µM) in primary T-cell cultures. Inhibition of p56^{Lck} was dose- and time-dependent and was irreversible, probably due to the –SH groups within the p56^{Lck} catalytic domain reacting with the isothiazolone ring, leading to ring opening and disulphide bond formation.

The most potent and selective inhibitors described in the literature to date are the pyrazolopyrimidines developed by Pfizer [153]. These inhibitors (PP1 and PP2) demonstrate specificity for the Src-family kinases (inhibiting p56^{Lck} and p59^{Fyn} (IC$_{50}$ 4–6 nM), p60^{Src} (PP1 170 nM) and p59^{Hck} (PP1 20 nM and PP2 5 nM)), over the epidermal growth factor (EGF) receptor kinase (250–480 nM), and are essentially inactive against Jak2 (> 50 µM), Zap-70 (> 100 µM) and PKA. PP1 significantly inhibited tyrosine phosphorylation of a number of proteins in human T cells stimulated with anti-CD3, an IC$_{50}$ of 0.5 µM was determined by quantifying the phosphorylation of a band of 70 kDa. PP1 and PP2 inhibited proliferation of human peripheral blood lymphocytes (PBLs) stimulated with anti-CD3 (IC$_{50}$ 0.5–0.6 µM), influenza virus vaccine (4–5.2 µM) and in the MLR (1.9–3.9 µM). Consistent with their point of action, PP1 and PP2 were less effective inhibitors of PMA/IL-2-dependent proliferation (IC$_{50}$ 18–26 µM), a stimulus that bypasses the T cell receptor complex. PP1 (1 µM) was also able to almost completely inhibit IL-2 mRNA induction by PHA/PMA, while slightly increasing IL-2 receptor mRNA. This result was confirmed by a reporter gene assay carried out in a human Jurkat T cell line, stimulated in the same way. PP1 inhibited the IL-2 reporter with an IC$_{50}$ of 1 µM, whilst having no effect on the IL-2 receptor reporter at 35 µM. These inhibitors have been reported to favour Th2 differentiation, by enhancing IL-4 and inhibiting IFNγ in cultures of splenocytes from ovalbumin-specific TCR-transgenic mice [154].

Table 1 - Kinase inhibition profile of p56^{Lck} inhibitors (IC$_{50}$ values in nM)

	Lck	Zap-70	PKC	EGFr	Csk	cdc2	Fyn	Lyn
PP2	171	>10000	8900	700	281	ND	ND	51
CT5215	3.5	>10000	1531	606	330	9036	61	21
CT5264	3.9	9209	1270	2938	126	>10000	24	15
CT5269	1.1	>10000	>10000	3155	86	8463	13	13

Table 2 - Effect of inhibitors on IL-2, IFNγ production and CD25 expression in OKT3-stimulated PBMC (IC$_{50}$ values in nM)

	IL-2	IFNγ	CD25
CsA	10	7	6
Staurosporine	2	ND	3
PP2	263	200	700
CT5264	100	400	500
CT5269	35	200	ND

We, at Celltech, have developed p56^{Lck} inhibitors that demonstrate selectivity over all the non-Src-family kinases examined (see Tab. 1). Comparison with the Src-family kinase inhibitor PP2 shows that, under the assay conditions employed, our compounds are more potent and selective than PP2. The compounds are nanomolar inhibitors of p56^{Lck} that demonstrate up to 10,000-fold selectivity over Zap-70, PKC and cdc2, 3,000-fold selectivity over EGFr and an 80-fold selectivity over Csk. Particularly important is the selectivity shown over inhibition of Csk, since this kinase is a negative regulator of Src-family kinases [5]. As expected from the very high sequence homology in the Src family there was incomplete specificity; however, we did develop compounds showing ~ 80-fold selectivity for p56Lck over the other Src-family kinases examined (data not shown).

The ability of a range of inhibitors, including CsA, staurosporine, PP2 and Celltech inhibitors, to inhibit the production of IL-2 and IFNγ from OKT3-stimulated peripheral blood mononuclear cells (PBMC) was tested (see Tab. 2). CsA and staurosporine were very potent inhibitors of both IL-2 and IFNγ production, and PP2, while equipotent against both cytokines, was very much weaker. The Celltech inhibitors were more potent inhibitors of IL-2 than IFNγ. These inhibitors were also capable of inhibiting the expression of CD25 (see Tab. 2) and proliferation in

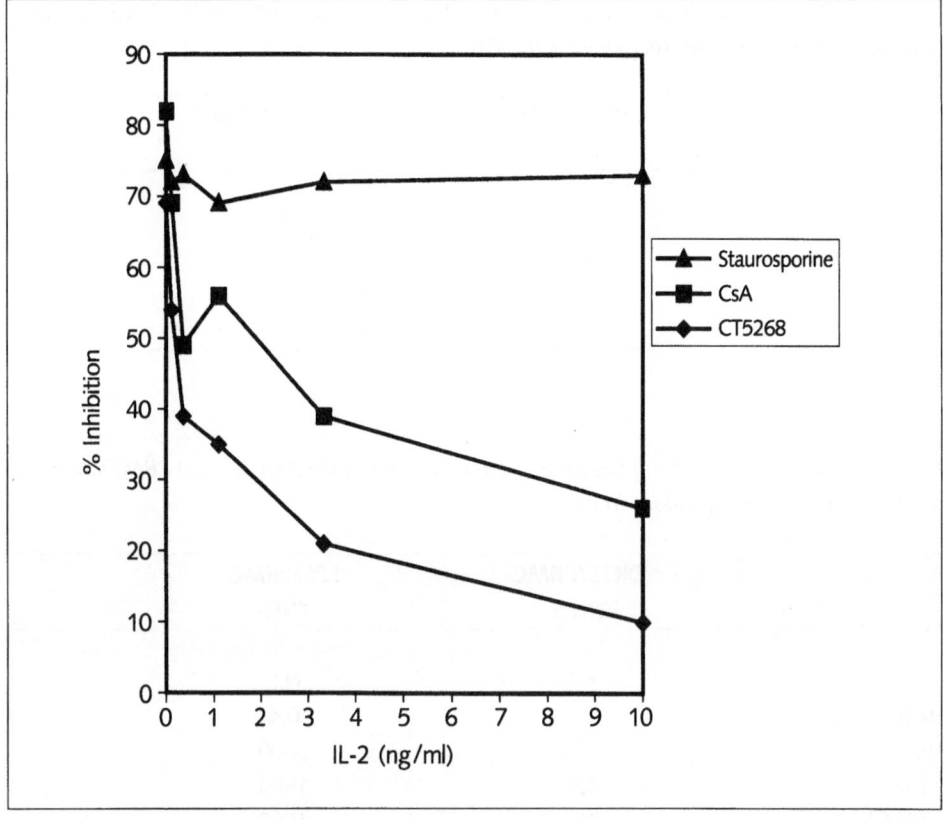

Figure 1
Reconstitution of the OKT3-stimulated proliferation of PBMC, by addition of IL-2, to over-
come the effect of various inhibitors.
The inhibitors were used at a concentration giving ~75% inhibition of proliferation (50 nM
staurosporine, 25 nM CsA and 500 nM CT5269).

OKT3-stimulated PBMC. The ability of exogenously added IL-2 to reconstitute the proliferative response was utilised to assess the specificity of inhibition of proliferation, and to ascertain whether the inhibition was up- or downstream of the IL-2 receptor. The results are shown in Figure 1. Reconstitution of the proliferative response by addition of IL-2 was achieved with CT5269 and CsA, but not with staurosporine as would be expected with a non-selective kinase inhibitor. This demonstrates that the inhibitory block on proliferation is upstream of the IL-2 receptor in the case of CT5269 and CsA. This raises an interesting point as there are a number of reports that the IL-2 receptor interacts both physically and functionally with p56[Lck] [155–159], but CT5269, a potent inhibitor of p56[Lck], does not block

Table 3 - Effect of inhibitors on IL-2 production via the T-cell receptor (OKT3) or PKC and calcium (PMA+ionomycin) (IC_{50} values in nM).

	OKT3/PBMC	PMA + ION/PBMC
	IL-2	IL-2
CsA	10	8
Staurosporine	2	3
PP2	263	3800
CT5215	78	2250

Table 4 - Effect of inhibitors on the production of TNFα in OKT3-stimulated PBMCs or LPS-stimulated PBMCs (IC_{50} values in nM).

	OKT3/PBMC	LPS/PBMC
	TNFα	TNFα
CsA	4.5	N.I.
Staurosporine	2.3	0.4
PP2	450	3000
CT5215	425	1800
CT5264	400	1500

the IL-2 signal leading to proliferation. Redundancy in the kinase signalling pathways may be an explanation, as two of these reports state that there are three protein tyrosine kinases, in addition to p56^{Lck}, that are physically associated with the IL-2 receptor, including Syk, Jak1 and Jak3 [157, 159]. In addition there is a recent report that states that p56^{Lck} is not required for IL-2 receptor-attenuation upon CD4-ligation [160]. To further define the point of action of the inhibitors, the inhibition of IL-2 production by two stimuli, one *via* the T-cell receptor (OKT3) and one acting at a point below the receptor (PMA/ionomycin), was compared. The results are shown in Table 3. Both CsA and staurosporine inhibited the IL-2 production equally whether the stimulus was *via* the T-cell receptor or bypassing it. This is because CsA acts at the level of calcineurin, which is at or below the two stimuli, and staurosporine is a non-specific kinase inhibitor exerting its effects at numerous points on the signalling cascade. The two p56^{Lck} inhibitors, CT5215 and PP2, demonstrated selective inhibition of the T-cell receptor signal over the PMA+iono-

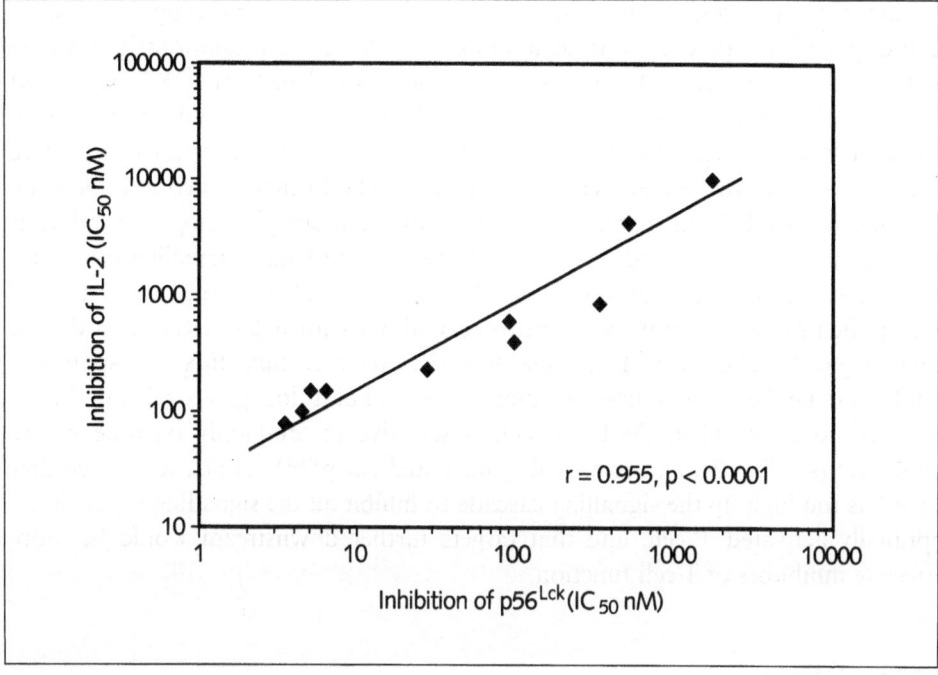

Figure 2
Correlation of inhibition of IL-2 production with inhibition of p56^{Lck}.

mycin signal, by ~30-fold and ~15-fold, respectively. The relationship of the inhibition of p56^{Lck} and inhibition of IL-2 production is shown by the highly significant correlation ($r = 0.955$, $p < 0.0001$) obtained when these two parameters are plotted for a series of structurally related p56^{Lck} inhibitors, as shown in Figure 2.

Comparing the inhibition of TNFα production from either OKT3-stimulated PBMCs or LPS-stimulated PBMCs further assessed the selectivity of the inhibitors. The results are shown in Table 4. CsA was a potent inhibitor of TNFα production in OKT3-stimulated PBMCs whilst it did not inhibit the LPS-stimulated TNFα production. Staurosporine was a potent inhibitor of TNFα production from both cell types, while the three p56^{Lck} inhibitors demonstrated cellular selectivity, inhibiting T cell- over monocyte-derived TNFα production more potently, by a factor of 4 to 6-fold.

The inhibition of p56^{Lck}, by a series of 4-phenylthiophenylaminopyrimidines, correlates with inhibition of IL-2 production (OKT3/PBMC), calcium flux (OKT3/E6.1), TNFα production (OKT3/PBMC), OKT3-mediated PBMC proliferation and the MLR (PBMC), but not with TNFα production (LPS/PBMC) or JY cell proliferation (a B-cell line). This leads to the conclusion that p56^{Lck} plays a central role in the signalling pathway(s) utilised for calcium flux, cytokine production and proliferation. However, despite inhibition of these cell responses correlating with inhibition of p56^{Lck}, the potency of these inhibitory effects reduces significantly from the isolated enzyme (2–10 nM), to IL-2 production (~ 100 nM), to OKT3-mediated PBMC proliferation (0.4–0.7 μM) to the MLR (0.75–2 μM). The drop in potency going from isolated enzyme to a cell system is to be expected due to the competitive effects of a high intracellular ATP concentration. The further reduction in potency in going from OKT3-mediated proliferation to the more physiologically relevant MLR may be due to redundancy amongst the tyrosine kinase signalling molecules that are triggered in the more complex MLR system. This redundancy is further exemplified by calcium flux experiments carried out with JCaM-1 cells that do not contain p56^{Lck}. The JCaM-1 cell line does not give a calcium flux in response to OKT3 (unlike the parental line containing p56^{Lck}), but it does give a calcium flux in response to cross-linked OKT3, which is sensitive to Src-family tyrosine kinase inhibitors (see Fig. 3), and is probably mediated *via* p59^{Fyn} [161]. It may be that p56^{Lck} is too high up the signalling cascade to inhibit all the signalling inputs in an optimally activated T cell, and that targets further downstream would be more effective inhibitors of T-cell function.

PDE4 inhibitors

Phosphodiesterases (PDEs) control the level of cyclic nucleotides in the cell by degrading them to inactive metabolites [162, 163], hence playing an essential role in terminating signal transduction processes mediated by the second messengers cAMP and cGMP. Nine PDE families have been described, encoded by distinct genes [162, 164–166], and differing in their substrate preference, sensitivity to endogenous activators and inhibitors, and in their tissue distribution [162–166]. Attention has recently focused on the therapeutic potential of inhibitors of the low K_m cAMP-specific PDE (PDE4) for the treatment of immune and inflammatory diseases. This is largely due to the cellular distribution of the PDE4 enzyme, being expressed in mast cells, monocytes, macrophages, eosinophils, neutrophils and T lymphocytes.

If we focus initially on the effects of PDE inhibitors in T-cell function, it is clear that both antigen- and mitogen-mediated T-cell proliferation is inhibited by non-specific PDE inhibitors, such as isobutylmethylxanthine (IBMX), pentoxyfylline, theophylline and enprofylline [167–169] and PDE4 inhibitors, such as rolipram, RP73401 and Ro20-1724 [168–174]. T-cell proliferation is unaffected by PDE3 inhibitors, such as siguazodan, CI-930 and SK&F 95654 [168, 170, 172–175] and the PDE5 inhibitor zaprinast [168, 176]. However, there is a synergistic effect with PDE3 and PDE4 inhibitors in combination [170, 172, 173]. The situation with respect to IL-2 is very similar, in that non-specific inhibitors of PDE inhibit IL-2 release from both antigen- and mitogen-stimulated T cells and PBMC [167, 175]. PDE4 inhibitors have the same effect in both murine splenocytes [173] and human

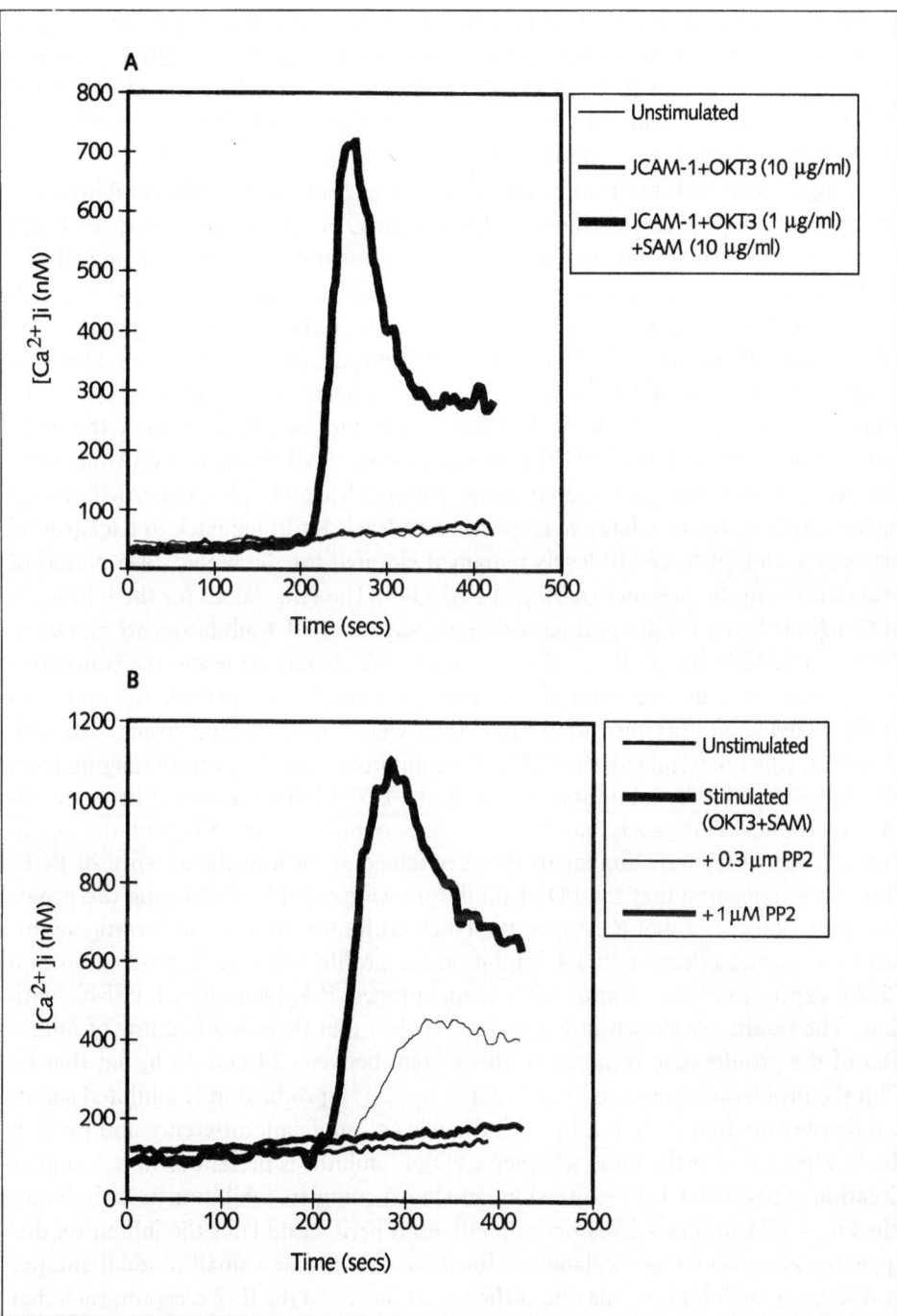

Figure 3
JCaM-1 cells give calcium flux in response to crosslinked OKT3: Sensitivity to PP2.

T cells and PBMC [170, 175], although one study carried out in Jurkat cells concluded that there was no effect on the IL-2 mRNA levels [175]. PDE3 inhibitors alone have no effect on IL-2 [170, 173, 175], while the combination of PDE3 and PDE4 inhibitors is claimed to be synergistic in CD4+ and CD8+ T lymphocytes [170], but no synergy was demonstrated in murine splenocytes [173].

In agreement with the results above, we have shown that PDE4 inhibitors can inhibit both antigen- and mitogen-mediated PBMC proliferation. A series of PDE4 inhibitors inhibited tetanus toxoid- and house dust mite antigen-mediated PBMC proliferation equivalently, and the IC_{50} values correlated with binding at the high affinity rolipram-binding site (r = 0.915, p < 0.001) rather than binding at the low-affinity catalytic site (r = -0.019) of the PDE4 enzyme (data not shown). This is in contrast to inhibition of SAg-stimulated murine splenocyte proliferation, where the inhibition correlated with the low-affinity conformer of PDE4 and not the high-affinity rolipram binding site [173]. In monitoring cAMP levels in ConA-mediated proliferation of PBMC, it was found that 100 nM RS25344 elevated cAMP during the first hour of the proliferative response with levels returning back to background between 1 and 24 h. cAMP levels remained elevated for the whole 24-h period of proliferation in the presence of 10 μM RS25344. The IC_{50} values for the inhibition of ConA-mediated PBMC proliferation by a series of PDE4 inhibitors are shown in Table 5. RS25344 has an IC_{50} value of 1.8 μM which may represent the concentration required to cause elevation of cAMP over the whole 24-h period. The potencies of the PDE4 inhibitors measured after 48 h were weak, ranging from 1–20 μM. However, when determined after 24 h, the inhibitors gave IC_{50} values ranging from 273 nM–4.3 μM. The inhibition seen with the PDE4 inhibitors could be augmented with 100 nM PGE_2 such that the IC_{50} values obtained after 48 h, with the exception of SB207499, were similar to those obtained at 24 h in the absence of PGE_2 (Tab. 5). It appeared that the PDE4 inhibitors were capable of delaying the proliferative response, but that it escaped the block with time. In order to investigate this observation, the effect of PDE4 inhibition on proliferation, IL-2 production and CD25 expression was monitored in superantigen (SAg)-stimulated PBMC with time. The results are shown in Figure 4. It is clear that there is a window of inhibition of the proliferative response in this system between 24 and 48 h, but that by 72 h the proliferative response has "caught up". IL-2 production is inhibited significantly over the first 24 h, but by 36 h there is no significant difference and by 48 h the IL-2 produced is the same whether a PDE4 inhibitor is present or not. A similar situation exists with CD25 expression; an almost complete inhibition at 20 h diminished to ~33% and to ~25% over the 40–60 h period and later the inhibition disappeared altogether. One explanation for these data is that a small residual amount of IL-2 escaping inhibition may be sufficient to signal *via* the IL-2 receptor, such that more IL-2 is produced and the block is overcome with time. Another possibility from these data and those of others is that proliferation of T lymphocytes may be governed by mitogenic factors in addition to IL-2 [170, 173].

Table 5 - Effect of PDE4 inhibitors on ConA-mediated PBMC proliferation. PGE_2 was used at 100 nM (data are n = 3).

	ConA (48 h) IC_{50} (nM)	ConA (24 h) IC_{50} (nM)	ConA + PGE_2 (48 h) IC_{50} (nM)
RP73401	2,300	584	525
RS25344	5,300	1,800	2,300
SB207499	>20,000	767	>15,000
CDP840	1, 200	273	470
R-rolipram	15,700	1,000	1,800
S-rolipram	17,500	4,300	4,500

As well as inhibiting IL-2, PDE4 inhibitors were able to potently inhibit TNFα generated from αCD3-stimulated PBMC (S.E. Rapecki and R.A. Allen, unpublished data) and activated T cells [167], and IFNγ gene expression in human PBMC [171, 175], as well as isolated lymphocytes [171]. In addition, there is one report that IL-13 gene expression and protein secretion were inhibited by rolipram in ragweed-specific T-cell clones derived from a ragweed allergic, asthmatic subject [172]. When the effects of PDE inhibitors are assessed on IL-4 and IL-5, there are distinct differences between the human and murine systems. IL-4 and IL-5 secretion is inhibited by the non-specific PDE inhibitor theophylline, and the PDE4 specific inhibitor Ro20-1724 in anti-CD3-stimulated Th2-like cell lines [169], and by rolipram in PBMC and isolated lymphocytes [171]. However, a study with an anti-CD3-stimulated murine Th2 T-cell clone (D10.G4.1) found the non-specific PDE inhibitors IBMX, theophylline and enprofylline, and the PDE4 inhibitor rolipram to have no effect on IL-4 production, whilst stimulating IL-5 production [176].

Despite potent effects of PDE4 inhibitors on a range of cytokines including IL-2, IL-4, IL-5, IFNγ, TNFα and IL-13, the inhibitory effects on T-cell proliferation are less potent, in particular in the MLR, where potent PDE4 inhibitors have IC_{50} values of ~ 20 μM. Synergy between PDE3 and PDE4 inhibitors holds some promise for meaningful inhibition of T-cell proliferation, and the functional role of PDE7 in the T cell awaits elucidation.

PDE4 inhibitors are showing great promise as anti-inflammatory agents, as PDE4 plays a role in modulating the activity of virtually all cells involved in the inflammatory process. The elevation of intracellular cAMP, *via* PDE4 inhibition, has been associated with inhibition of the respiratory burst, cytokine and lipid mediator production, degranulation, phagocytosis, chemotaxis, and adhesion in a wide range of cells including neutrophils [177–180], eosinophils [181–183], mast cells [184],

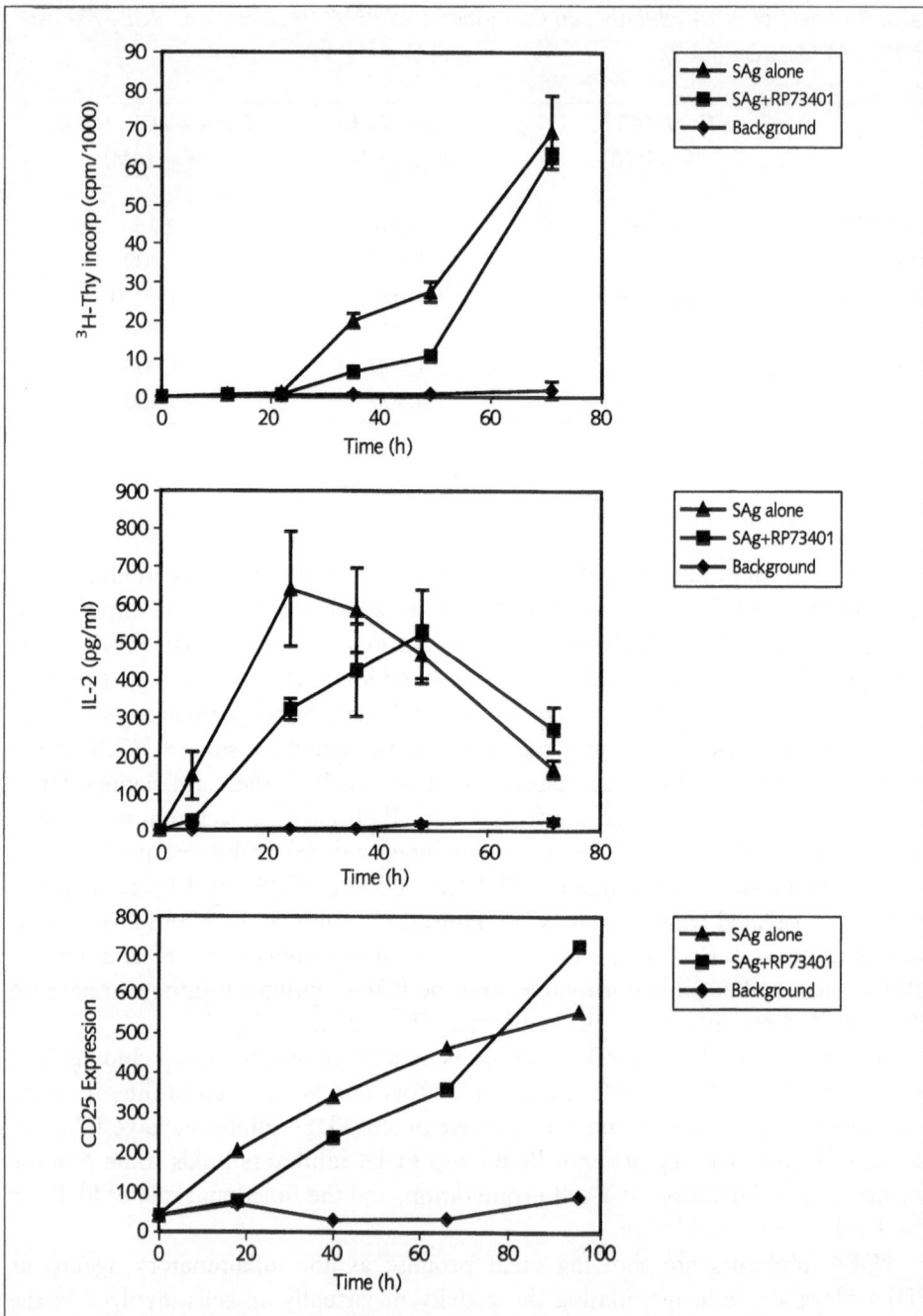

Figure 4
Time-course of proliferation, IL-2 production and CD25 expression in SAg-stimulated
PBMC: Effect of 1 µM RP73401.

basophils [185, 186], lymphocytes [170, 187, 188], monocytes/macrophages [189–191], mesangial cells [192, 193] and endothelial cells [194, 195].

PDE4 inhibitors are potent inhibitors of TNFα release from LPS-stimulated human monocytes [196–201]. This inhibition is correlated with the elevation of cAMP [199] and can be achieved by the addition of non-hydrolysable analogues of cAMP [200]. It is reported that rolipram causes a reduction in TNFα mRNA [197, 198], indicating that cAMP regulates TNFα at the transcriptional level. However, rolipram is also active when added post-LPS stimulation, indicating that rolipram may also have effects at the translational level [197]. There are reports that PDE4 inhibitors can also inhibit IL-1β release from LPS-stimulated human monocytes *via* an elevation of cAMP [198, 200]; however, inhibition of IL-1β is not a universal finding [196, 197]. An inhibition of IL-6 release is reported for the PDE4 inhibitors rolipram and Ro20-1724 in LPS-stimulated murine macrophages [202], but this inhibition is not seen in an LPS-stimulated human monocyte system, utilising the same inhibitors [197]. Work carried out in our lab shows a clear and complete inhibition of TNFα release from LPS-stimulated PBMC with a wide range of PDE4 inhibitors. In the same system we also demonstrated an inhibition of IL-1α and IL-1β (~80%) and a more modest inhibitory effect on IL-6 (~60%). IL-10 plays a major role in regulating TNFα levels. Addition of human recombinant IL-10 into the LPS-stimulated PBMC system caused inhibition of TNFα release [203–205], while addition of anti-IL-10 or anti-IL-10 receptor antibodies leads to increased TNFα production [203]. However, the addition of anti-IL-10 had no effect on the ability of a PDE4 inhibitor, or other cAMP-elevating drugs, to inhibit TNFα in human monocytes [205], but was reported to significantly reverse the inhibitory effect of rolipram on TNFα in murine macrophages [202]. The difference between these results may be due to species differences [205] or be related to the differentiation of monocytes to macrophages [205], when there is a change in the PDE profile and its relationship to suppression of TNFα release [206]. IFNγ, at 1ng/ml, is synergistic with LPS-stimulation of PBMC leading to increased TNFα production, and the release of IL-12, both of which are sensitive to PDE4 inhibitors, and in this system PDE4 inhibitors elevate IL-10 (S.E. Rapecki and R.A. Allen, unpublished observations).

PDE4 inhibitors have been shown to have effects in a wide range of experimental models, including asthma [207–213], rheumatoid arthritis [214–216], glomerulonephritis [217], septic shock [218–220], acute respiratory distress syndrome [221, 222], multiple sclerosis [223–225], ischemia-reperfusion injury [226, 227] and Parkinson's disease [228]. The greatest progress, in terms of advancing a PDE4 inhibitor to the clinic, has been made in rheumatoid arthritis and asthma. Analysis of cytokine mRNA and protein in rheumatoid arthritis tissue revealed that many proinflammatory cytokines, such as TNFα, IL-1, IL-6, granulocyte macrophage colony-stimulating factor (GM-CSF) and IL-8 are abundant in all patients regardless of therapy [229]. This is compensated to some degree by increased production of anti-inflammatory cytokines, such as IL-10 and TGFβ, and cytokine inhibitors, such as IL-1ra and soluble TNF-receptor [229], but the disease progresses when the cytokines are in an imbalance, favouring inflammation. It is believed that there is a network of proinflammatory cytokines with TNFα at the apex [229], and it has been shown that blockade of TNFα ameliorates the disease [230, 231]. Good clinical results have been obtained for a recombinant human TNF receptor (p75)-Fc fusion protein termed Enbrel or Etanercept [232–236] and a chimeric anti-TNFα antibody termed Infliximab or Remicade [237, 238]. However, it is not clear whether anti-TNFα therapies will protect against the erosion of cartilage and bone which is an important component of the disease. If they do not protect, then there is an opportunity for anti-IL-1 therapies [239], which animal studies suggest will be anti-erosive [240]. Inhibitors of p38 MAP kinase, as well as PDE4 inhibitors, have the potential to block both TNFα and IL-1. The anti-inflammatory activity of rolipram has been demonstrated in animal models of carrageenan-induced paw oedema and adjuvant arthritis [215] and has been shown to ameliorate collagen II-induced arthritis (CIA) in mice [216]. RP73401 (Fig. 5) has shown excellent anti-arthritic activities in both the CIA and Streptococcal cell wall (SCW)-induced arthritis models, significantly blocking joint destruction, synovitis, erosion and fibrosis, and decreasing TNFα mRNA expression at the pannus/cartilage interface of paw joints of collagen II-treated mice [241]. In the first clinical study with a PDE4 inhibitor (RP73401) in rheumatoid arthritis, patients reported an improved sense of well-being (pain relief, improved mobility, etc.), despite the TNFα and IL-1 levels being unaffected [242]. The administration of higher doses of RP73401 was prohibited due to side-effects, making it impossible to test the hypothesis that suppressing TNFα with a PDE4 inhibitor would confer benefits in the treatment of arthritis. Thus PDE4 inhibitors with improved therapeutic ratios are required.

There are many examples of PDE4 inhibitors demonstrating activity in models of asthma [207–213]. CDP840 has been shown to cause a dose-dependent reduction of IL-5-induced pleural eosinophilia and increase eosinophil stabilisation in rats, as well as reduce antigen-induced bronchoconstriction and pulmonary eosinophilic inflammation in guinea pigs [210]. CDP840 also inhibited ozone-induced airway hyper-responsiveness (AHR) to inhaled histamine in the guinea pig [211] and antigen-induced AHR and eosinophilia in neonatally immunised rabbits [209]. CDP840 was investigated in a series of phase IIa studies on the allergen-induced asthmatic response [243]. CDP840 was well tolerated, with no patients reporting nausea, one of the main side-effects limiting development of PDE4 inhibitors. CDP840 significantly attenuated the late phase asthmatic response to allergen challenge in the absence of any bronchodilatory or histamine antagonist effect [243]. The development of CDP840 was not pursued after a further study, in which twice the dose was administered, did not achieve a sufficiently significant clinical effect. The development of second-generation inhibitors is ongoing.

RP73401 has also been reported to inhibit antigen-induced bronchospasm in guinea pigs and rats, and reduce eosinophil, as well as total inflammatory cell, infil-

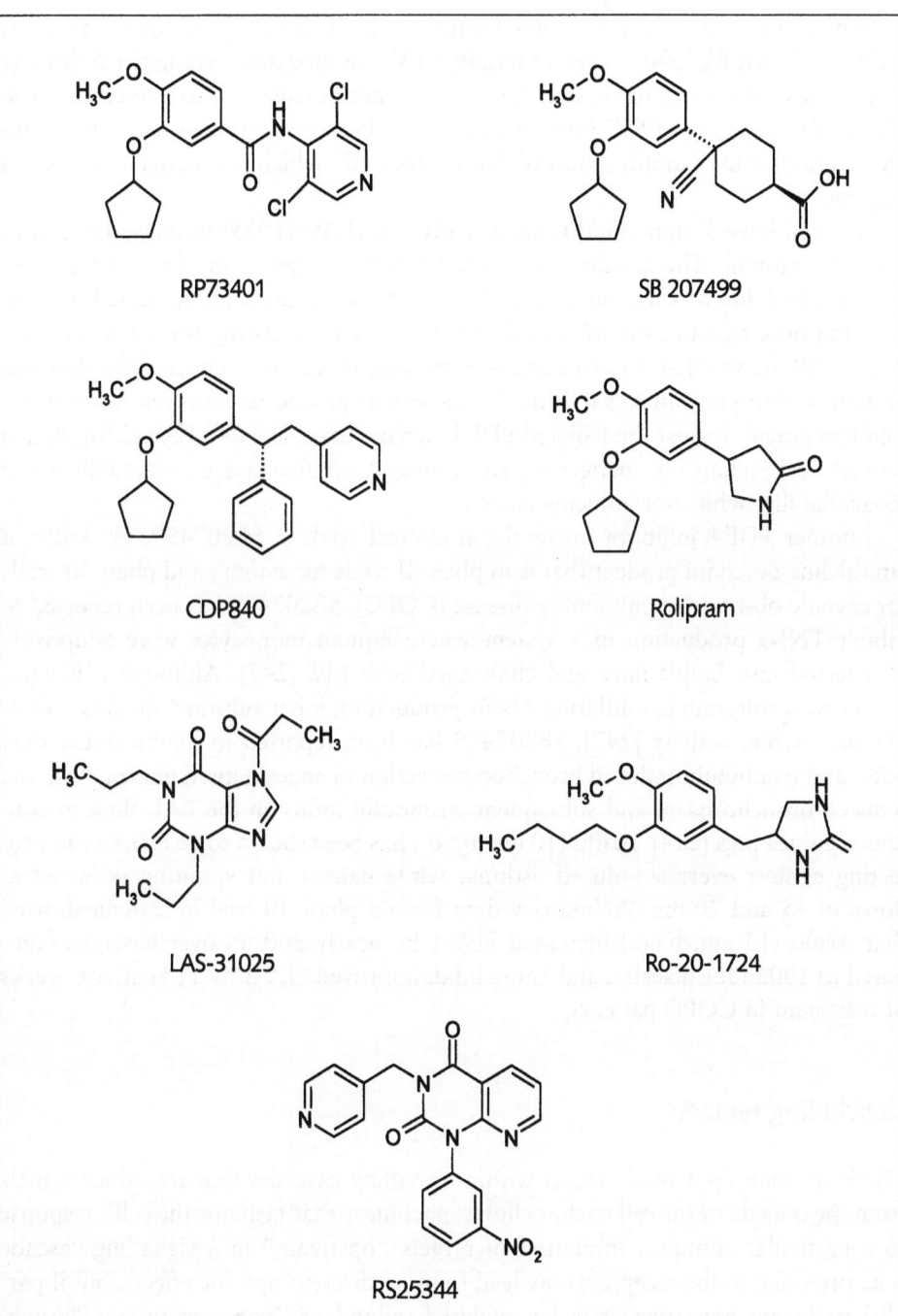

Figure 5
Structures of PDE4 inhibitors.

trate in guinea pig bronchoalveolar lavage (BAL) fluid [207]. In clinical trials in asthma, inhaled RP73401 failed to modify FEV-1 in moderate asthmatics following a single dose of 0.8 mg; increasing this dose was not possible due to side-effects. The clinical development of RP73401 in asthma has been discontinued, but once again the development of orally bioavailable compounds which are better tolerated is ongoing.

Almirall have a compound termed Arofylline (LAS-31025) in phase III clinical trials for asthma. The affinity for isolated PDE4 is significantly lower than most other PDE4 inhibitors having an IC_{50} value of 5.3 µM, although in phase II studies a 20 mg dose significantly increased FEV-1 with effects lasting for ~ 8 h. In these studies, 10 mg was less effective and 4mg produced negligible effects. The shortage of data on Arofylline makes it difficult to assess its prospects; however, it is ~ 1000-fold less potent against the isolated PDE4 enzyme than CDP840 [244–246], which proved to be clinically ineffective, suggesting Arofylline has exceptionally good bioavailability while not inducing emesis.

Another PDE4 inhibitor currently in clinical trials is SB207499, or Ariflo, a SmithKline Beecham product that is in phase II trials for asthma and phase III trials for chronic obstructive pulmonary disease (COPD). SB207499 has been reported to inhibit TNFα production in a system where human monocytes were adoptively transferred into Balb/c mice and challenged with LPS [247]. Although it is equipotent with rolipram in inhibiting TNFα production, it has substantially less central nervous system activity [247]. SB207499 has been reported to inhibit house dust mite- and ovalbumin-induced bronchoconstriction in anaesthetised guinea pigs, and reduced bronchospasm and subsequent eosinophil influx in the BAL fluid in conscious guinea pigs [214]. Ariflo (10 mg, b.i.d.) has been shown to be effective in protecting against exercise-induced asthma, while nausea and vomiting occurred at doses of 15 and 20 mg. Preliminary data from a phase III trial in asthma showed that Ariflo (15 mg, b.i.d.) increased FEV-1 by nearly 300 ml over baseline, compared to 100ml for placebo, and 15mg b.i.d. improved FEV-1 by 11% after 6 weeks of treatment in COPD patients.

Concluding remarks

There are many potential targets within signalling cascades that transduce signals from the outside of the cell to the cellular machinery that fashions the cell's response to a particular stimulus. Inhibition of targets "upstream" in a signalling cascade (i.e., proximal to the receptor) may lead to more discrete, specific effects, but if parallel pathways exist they may be rendered redundant. However, targets "downstream" in the cascade (i.e., distal to the receptor), at, or below, the convergence of two or more pathways, would be non-selective if it was undesirable to inhibit the response of both or all the stimuli. Alternatively, if the multiple signals converging gave rise to the same cell response, and it was desirable to inhibit all these stimuli, this would lead to a more complete and therefore more effective inhibition.

This review has focused primarily on IL-2 and TNFα produced by two cell types, the T cell and the monocyte, and two classes of inhibitors, p56^{Lck} inhibitors and PDE4 inhibitors; however, there are, as alluded to above, many other approaches to inhibiting these cytokines. Simulect (basiliximab), a chimeric anti-IL-2 receptor monoclonal antibody, has been shown to reduce the occurrence of acute cellular rejection among renal allograft recipients. RAD-001, an analogue of rapamycin, is an oral immunosuppressant that blocks IL-2 receptor-driven cell signalling, that is in phase III clinical trials for transplant rejection.

TNFα is a key cytokine that is a validated target with wide-ranging therapeutic potential and, as such, has attracted a range of approaches to inhibit it. At various stages of development there are anti-TNF antibodies, soluble TNF receptor fusion proteins, PDE4 inhibitors and antagonists of TNFα-converting enzyme (TACE). Other more general approaches to inhibiting IL-2 and TNFα, include targeting nuclear transcription factors, such as AP-1, NFAT and NF-κB, and MAP kinase pathways, such as Jnk (c-Jun N-terminal kinase) and p38. Antisense oligonucleotide and gene therapy approaches are also being developed.

With the plethora of approaches being pursued, the chance of success seems almost guaranteed, which will lead to advances in the treatment of a wide range of immune and inflammatory diseases.

References

1   Weiss A, Kadlecek T, Iwashima M, Chan A, van Oers N (1995) Molecular and genetic insights into T-cell antigen receptor signalling. *Annals New York Acad Sci* 766: 149–156

2   Wange RL, Samelson LE (1996) Complex complexes: signalling at the TCR. *Immunity* 5: 197–205

3   Ashwell JD, Klausner RD (1990) Genetic and mutational analysis of the T-cell antigen receptor. *Ann Rev Immunol* 8: 139–167

4   Reth M (1989) Antigen receptor tail clue [letter]. *Nature* 338: 383–384

5   Weiss A, Littman DR (1994) Signal transduction by lymphocyte antigen receptors. *Cell* 76: 263–274

6   Cambier JC (1995) Antigen and Fc receptor signalling: the awesome power of the immunoreceptor tyrosine-based activation motif (ITAM). *J Immunol* 155: 3281–3285

7   Irving BA, Chan AC, Weiss A (1993) Functional characterization of a signal transducing motif present in the T cell antigen receptor ζ chain. *J Exp Med* 177: 1093–1103

8   Exley M, Varticovsky L, Peter M, Sancho J, Terhorst C (1994) Association of phosphatidylinositol-3-kinase with a specific sequence of the T cell receptor zeta chain is dependent on T cell activation. *J Biol Chem* 269: 15140–15146

9   Cambier JC, Johnson SA (1995) Differential binding activity of ARH1/TAM motifs. *Immunol Lett* 44: 77–80

10   Isakov N, Wange RL, Burgess WH, Watts JD, Aebersold R, Samelson LE. ZAP-70 binding specificity to T cell based tyrosine-based activation motifs: the tandem SH2 domains of ZAP-70 bind distinct tyrosine-based activation motifs with varying affinity. *J Exp Med* 181: 375–380

11   Osman N, Turner H, Lucas S, Reif K, Cantrell DA (1996) The protein interactions of the immunoglobulin receptor family tyrosine-based activation motifs present in the T cell receptor ζ subunits and the CD3 γ, δ and ε chains. *Eur J Immunol* 26: 1063–1068

12   Howe LR, Weiss A (1995) Multiple kinases mediate T-cell-receptor signaling. *Trends Biochem Sci* 20: 59–64

13   Van Oers NSC, Lowin-Kropf B, Finlay D, Connolly K, Weiss A (1996) $\alpha\beta$ T cell development is abolished in mice lacking both Lck and Fyn protein tyrosine kinases. *Immunity* 5: 429–436

14   Straus DB, Weiss A (1992) Genetic evidence for the involvement of lck tyrosine kinase in signal transduction through the T cell antigen receptor. *Cell* 70: 585–593

15   Molina TJ, Kishihara K, Siderovski DP, van Ewijk W, Narendran A, Timms E, Wakeham A, Paige CJ, Hartmann K-U, Veillette A, Davidson D, Mak TW (1992) Profound block in thymocyte development in mice lacking p56^{lck}. *Nature* 357: 161–164

16   Nakamura K, Koga Y, Yoshida H, Tanaka K, Sasaki M, Kimura G, Nomoto K (1994) Inhibition of the T cell receptor-mediated signal transduction by microinjection of anti-Lck monoclonal antibody into T-cells. *Biochim Biophys Acta* 1224: 495–505

17   Samelson LE, Phillips AF, Luong ET, Klausner RD (1990) Association of the Fyn protein-tyrosine kinase with the T-cell antigen receptor. *Proc Natl Acad Sci USA* 87: 4358–4362

18   Timson Gauen LK, Kong AN, Samelson LE, Shaw AS (1992) p59fyn tyrosine kinase associates with multiple T-cell receptor subunits through its unique amino-terminal domain. *Mol Cell Biol* 12: 5438–5446

19   Tsygankov AY, Broker BM, Fargnoli J, Ledbetter JA, Bolen JB (1992) Activation of tyrosine kinase p60fyn following T cell antigen receptor cross-linking. *J Biol Chem* 267: 18259–18262

20   Davidson D, Chow LML, Fournel M, Veillette A (1992) Differential regulation of T cell antigen responsiveness by isoforms of the src-related tyrosine protein kinase p59fyn. *J Exp Med* 175: 1483–1492

21   Cooke MP, Abraham KM, Forbush KA, Perlmutter RM (1991) Regulation of T cell receptor signaling by a src family protein-tyrosine kinase (p59fyn). *Cell* 65: 281–291

22   Arpaia E, Shahar M, Dadi H, Cohen A, Roifman CM (1994) Defective T cell receptor signaling and CD8+ thymic selection in humans lacking ZAP-70 kinase. *Cell* 76: 947–958

23   Chan AC, Kadlecek TA, Elder ME, Filipovich AH, Kuo WL, Iwashima M, Parslow TG, Weiss A (1994) ZAP-70 deficiency in an autosomal recessive form of severe combined immunodeficiency. *Science* 264: 1599–1601

24   Elder ME, Clever J, Chan AC, Hope TJ, Weiss A, Parslow T (1994) Human severe combined immunodeficiency due to a defect in ZAP-70, a T cell tyrosine kinase. *Science* 264: 1596–1599

25   Gelfand E, Mazer B, Kadelecek T, Weinberg K, Weiss A (1995) Absence of ZAP-70 prevents signaling through the antigen receptor on peripheral blood T cells but not thymocytes. *J Exp Med* 182: 1057–1066

26   Negishi I, Motoyama N, Nakayama K, Nakayama K, Senju S, Hatakyama S, Zhang Q, Chan AC, Loh DY (1995) Essential role for ZAP-70 in both positive and negative selection of thymocytes. *Nature* 376: 435–438

27   Chan AC, Iwashima M, Turck CW, Weiss A (1992) ZAP-70: a 70 kd protein-tyrosine kinase that associates with the TCR zeta chain. *Cell* 71: 649–662

28   Iwashima M, Irving BA, van Oers NSC, Chan AC, Weiss A (1994) Sequential interactions of the TCR with two distinct cytoplasmic tyrosine kinases. *Science* 263: 1136–1139

29   Latour S, Fournel M, Veillette A (1997) Regulation of T-cell antigen receptor signalling by Syk tyrosine protein kinase. *Mol Cell Biol* 17: 4434–4441

30   Williams BL, Schreiber KL, Zhang W, Wange RL, Samelson LE, Leibson PJ, Abraham RT (1998) Genetic evidence for differential coupling of Syk family kinases to the T-cell receptor: reconstitution studies in a ZAP-70-deficient Jurkat T-cell line. *Mol Cell Biol* 18: 1388–1399

31   Chu DH, Morita CT, Weiss A (1998) The Syk family of protein tyrosine kinases in T-cell activation and development. *Immunol Rev* 165: 167–180

32   Desiderio S, Siliciano JD (1994) The Itk/Btk/Tec family of protein-tyrosine kinases. *Chem Immunol* 59: 191–208

33   Gibson S, Truitt K, Lapushin R, Khan H, Imboden JB, Mills GB (1998) Efficient CD28 signalling leads to increases in the kinase activities of the TEC family tyrosine kinase EMT/ITK/TSK and the SRC family tyrosine kinase LCK Biochem J 15: 1123–1128

34   Lu Y, Cuevas B, Gibson S, Khan H, LaPushin R, Imboden J, Mills GB (1998) Phospatidylinositol 3-kinase is required for CD28 but not CD3 regulation of the TEC family tyrosine kinase EMT/ITK/TSK: functional and physical interaction of EMT with phosphatidylinositol 3-kinase. *J Immunol* 161: 5404–5412

35   Liao XC, Littman DR, Weiss A (1997) Itk and Fyn make independent contributions to T cell activation. *J Exp Med* 186: 2069–2073

36   Lui KQ, Bunnell SC, Gurniak CB, Berg LJ (1998) T cell receptor-initiated calcium release is uncoupled from capacitative calcium entry in Itk-deficient T cells. *J Exp Med* 187: 1721–1727

37   Liao XC, Littman DR (1995) Altered T cell receptor signalling and disrupted T cell development in mice lacking Itk. *Immunity* 3: 757–769

38   Yang WC, Ghiotto M, Barbarat B, Olive D (1999) The role of Tec protein-tyrosine kinase in T cell signalling. *J Biol Chem* 274: 607–617

39   Weiss A, Kadlecek T, Iwashima M, Chan A, van Oers N (1995) Molecular and genetic insights into T-cell antigen receptor signaling. *Annals NY Acad Sci* 766: 149–156

40 Samelson LE, Donovan JA, Isakov N, Ota Y, Wange RL (1995) Signal transduction mediated by the T-cell antigen receptor. *Annals NY Acad Sci* 766: 157–172

41 Peterson EJ, Koretzky GA (1999) Signal transduction in T lymphocytes. *Clin Exp Rheumatol* 17: 107–114

42 Qian D, Weiss A (1997) T cell antigen receptor signal transduction. *Curr Opin in Cell Biol* 9: 205–212

43 Alberola-Ila J, Takaki S, Kerner JD, Perlmutter RM (1997) Differential signaling by lymphocyte antigen receptors. *Ann Rev Immunol* 15: 125–154

44 Peri KG, Veillette A (1994) Tyrosine protein kinases in T lymphocytes. *Chem Immunol* 59: 19–39

45 Jayaraman T, Ondrias K, Ondriasova E, Marks AR (1996) Regulation of the inositol 1, 4, 5-triphosphate receptor by tyrosine phosphorylation. *Science* 272: 1492–1494

46 Wardenburg JB, Fu C, Jackman JK, Flotow H, Wilkinson SE, Williams DH, Johnson R, Kong G, Chan AC, Findell PR (1996) Phosphorylation of SLP-76 by the ZAP-70 protein tyrosine kinase is required for T cell receptor function. *J Biol Chem* 271: 19641–19644

47 Wu J, Motto DG, Koretzky GA, Weiss A (1996) Vav and SLP-76 interact and functionally cooperate in IL-2 gene activation. *Immunity* 4: 593–602

48 Onodera H, Motto DG, Koretzky GA, Rothstein DM (1996) Differential regulation of activation-induced tyrosine phosphorylation and recruitment of SLP-76 to Vav by distinct isoforms of the CD45 protein-tyrosine phosphatase. *J Biol Chem* 271: 22225–22230

49 Tuosto L, Michel F, Acuto O (1996) p95vav associates with tyrosine phosphorylated SLP-76 in antigen-stimulated T cells. *J Exp Med* 184: 1161–1166

50 Wu J, Katzav S, Weiss A (1995) A functional T cell receptor signaling pathway is required for p95vav activity. *Mol Cell Biol* 15: 4337–4346

51 Donovan JA, Wange RL, Langdon WY, Samelson LE (1994) The protein product of the c-Cbl protooncogene is the 120kDa tyrosine-phosphorylated protein in Jurkat cells activated *via* the T cell antigen receptor. *J Biol Chem* 271: 19641–19644

52 Motto DG, Ross SE, Wu J, Hendricks-Taylor LR, Koretzky GA (1996) Implication of the GRB2-associated phosphoprotein SLP-76 in T cell receptor-mediated interleukin 2 production. *J Exp Med* 183: 1937–1943

53 Cantrell D (1996) T cell antigen receptor signal transduction pathways. *Ann Rev Immunol* 14: 259–274

54 Fukazawa T, Reedquist KA, Trub T, Soltoff S, Panchamoorthy G, Druker B, Cantly L, Shoelson SE, Band H (1995) The SH3 domain-binding T cell tyrosyl phosphoprotein p120. *J Biol Chem* 270: 19141–19150

55 Meisner H, Conway BR, Hartley D, Czech MP (1995) Interaction of Cbl with Grb2 and phosphatidylinositol-3'-kinase in activated Jurkat cells. *Mol Cell Biol* 15: 3571–3578

56 Jackman JK, Motto DG, Sun Q, Tanemoto M, Turck CW, Peltz GA, Koretzky GA, Findell PR (1995) Molecular cloning of SLP-76, a 76-kDa tyrosine phosphoprotein associated with Grb2 in T cells. *J Biol Chem* 270: 7029–7032

57 Downward J (1994) The GRB2/Sem-5 adaptor protein. *FEBS Lett* 338: 113–117

58 Holsinger LJ, Spencer DM, Austin DJ, Schreiber SL, Crabtree GR (1995) Signal transduction in T lymphocytes using a conditional allele of Sos. *Proc Natl Acad Sci USA* 92: 9810–9814

59 Fields PE, Gajewski TF, Fitch FW (1996) Blocked Ras activation in anergic CD4⁺ T cells. *Science* 271: 1276–1278

60 Li W, Whaley CD, Mondino A, Mueller DL (1996) Blocked signal transduction to the ERK and JNK protein kinases in anergic CD4⁺ T cells. *Science* 271: 1272–1276

61 Denny MF, Kaufman HC, Chan AC, Straus DB (1999) The Lck SH3 domain is required for activation of the mitogen-activated protein kinase pathway but not the initiation of T-cell antigen receptor signaling. *J Biol Chem* 274: 5146–5152

62 Wright SD, Ramos RA, Tobias PS, Ulevitch RJ, Mathison JC (1990) CD14, a receptor for complexes of LPS and LPS binding protein. *Science* 249: 1431–1433

63 Leturcq DJ, Moriarty AM, Talbott G, Winn RK, Martin TR, Ulevitch RJ (1996) Antibodies against CD14 protect primates from endotoxin-induced shock. *J Clin Invest* 98: 1533–1538

64 Lee J-D, Kato K, Tobias PS, Kirkland TN, Ulevitch RJ (1992) Transfection of CD14 into 70Z/3 cells dramatically enhances the sensitivity to complexes of lipopolysaccharide (LPS) and LPS binding protein. *J Exp Med* 175: 1697–1705

65 Golenbock DT, Liu Y, Millham FH, Freeman MW, Zoeller RA (1993) Surface expression of human CD14 in Chinese hamster ovary fibroblasts imparts macrophage-like responsiveness to bacterial endotoxin. *J Biol Chem* 268: 22055–22059

66 Ferrero E, Jiao D, Tsuberi BZ, Tesio L, Rong GW, Haziot A, Goyert SA (1993) Transgenic mice expressing human CD14 are hypersensitive to lipopolysaccharide. *Proc Natl Acad Sci USA* 90: 2380–2384

67 Ulevitch RJ, Tobias PS (1995) Receptor-dependent mechanisms of cell stimulation by bacterial endotoxin. *Ann Rev Immunol* 13: 437–457

68 Ulevitch RJ, Tobias PS (1994) Recognition of endotoxin by cells leading to transmembrane signaling. *Curr Opin in Immunol* 6: 125–130

69 Jack RS, Fan W, Bernheiden M, Rune G, Ehlers M, Weber A, Kirsch G, Mentel R, Furll B, Freudenberg M, Schmitz G, Stelter F, Schutt C (1997) Lipopolysaccharide-binding protein is required to combat a murine gram-negative bacterial infection. *Nature* 389: 742–745

70 Wurfel MM, Monks BG, Ingalls RR, Dedrick RL, Delude R, Zhou D, Lamping N, Schumann RR, Thieringer R, Fenton MJ, Wright SD, Golenbock D (1997) Targetted deletion of the lipopolysaccharide (LPS)-binding protein gene leads to profound suppression of LPS responses *ex vivo*, whereas *in vivo* responses remain intact. *J Exp Med* 186: 2051–2056

71 Elsbach P, Weiss J (1998) Role of the bacterial/permeability-increasing protein in host defence. *Curr Opin in Immunol* 10: 45–49

72 Haziot A, Ferrero E, Kontgen F, Hijiya N, Yamamoto S, Silver J, Stewart CL, Goyert SM

(1996) Resistance to endotoxin shock and reduced dissemination of gram-negative bacteria in CD14-deficient mice. *Immunity* 4: 407–414

73 Perera P-Y, Vogel SN, Detore GR, Haziot A, Goyert SM (1997) CD14-dependent and CD14-independent signaling pathways in murine macrophages from normal and CD14 knockout mice stimulated with lipopolysaccharide or taxol. *J Immunol* 158: 4422–4429

74 Haziot A, Lin XY, Zhang F, Goyert SM (1998) The induction of acute phase proteins by lipopolysaccharide uses a novel pathway that is CD14-independent. *J Immunol* 160: 2570–2572

75 Weinstein SL, Gold MR, DeFranco AL (1991) Bacterial lipopolysaccharide stimulates protein tyrosine phosphorylation in macrophages. *Proc Natl Acad Sci USA* 88: 4148–4152

76 DeFranco AL, Crowley MT, Finn A, Hambleton J, Weinstein SL (1998) The role of tyrosine kinases and MAP kinases in LPS-induced signaling. *Prog Clin Biol Res* 397: 119–136

77 Weinstein SL, Sanghera JS, Lemke K, DeFranco AL, Pelech SL (1992) Bacterial lipopolysaccharide induces tyrosine phosphorylation and activation of mitogen-activated protein kinases in macrophages. *J Biol Chem* 276: 14955–14962

78 Novogrodsky A, Vanichkin A, Patya M, Gazit A, Osherov N, Levitzki A (1994) Prevention of lipopolysaccharide-induced lethal toxicity by tyrosine kinase inhibitors. *Science* 264: 1319–1322

79 Ding A, Sanchez E, Nathan CF (1993) Taxol shares the ability of bacterial lipopolysaccharide to induce tyrosine phosphorylation of microtubule-associated protein kinase. *J Immunol* 151: 5596–5602

80 Shapira L, Takashiba S, Champagne C, Amar S, Van Dyke TE (1994) Involvement of protein kinase C and protein tyrosine kinase in lipopolysaccharide-induced TNFα and IL-1β production by human monocytes. *J Immunol* 153: 1818–1824

81 Geng Y, Gulbins E, Altman A, Lotz M (1994) Monocyte deactivation by interleukin 10 *via* inhibition of tyrosine kinase activity and the ras signaling pathway. *Proc Natl Acad Sci USA* 91: 8602–8606

82 Beaty CD, Franklin TL, Uehara Y, Wilson CB (1994) Lipopolysaccharide-induced cytokine production in human monocytes: Role of tyrosine phosphorylation in transmembrane signal transduction Eur. *J Immunol* 24: 1278–1284

83 Stefanova I, Corcoran ML, Horak EM, Wahl LM, Bolen JB, Horak ID (1993) Lipopolysaccharide induces activation of CD14-associated protein tyrosine kinase p53/56lyn. *J Biol Chem* 268: 20725–20728

84 Meng F, Lowell CA (1997) Lipopolysaccharide (LPS)-induced macrophage activation and signal transduction in the absence of Src-family kinases Hck, Fgr, Lyn. *J Exp Med* 185: 1661–1670

85 Crowley MT, Harmer SL, DeFranco AL (1996) Activation-induced association of a 145-kDa tyrosine-phosphorylated protein with Shc and Syk in B lymphocytes and macrophages. *J Biol Chem* 271: 1145–1152

86 Crowley MT, Costello PS, Fitzer-Attas CJ, Turner M, Meng F, Lowell C, Tybulewicz

VLJ, DeFranco AL (1997) A critical role for Syk in signal transduction and phagocytosis mediated by Fc-γ receptors on macrophages. *J Exp Med* 186: 1027–1039

87   Dong Z, Qi X, Fidler IJ (1993) Tyrosine phosphorylation of mitogen-activated protein kinases is necessary for activation of murine macrophages by natural and synthetic bacterial products. *J Exp Med* 177: 1071–1077

88   Hambleton J, Weinstein SL, Lem L, DeFranco AL (1996) Activation of c-Jun N-terminal kinase in bacterial lipopolysaccharide-stimulated macrophages. *Proc Natl Acad Sci USA* 93: 2774–2778

89   Sanghera JS, Weinstein SL, Aluwalia M, Girn J, Pelech SL (1996) Activation of multiple proline-directed kinases by bacterial lipopolysaccharide in murine macrophages. *J Immunol* 156: 4457–4465

90   Han J, Lee JD, Bibbs L, Ulevitch RJ (1994) A MAP kinase targetted by endotoxin and hyperosmolarity in mammalian cells. *Science* 265: 808–811

91   Foey AD, Parry SL, Williams LM, Feldmann M, Foxwell BMJ, Brennan FM (1998) Regulation of monocyte IL-10 synthesis by endogenous IL-1 and TNFα: Role of the p38 and p42/44 mitogen-activated protein kinases. *J Immunol* 160: 920–928

92   Lee JC, Laydon JT, McDonnell PC, Gallagher TF, Kumar S, Green D, McNulty D, Blumenthal MJ, Heys JR, Landvatter SW et al (1994) A protein kinase involved in the regulation of inflammatory cytokine biosynthesis. *Nature* 372: 739–746

93   Beyaert R, Cuenda A, Berghe WV, Plaisance S, Lee JC, Haegeman G, Cohen P, Fiers W (1996) The p38/RK mitogen-activated protein kinase pathway regulates interleukin-6 synthesis in response to tumor necrosis factor. *EMBO J* 15: 1914–1923

94   Su B, Karin M (1996) Mitogen-activated protein kinase cascades and regulation of gene expression. *Curr Opin in Immunol* 8: 402–411

95   Karin M, Liu Z-G, Zandi E (1997) AP-1 function and regulation. *Curr Opin in Cell Biol* 9: 240–246

96   Whitmarsh AJ, Davis RJ (1996) Transcription factor AP-1 regulation by mitogen-activated protein kinase signal transduction pathways. *J Mol Med* 74: 589–607

97   Foletta VC (1996) Transcription factor AP-1, and the role of Fra-2. *Immunol Cell Biol* 74: 121–133

98   Bretscher PA (1992) The two-signal model of lymphocyte activation twenty-one years later. *Immunol Today* 13: 74–76

99   Linsley PS, Ledbetter JA (1993) The role of CD28 receptor during T cell responses to antigen. *Ann Rev Immunol* 11: 191–212

100   Ward SG, Westwick J, Hall ND, Sansom DM (1993) Ligation of CD28 receptor by B7 induces formation of D-3 phosphoinositides in T lymphocytes independently of T cell receptor/CD3 activation. *Eur J Immunol* 23: 2572–2577

101   Pages F, Ragueneau M, Rottapel R, Truneh A, Nunes J, Imbert J, Olive D (1994) Binding of phosphatidyl-inositol-3-OH kinase to CD28 is required for T-cell signalling. *Nature* 36: 327–330

102   Boucher LM, Wiegmann K, Futterer A, Pfeffer K, Machleidt T, Schutze S, Mak TW,

Kronke M (1995) CD28 signals through acidic sphingomyelinase. *J Exp Med* 181: 2059–2068

103 Su B, Jacinto E, Hibi M, Kallunki T, Karin M, Ben-Neriah Y (1994) JNK is involved in signal integration during costimulation of T lymphocytes Cell 77: 727–736

104 Edmead CE, Patel YI, Wilson A, Boulougouris G, Hall ND, Ward SG, Sansom DM (1996) Induction of activator protein (AP)-1 and nuclear factor-κB by CD28 stimulation involves both phosphatidyl-inositol 3-kinase and acidic sphingomyelinase signals. *J Immunol* 157: 3290–3297

105 Granelli-Piperno A, Nolan P (1991) Nuclear transcription factors that bind to elements of the IL-2 promoter Induction requirements in primary T cells. *J Immunol* 147: 2734–2739

106 Rincon M, Flavell RA (1994) AP-1 transcriptional activity requires both T-cell receptor-mediated and co-stimulatory signals in primary T lymphocytes. *EMBO J* 13: 4370–4381

107 Angel P, Imagawa M, Chiu R, Stein B, Imbra RJ, Rahmsdorf HJ, Jonat C, Herrlich P, Karin M (1987) Phorbol ester-inducible genes contain a common cis element recognized by a TPA-modulated trans-acting factor. *Cell* 49: 729–739

108 Crabtree GR, Clipstone NA (1994) Signal transmission between the plasma membrane and nucleus of T-lymphocytes. *Ann Rev Biochem* 63: 1045–1083

109 Serfling E, Avots A, Neumann M (1995) The architecture of the interleukin-2 promoter: a reflection of T lymphocyte activation. *Biochim Biophys Acta* 1263: 181–200

110 Jain J, Loh C, Rao A (1995) Transcriptional regulation of the IL-2 gene. *Curr Opin in Immunol* 7: 333–342

111 Bierer BE, Hollander G, Fruman D, Burakoff SJ (1993) Cyclosporin A and FK506: molecular mechanisms of immunosuppression and probes for transplantation biology. *Curr Opin in Immunol 5*: 763–773

112 Bram RJ, Hung DT, Martin PK, Schreiber SL, Crabtree GR (1993) Identification of the immunophilins capable of mediating inhibition of signal transduction by cyclosporin A and FK506: roles of calcineurin binding and cellular location. *Mol Cell Biol* 13: 4760–4769

113 Sen R, Baltimore D (1986) Multiple nuclear factors interact with the immunoglobulin enhancer sequences. *Cell* 46: 705–716

114 Baldwin AS Jr (1996) The NF-B and IB proteins: new discoveries and insights. *Ann Rev Immunol* 14: 649–681

115 Siebenlist U, Franzoso G, Brown K (1994) Structure, regulation and function of NF-κB Ann Rev Cell Biol 10: 405–455

116 Brockman JA, Scherer DC, Hall SM, McKinsey TA, Qi X, Lee WY, Ballard DW (1995) Coupling of a signal-response domain in IκBα to mutiple pathways for NF-κB activation. *Mol Cell Biol* 15: 2809–2818

117 Chen Z, Parent L, Maniatis T (1996) Site-specific phosphorylation of IκBα by a novel ubiquitin-dependent protein kinase activity. *Cell* 84: 853–862

118 Chen Z, Hagler J, Palombella VJ, Melandri F, Scherer D, Ballard D, Maniatis T (1995)

Signal-induced site-specific phosphorylation targets IκBα to the ubiquitin-proteasome pathway. *Genes Dev* 9: 1586–1597

119 Sassone-Corsi P (1998) Coupling gene expression to cAMP signalling: role of CREB and CREM. *Int J Biochem Cell Biol* 30: 27–38

120 Sassone-Corsi P (1995) Transcription factors responsive to cAMP. *Ann Rev Cell Dev Biol* 11: 355–377

121 Bohm M, Moellmann G, Cheng E, Alvarez-Franco M, Wagner S, Sassone-Corsi P, Halaban R (1995) Identification of p90rsk as the probable CREB-Ser133 kinase in human melanocytes. *Cell Growth Differ* 6: 291–302

122 Ginty DD, Bonni A, Greenberg ME (1994) Nerve growth factor activates a Ras-dependent protein kinase that stimulates c-fos transcription *via* phosphorylation of CREB. *Cell* 77: 713–725

123 Cambier JC, Newell MK, Justement LB, McGuire JC, Leach KL, Chen ZZ (1987) Ia binding ligands and cAMP stimulate nuclear translocation of PKC in B lymphocytes. *Nature* 327: 629–632

124 Yoshimasa T, Sibley DR, Bouvier M, Lefkowitz RJ, Caron MG (1987) Cross-talk between cellular signalling pathways suggested by phorbol ester adenylate cyclase phosphorylation. *Nature* 327: 67–70

125 Frodin M, Peraldi P, Van Obberghen E (1994) Cyclic AMP activates the mitogen-activated protein cascade in PC12 cells. *J Biol Chem* 269: 6207–6214

126 Masquilier D, Sassone-Corsi P (1992) Transcriptional cross-talk: nuclear factors CREM and CREB bind to AP-1 sites and inhibit activation by Jun. *J Biol Chem* 267: 22460–22466

127 Gothel SF, Marahiel MA (1999) Peptidyl-prolyl cis-trans isomerases, a superfamily of ubiquitous folding catalysts. *Cell Mol Life Sci* 55: 423–436

128 Siekierka JJ (1994) Probing T-cell signal transduction pathways with the immunosuppressive drugs, FK-506 and rapamycin. *Immunol Res* 13: 110–116

129 Friedman J, Weissman I (1991) Two cytoplasmic candidates for immunophilin action are revealed by affinity for a new cyclophilin: one in the presence and one in the absence of CsA. *Cell* 66: 799–806

130 Liu J, Farmer JD, Lane WS, Friedman J, Weissman I, Schreiber SL (1991) Calcineurin is a common target of cyclophilin-cyclosporin A and FKBP-FK-506 complexes. *Cell* 66: 807–815

131 Wiederrecht G, Hung S, Chan KH, Marcy A, Martin M, Calaycay J, Boulton D, Signal N, Kincaid RL, Siekierka JJ (1992) Characterization of high molecular weight FK-506 binding activities reveals a novel FK-506-binding protein as well as a protein complex. *J Biol Chem* 267: 21753– 21760

132 Sewell TJ, Lam E, Martin MM, Leszyk J, Weidner J, Calaycay J, Griffin P, Williams H, Hung S, Cryan J et al (1994) Inhibition of calcineurin by a novel FK-506-binding protein. *J Biol Chem* 269: 21094–21102

133 Dennis PB, Fumagalli S, Thomas G (1999) Target of rapamycin (TOR): balancing the opposing forces of protein synthesis and degradation. *Curr Opin Genet Dev* 9: 49–54

134 Woerly G, Brooks N, Ryffel B (1996) Effect of rapamycin on the expression of the IL-2 receptor (CD25). *Clin Exp Immunol* 103: 322–327

135 Bennett WM (1998) The nephrotoxicity of new and old immunosuppressive drugs. *Ren Fail* 20: 687–690

136 Zachariae H (1999) Renal toxicity of long-term cyclosporin. *Scand J Rheumatol* 28: 65–68

137 Hanks SK, Quinn AM, Hunter T (1988) The protein kinase family: conserved features and deduced phylogeny of the catalytic domains. *Science* 241: 42–52

138 Carrera AC, Alexandrov K, Roberts TM (1993) The conserved lysine of the catalytic domain of protein kinases is actively involved in the phosphotransfer reaction and not required for anchoring ATP. *Proc Natl Acad Sci USA* 90: 442–446

139 Levitzki A (1992) Tyrphostins: tyrosine kinase blockers as novel antiproliferative agents and dissectors of signal transduction. *FASEB J* 6: 3275–3282

140 Akiyama T, Ishida J, Nakagawa J, Ogawara H, Watanabe S, Itoh N, Shibuya M, Fukami Y (1987) Genestein, a specific inhibitor of tyrosine-specific protein kinases. *J Biol Chem* 262: 5592–5595

141 Uehara Y, Murakami Y, Suzukake-Tsuchiya K, Moriya Y, Sano H, Shibata K, Omura S (1988) Effects of herbimycin derivatives on src oncogene function in relation to antitumor activity. *J Antibiot (Tokyo)* 41: 831–834

142 June CH, Fletcher MC, Ledbetter JA, Schieven GL, Siegel JN, Phillips AF, Samelson LE (1990) Inhibition of tyrosine phosphorylation prevents T-cell receptor-mediated signal transduction. *Proc Natl Acad Sci USA* 87: 7722–7726

143 Trevillyan JM, Lu Y, Atluru D, Phillips CA, Bjorndahl JM (1990) Differential inhibition of T cell receptor signal transduction and early activation events by a selective inhibitor of protein-tyrosine kinase. *J Immunol* 145: 3223–3230

144 Mustelin T, Coggeshall KM, Isakov N, Altman A (1990) T cell antigen receptor-mediated activation of phospholipase C requires tyrosine phosphorylation. *Science* 247: 1584–1587

145 Graber M, June CH, Samelson LE, Weiss A (1992) The protein tyrosine kinase inhibitor herbimycin A, but not genistein, specifically inhibits signal transduction by the T cell antigen receptor. *Int Immunol* 4: 1201–1210

146 Baldari CT, Telford JL (1994) Dissection of T cell antigen receptor signaling using protein tyrosine kinase inhibitors. *Eur J Immunol* 24: 1046–1052

147 Hanke JH, Pollok BA, Changelian PS (1995) Role of tyrosine kinases in lymphocyte activation: Targets for drug intervention. *Inflamm Res* 44: 357–371

148 Hanson GJ (1997) Inhibitors of p38 kinase. *Exp Opin Ther Patents* 7: 729–733

149 Lee JC, Laydon JT, McDonnell PC, Gallagher TF, Kumar S, Green D, McNulty D, Blumenthal MJ, Heys JR, Landvatter SW et al (1994) A protein kinase involved in the regulation of inflammatory cytokine biosynthesis. *Nature* 372: 739–746

150 Ward SG, Parry RV, Matthews J, O'Neill L (1997) A p38 MAP kinase inhibitor SB203580 inhibits CD-28-dependent T cell proliferation and IL-2 production. *Biochem Soc Trans* 25: 304S

151 Faltynek CR, Wang S, Miller D, Mauvais P, Gauvin B, Reid J, Xie W, Hoekstra S, Juniewicz P, Sarup J et al (1995) Inhibition of T lymphocyte activation by a novel p56^{lck} tyrosine kinase inhibitor. *J Enzyme Inhib* 9: 111–122

152 Trevillyan JM, Chiou XG, Ballaron SJ, Tang QM, Buko A, Sheets MP, Smith ML, Putman CB, Wiedeman P, Tu N et al (1999) Inhibition of p56(lck) tyrosine kinase by isothiazolones. *Arch Biochem Biophys* 362: 19–29

153 Hanke JH, Gardner JP, Dow RL, Changelian PS, Brissette WH, Weringer EJ, Pollok BA, Connelly PA (1996) Discovery of a novel, potent and Src family-selective tyrosine kinase inhibitor. Study of Lck- and FynT-dependent T cell activation. *J Biol Chem* 271: 695–701

154 Gimsa U, Mitchison A, Allen R (1999) Inhibitors of Src-family tyrosine kinases favour Th2 differentiation. *Cytokine* 11: 208–215

155 Shibuya H, Kohu K, Yamada K, Barsoumian EL, Perlmutter RM, Taniguchi T (1994) Functional dissection of p56lck, a protein tyrosine kinase which mediates interleukin-2-induced activation of the c-fos gene. *Mol Cell Biol* 14: 5812– 5819

156 Nishio K, Miura K, Ohira T, Heike Y, Saijo N (1994) Genistein, a tyrosine kinase inhibitor, decreased the affinity of p56lck to beta-chain of interleukin-2 receptor in human natural killer (NK)-rich cells and decreased NK-mediated cytotoxicity. *Proc Soc Exp Biol Med* 207: 227–233

157 Taniguchi T, Miyazaki T, Minami Y, Kawahara A, Fujii H, Nakagawa Y, Hatakeyama M, Liu ZJ (1995) IL-2 signaling involves recruitment and activation of multiple tyrosine kinases by the IL-2 receptor. *Ann NY Acad Sci* 766: 235-244

158 Taieb J, Blanchard DA, Auffredou MT, Chaouchi N, Vazquez A (1995) *In vivo* association between p56lck and MAP kinase during IL-2-mediated lymphocyte proliferation. *J Immunol* 155: 5623–5630

159 Miyazaki T, Taniguchi T (1996) Coupling of the IL2 receptor complex with non-receptor protein tyrosine kinases. *Cancer Surv* 27: 25–40

160 Goebel J, Franks A, Robey F, Mikovits J, Lowry RP (1999) Attenuation of IL-2 receptor signaling by CD4-ligation requires polymerized cytoskeletal actin but not P56LCK. *Transplant Proc* 31: 822–824

161 Jayaraman T, Ondrias K, Ondriasova E, Marks AR (1996) Regulation of the inositol 1, 4, 5-triphosphate receptor by tyrosine phosphorylation. *Science* 272: 1492–1494

162 Beavo JA, Reifsnyder DH (1990) Primary sequence of cyclic nucleotide phosphodiesterase isozymes and the design of selective inhibitors. *Trends Pharmacol Sci* 11: 150–155

163 Thompson WJ (1991) Cyclic nucleotide phosphodiesterases: Pharmacology, biochemistry and function. *Pharmacol Ther* 51: 13–33

164 Fisher DA, Smith JF, Pillar JS, St Denis SH, Cheng JB (1998) Isolation and characterization of PDE8A, a novel human cAMP-specific phosphodiesterase. *Biochem Biophys Res Comm* 246: 570–577

165 Soderling SH, Bayuga SJ, Beavo JA (1998) Identification and characterization of a novel family of cyclic nucleotide phosphodiesterases. *J Biol Chem* 273: 15553–15558

166 Fisher DA, Smith JF, Pillar JS, St Denis SH, Cheng JB (1998) Isolation and characterization of PDE9A, a novel human cGMP-specific phosphodiesterase. *J Biol Chem* 273: 15559–15564

167 Rott O, Cash E, Fleischer B (1993) Phosphodiesterase inhibitor pentoxyfylline, a selective suppressor of T helper type 1- but not type 2-associated lymphokine production, prevents induction of experimental autoimmune encephalomyelitis in Lewis rats. *Eur J Immunol* 23: 1745–1751

168 Essayan DM, Huang SK, Undem BJ, Kagey-Sobotka A, Lichtenstein LM (1994) Modulation of antigen- and mitogen-induced proliferative responses of peripheral blood mononuclear cells by nonselective and isozyme selective cyclic nucleotide phosphodiesterase inhibitors. *J Immunol* 153: 3408–3416

169 Crocker IC, Townley RG, Khan MM (1996) Phosphodiesterase inhibitors suppress proliferation of peripheral blood mononuclear cells and interleukin-4 and -5 secretion by human T-helper type 2 cells. *Immunopharmacol* 31: 223–235

170 Giembycz MA, Corrigan CJ, Seybold J, Newton R, Barnes PJ (1996) Identification of cyclic AMP phosphodiesterases 3, 4 and 7 in human CD4+ and CD8+ T-lymphocytes: role in regulating proliferation and the biosynthesis of interleukin-2. *Brit J Pharmacol* 118: 1945–1958

171 Essayan DM, Huang SK, Kagey-Sobotka A, Lichtenstein LM (1997) Differential efficacy of lymphocyte- and monocyte-selective pretreatment with a type 4 phosphodiesterase inhibitor on antigen-driven proliferation and cytokine gene expression. *J Allergy Clin Immunol* 99: 28–37

172 Essayan DM, Kagey-Sobotka A, Lichtenstein LM, Huang SK (1997) Regulation of interleukin-13 by type 4 cyclic nucleotide phosphodiesterase (PDE) inhibitors in allergen-specific human T lymphocyte clones. *Biochem Pharmacol* 53: 1055–1060

173 Souness JE, Houghton C, Sardar N, Withnall MT (1997) Evidence that cyclic AMP phosphodiesterase inhibitors suppress interleukin-2 release from murine splenocytes by interacting with a 'low-affinity' phosphodiesterase 4 conformer. *Br J Pharmacol* 121: 743–750

174 Essayan DM, Kagey-Sobotka A, Lichtenstein LM, Huang SK (1997) Differential regulation of human antigen-specific Th1 and Th2 lymphocyte responses by isozyme selective cyclic nucleotide phosphodiesterase inhibitors. *J Pharmacol Exp Ther* 282: 505–512

175 Lewis GM, Caccese RG, Heaslip RJ, Bansbach CC (1993) Effects of rolipram and CI-930 on IL-2 mRNA transcription in human Jurkat cells. *Agents Actions* 39: C89–92

176 Schmidt J, Hatzelmann A, Fleissner S, Heimann-Weitschat I, Lindstaedt R, Szelenyi I (1995) Effect of phosphodiesterase inhibition on IL-4 and IL-5 production of the murine TH2-type T cell clone D10G41. *Immunopharmacol* 30: 191–198

177 Derian CK, Santulli RJ, Rao PE, Solomon HF, Barrett JA (1995) Inhibition of chemotactic peptide-induced neutrophil adhesion to vascular endothelium by cAMP modulators. *J Immunol* 154: 308–317

178 Giembycz MA (1992) Could isozyme-selective phosphodiesterase inhibitors render

bronchodilator therapy redundant in the treatment of bronchial asthma? *Biochem Pharmacol* 43: 2041–2051

179 Simpson PJ, Schelm JA, Smallwood JK, Clay MP, Lindstrom TD (1992) Inhibition of granulocyte cAMP-phosphodiesterase by rolipram *in vivo* is not sufficient to protect the canine myocardium from reperfusion injury. *J Cardiovasc Pharmacol* 19: 987–995

180 Nielson CP, Vestal RE, Sturm RJ, Heaslip R (1990) Effects of selective phosphodiesterase inhibitors on the polymorphonuclear leukocyte respiratory burst. *J Allergy Clin Immunol* 86: 801–808

181 Dent G, Giembycz MA, Rabe KF, Barnes PJ (1991) Inhibition of eosinophil cyclic nucleotide PDE activity and opsonised zymosan-stimulated respiratory burst by 'type IV'-selective PDE inhibitors. *Brit J Pharmacol* 103: 1339–1346

182 Hatzelmann A, Tenor H, Schudt C (1995) Differential effects of non-selective and selective phosphodiesterase inhibitors on human eosinophil functions. *Brit J Pharmacol* 114: 821–831

183 Souness JE, Carter CM, Diocee BK, Hassall GA, Wood LJ, Turner NC (1991) Characterisation of guinea-pig eosinophil phosphodiesterase activity Assessment of its involvement in regulating superoxide generation. *Biochem Pharmacol* 42: 937–945

184 Torphy TJ, Livi GP (1992) Therapeutic potential of isozyme-selective phosphodiesterase inhibitors in the treatment of asthma. *Adv Second Messenger Phosphoprotein Res* 25: 289–305

185 Peachell PT, Undem BJ, Schleimer RP, MacGlashan DW Jr, Lichtenstein LM, Cieslinski LB, Torphy TJ (1992) Preliminary identification and role of phosphodiesterase isoenzymes in human basophils. *J Immunol* 148: 2503–2510

186 Kleine-Tebbe J, Wicht L, Gagne H, Friese A, Schunack W, Schudt C, Kunkel G (1992) Inhibition of IgE-mediated histamine release from human peripheral leukocytes by selective phosphodiesterase inhibitors. *Agents Actions* 36: 200–206

187 Sommer N, Loschmann PA, Northoff GH, Weller M, Steinbrecher A, Steinbach JP, Lichtenfels R, Meyermann R, Riethmuller A, Fontana A et al (1995) The antidepressant rolipram suppresses cytokine production and prevents autoimmune encephalomyelitis. *Nature* Med 1: 244–248

188 Foissier L, Lonchampt M, Coge F, Canet E (1996) *In vitro* down-regulation of antigeninduced IL-5 gene expression and protein production by cAMP-specific phosphodiesterase type 4 inhibitor. *J Pharmacol Exp Ther* 278: 1484–1490

189 Kambayashi T, Jacob CO, Zhou D, Mazurek N, Fong M, Strassmann G (1995) Cyclic nucleotide phosphodiesterase type IV participates in the regulation of IL-10 and in the subsequent inhibition of TNFα and IL-6 release by endotoxin-stimulated macrophages. *J Immunol* 155: 4909–4916

190 Turner NC, Wood LJ, Burns FM, Gueremy T, Souness JE (1993) The effect of cyclic AMP and cyclic GMP phosphodiesterase inhibitors on the superoxide burst of guineapig peritoneal macrophages. *Brit J Pharmacol* 108: 876–883

191 Verghese MW, McConnell RT, Strickland AB, Gooding RC, Stimpson SA, Yarnall DP, Taylor JD, Furdon PJ (1995) Differential regulation of human monocyte-derived TNFα

and IL-1β by type IV cAMP-phosphodiesterase (cAMP- PDE) inhibitors. *J Pharmacol Exp Ther* 272: 1313–1320

192 Chini CC, Chini EN, Williams JM, Matousovic K, Dousa TP (1994) Formation of reactive oxygen metabolites in glomeruli is suppressed by inhibition of cAMP phosphodiesterase isozyme type IV. *Kidney Int* 46: 28–36

193 Matousovic K, Grande JP, Chini CC, Chini EN, Dousa TP (1995) Inhibitors of cyclic nucleotide phosphodiesterase isozymes type-III and type-IV suppress mitogenesis of rat mesangial cells. *J Clin Invest* 96: 401–410

194 Morandini R, Ghanem G, Portier-Lemarie A, Robaye B, Renaud A, Boeynaems JM (1996) Action of cAMP on expression and release of adhesion molecules in human endothelial cells. *Am J Physiol* 270: H807–816

195 Suttorp N, Weber U, Welsch T, Schudt C (1993) Role of phosphodiesterases in the regulation of endothelial permeability *in vitro. J Clin Invest* 91: 1421–1428

196 Endres S, Fulle HJ, Sinha B, Stoll D, Dinarello CA, Gerzer R, Weber PC (1991) Cyclic nucleotides differentially regulate the synthesis of tumour necrosis factor-alpha and interleukin-1 beta by human mononuclear cells Immunol 72: 56–60

197 Prabhakar U, Lipshutz D, Bartus JO, Slivjak MJ, Smith EF 3rd, Lee JC, Esser KM (1994) Characterization of cAMP-dependent inhibition of LPS-induced TNF alpha production by rolipram, a specific phosphodiesterase IV (PDE IV) inhibitor. *Int J Immunopharmacol* 16: 805–816

198 Verghese MW, McConnell RT, Strickland AB, Gooding RC, Stimpson SA, Yarnall DP, Taylor JD, Furdon PJ (1995) Differential regulation of human monocyte-derived TNF alpha and IL-1 beta by type IV cAMP-phosphodiesterase (cAMP-PDE) inhibitors. *J Pharmacol Exp Ther* 272: 1313–1320

199 Souness JE, Griffin M, Maslen C, Ebsworth K, Scott LC, Pollock K, Palfreyman MN, Karlsson JA (1996) Evidence that cyclic AMP phosphodiesterase inhibitors suppress TNF alpha generation from human monocytes by interacting with a 'low-affinity' phosphodiesterase 4 conformer. *Brit J Pharmacol* 118: 649–658

200 Yoshimura T, Kurita C, Nagao T, Usami E, Nakao T, Watanabe S, Kobayashi J, Yamazaki F, Tanaka H, Nagai H (1997) Effects of cAMP-phosphodiesterase isozyme inhibitor on cytokine production by lipopolysaccharide-stimulated human peripheral blood mononuclear cells. *Gen Pharmacol* 29: 633–638

201 Barnette MS, Christensen SB, Essayan DM, Grous M, Prabhakar U, Rush JA, Kagey-Sobotka A, Torphy TJ (1998) SB 207499 (Ariflo), a potent and selective second-generation phosphodiesterase 4 inhibitor: *in vitro* anti-inflammatory actions. *J Pharmacol Exp Ther* 284: 420–426

202 Kambayashi T, Jacob CO, Zhou D, Mazurek N, Fong M, Strassmann G (1995) Cyclic nucleotide phosphodiesterase type IV participates in the regulation of IL-10 and in the subsequent inhibition of TNF-alpha and IL-6 release by endotoxin-stimulated macrophages. *J Immunol* 155: 4909–4916

203 Allen R, Rapecki S, Higgs G (1997) The role of IL-10 in the inhibition of LPS- mediat-

ed TNF release from human PBMCs by phosphodiesterase 4 (PDE4) inhibitors. *Inflamm Res* 46: S218, Abs P-1-3-11

204 Armstrong L, Jordan N, Millar A (1996) Interleukin 10 (IL-10) regulation of tumour necrosis factor alpha (TNF-alpha) from human alveolar macrophages and peripheral blood monocytes. *Thorax* 51: 143–149

205 Seldon PM, Barnes PJ, Giembycz MA (1998) Interleukin-10 does not mediate the inhibitory effect of PDE-4 inhibitors and other cAMP-elevating drugs on liposaccharide-induced tumour necrosis factor-alpha generation from human peripheral blood monocytes. *Cell Biochem Biophys* 29: 179–201

206 Gantner F, Kupferschmidt R, Schudt C, Wendel A, Hatzelmann A (1997) *In vitro* differentiation of human monocytes to macrophages: change of PDE profile and its relationship to suppression of tumour necrosis factor-alpha release by PDE inhibitors. *Brit J Pharmacol* 121: 221–231

207 Raeburn D, Underwood SL, Lewis SA, Woodman VR, Battram CH, Tomkinson A, Sharma S, Jordan R, Souness JE, Webber SE, Karlsson J-A (1994) Anti-inflammatory and bronchodilator properties of RP73401, a novel and selective phosphodiesterase type IV inhibitor. *Brit J Pharmacol* 113: 1423–1431

208 Banner KH, Marchini F, Buschi A, Moriggi E, Semeraro C, Page CP (1995) The effect of selective phosphodiesterase inhibitors in comparison with other anti-asthma drugs on allergen-induced eosinophilia in guinea-pig airways. *Pulm Pharmacol* 8: 37–42

209 Gozzard N, el-Hashim A, Herd CM, Blake SM, Holbrook M, Hughes B, Higgs GA, Page CP (1996) Effect of the glucocorticosteroid budesonide and a novel phosphodiesterase type 4 inhibitor CDP840 on antigen-induced airway responses in neonatally immunised rabbits. *Brit J Pharmacol* 118: 1201–1208

210 Hughes B, Howat D, Lisle H, Holbrook M, James T, Gozzard N, Blease K, Hughes P, Kingaby R, Warrellow G et al (1996) The inhibition of antigen- induced eosinophilia by CDP840, a novel stereo-selective inhibitor of phosphodiesterase type 4. *Brit J Pharmacol* 118: 1183–1191

211 Holbrook M, Gozzard N, James T, Higgs G, Hughes B (1996) Inhibition of bronchospasm and ozone-induced airway hyperresponsiveness in the guinea-pig by CDP840, a novel phosphodiesterase type 4 inhibitor. *Brit J Pharmacol* 118: 1192– 1200

212 Gozzard N, Herd CM, Blake SM, Holbrook M, Hughes B, Higgs G, Page CP (1996) Effects of theophylline and rolipram on antigen-induced airway responses in neonatally immunized rabbits. *Brit J Pharmacol* 117: 1405–1412

213 Turner CR, Cohan VL, Cheng JB, Showell HJ, Pazoles CJ, Watson JW (1996) The *in vivo* pharmacology of CP-80, 633, a selective inhibitor of phosphodiesterase 4. *J Pharmacol Exp Ther* 278: 1349–1355

214 Underwood DC, Bochnowicz S, Osborn RR, Kotzer CJ, Luttmann MA, Hay DW, Gorycki PD, Christensen SB, Torphy TJ (1998) Antiasthmatic activity of the second-generation phosphodiesterase 4 (PDE4) inhibitor SB 207499 (Ariflo) in the guinea-pig. *J Pharmacol Exp Ther* 287: 988–995

215 Sekut L, Yarnall D, Stimpson SA, Noel LS, Bateman-Fite R, Clark RL, Brackeen MF,

Menius JA Jr, Connolly KM (1995) Anti-inflammatory activity of phosphodiesterase (PDE)-IV inhibitors in acute and chronic models of inflammation. *Clin Exp Immunol* 100: 126–132

216 Nyman U, Mussener A, Larsson E, Lorentzen J, Klareskog L (1997) Amelioration of collagen II-induced arthritis in rats by the type IV phosphodiesterase inhibitor rolipram. *Clin Exp Immunol* 108: 415–419

217 Tsuboi Y, Shankland SJ, Grande JP, Walker HJ, Johnson RJ, Dousa TP (1996) Suppression of mesangial proliferative glomerulonephritis development in rats by inhibitors of cAMP phosphodiesterase isozymes types III and IV. *J Clin Invest* 98: 262–270

218 Badger AM, Olivera DL, Esser KM (1994) Beneficial effects of the phosphodiesterase inhibitors BRL61063, pentoxyfylline, and rolipram in a murine model of endotoxin shock. *Circ Shock* 44: 188–195

219 Fischer W, Schudt C, Wendel A (1993) Protection by phosphodiesterase inhibitors against endotoxin-induced liver injury in galactosamine-sensitized mice. *Biochem Pharmacol* 45: 2399–2404

220 Sekut L, Menius JA Jr, Brackeen MF, Connolly KM (1994) Evaluation of the significance of elevated levels of systemic and localized tumor necrosis factor in different animal models of inflammation. *J Lab Clin Med* 124: 813–820

221 Rabinovici R, Feuerstein G, Abdullah F, Whiteford M, Borboroglu P, Sheikh E, Phillip DR, Ovadia P, Bobroski L, Bagasra O, Neville LF (1996) Locally produced tumor necrosis factor-alpha mediates interleukin-2-induced lung injury. *Circ Res* 78: 329–336

222 Howell RE, Jenkins LP, Howell DE (1995) Inhibition of lipopolysaccharide- induced pulmonary edema by isozyme-selective phosphodiesterase inhibitors in guinea-pigs. *J Pharmacol Exp Ther* 275: 703–709

223 Sommer N, Loschmann PA, Northoff GH, Weller M, Steinbrecher A, Steinbach JP, Lichtenfels R, Meyermann R, Reithmuller A, Fontana A et al (1995) The antidepressant rolipram suppresses cytokine production and prevents autoimmune encephalomyelitis. *Nat Med* 1: 244–248

224 Genain CP, Roberts T, Davis RL, Nguyen MH, Uccelli A, Faulds D, Li Y, Hedgpeth J, Hauser SL (1995) Prevention of autoimmune demyelination in non-human primates by a cAMP-specific phosphodiesterase inhibitor. *Proc Natl Acad Sci USA* 92: 3601–3605

225 Jung S, Zielasek J, Kollner G, Donhauser T, Toyka K, Hartung HP (1996) Preventive but not therapeutic application of rolipram ameliorates experimental autoimmune encephalomyelitis in Lewis rats. *J Neuroimmunol* 68: 1–11

226 Kato H, Araki T, Itoyama Y, Kogure K (1995) Rolipram, a cyclic AMP- selective phosphodiesterase inhibitor, reduces neuronal damage following cerebral ischemia in the gerbil. *Eur J Pharmacol* 272: 107–110

227 Barnard JW, Siebert AF, Prasad VR, Smart DA, Strada SJ, Taylor AE, Thompson WJ (1994) Reversal of pulmonary capillary ischemia-reperfusion injury by rolipram, a cAMP phosphodiesterase inhibitor. *J Appl Physiol* 77: 774–781

228 Hulley P, Hartikka J, Abdel'Al S, Engels P, Buerki HR, Wiederhold KH, Muller T, Kelly

P, Lowe D, Lubbert H (1995) Inhibitors of type IV phosphodiesterases reduce the toxicity of MPTP in *substantia nigra* neurons *in vivo*. *Eur J Neurosci* 7: 2431–2440

229 Feldmann M, Brennan FM, Maini RN (1996) Role of cytokines in rheumatoid arthritis. *Ann Rev Immunol* 14: 397–440

230 Feldmann M, Brennan FM, Elliott MJ, Williams RO, Maini RN (1995) TNF alpha is an effective therapeutic target for rheumatoid arthritis. *Ann NY Acad Sci* 766: 272–278

231 Camussi G, Lupia E (1998) The future role of anti-tumour necrosis factor (TNF) products in the treatment of rheumatoid arthritis Drugs 55: 613–620

232 Moreland LW, Baumgartner SW, Schiff MH, Tindall EA, Fleischmann RM, Weaver AL, Ettlinger RE, Cohen S, Koopman WJ, Mohler K et al (1997) Treatment of rheumatoid arthritis with a recombinant human tumor necrosis factor receptor (p75)-Fc fusion protein. *N Eng J Med* 337: 141–147

233 Moreland LW (1998) Soluble tumor necrosis factor receptor (p75) fusion protein (ENBREL) as a therapy for rheumatoid arthritis. *Rheum Dis Clin North Am* 24: 579–591

234 Goldenberg MM (1999) Etanercept, a novel drug for the treatment of patients with severe, active rheumatoid arthritis. *Clin Ther* 21: 75–87

235 Weinblatt ME, Kremer JM, Bankhurst AD, Bulpitt KJ, Fleischmann RM, Fox RI, Jackson CG, Lange M, Burge DJ (1999) A trial of etanercept, a recombinant tumor necrosis factor receptor:Fc fusion protein, in patients with rheumatoid arthritis receiving methotrexate. *N Eng J Med* 340: 253–259

236 Moreland LW, Schiff MH, Baumgartner SW, Tindall EA, Fleischmann RM, Bulpitt KJ, Weaver AL, Keystone EC, Furst DE, Mease PJ et al (1999) Etanercept therapy in rheumatoid arthritis A randomized, controlled trial. *Ann Intern Med* 130: 478–486

237 Maini RN, Breeveld FC, Kalden JR, Smolen JS, Davis D, Macfarlane JD, Antoni C, Leeb B, Elliott MJ, Woody JN et al (1998) Therapeutic efficacy of multiple intravenous infusions of anti-tumor necrosis factor alpha monoclonal antibody combined with low-dose weekly methotrexate in rheumatoid arthritis. *Arthritis Rheum* 41: 1552–1563

238 Moreland LW (1999) Inhibitors of tumor necrosis factor for rheumatoid arthritis. *J Rheumatol* 26 (Suppl 57): 7–15

239 Dower SK, Franslow W, Jacobs C, Waugh S, Sims JE, Widmer MB (1994) Interleukin-1 antagonists. *Ther Immunol* 1: 113–122

240 Jones RE, Moreland LW (1999) Tumor necrosis factor inhibitors for rheumatoid arthritis. *Bull Rheum Dis* 48: 1–4

241 Souness JE, Foster M (1998) Potential of phosphodiesterase type IV inhibitors in the treatment of rheumatoid arthritis. *Invest Drugs* 1: 541–553

242 Chikanza IC, Jawed SJ, Blake DR, Perrot S, Menkes CJ, Barnes CG, Perry JD, Wright MG (1996) Treatment of patients with rheumatoid arthritis with RP 73401 phosphodiesterase type IV inhibitor. *Arthritis Rheum* 39: S282

243 Harbinson PL, MacLeod D, Hawksworth R, O'Toole S, Sullivan PJ, Heath P, Kilfeather S, Page CP, Costello J, Holgate ST, Lee TH (1997) The effect of a novel orally active

selective PDE4 isoenzyme inhibitor (CDP840) on allergen-induced responses in asthmatic subjects. *Eur Resp J* 10: 1008–1014

244 Hughes B, Owens R, Perry M, Warrellow G, Allen R (1997) PDE 4 inhibitors: the use of molecular cloning in the design and development of novel drugs. *Drug Disc Today* 2: 89–101

245 Perry MJ, O'Connell J, Walker C, Crabbe T, Baldock D, Russell A, Lumb S, Huang Z, Howat D, Allen R et al (1998) CDP840 A novel inhibitor of PDE-4. *Cell Biochem Biophys* 29: 113–132

246 Allen RA, Merriman MW, Perry MJ, Owens RJ (1999) Development of a recombinant cell-based system for the characterisation of phosphodiesterase 4 isoforms and evaluation of inhibitors. *Biochem Pharmacol* 57: 1375–1382

247 Griswold DE, Webb EF, Badger AM, Gorycki PD, Levandoski PA, Barnette MA, Grous M, Christensen S, Torphy TJ (1998) SB 207499 (Ariflo), a second generation phosphodiesterase 4 inhibitor, reduces tumor necrosis factor alpha and interleukin-4 production *in vivo. J Pharmacol Exp Ther* 287: 705–711

Oligonucleotide-based drugs in the control of cytokine synthesis

Stanley T. Crooke

Isis Pharmaceuticals, Inc., 2292 Faraday Avenue, Carlsbad, CA 92008, USA

Introduction

As cytokines play key roles in physiological and pathophysiological processes, substantial research has been directed to identifying and elucidating the structures and functions of various cytokines and to understanding the processes that regulate them. Significant efforts have also been directed toward the identification of selective inhibitors of many of the cytokines (for review, see [1]).

Antisense oligonucleotides that inhibit a number of cytokines have also been reported (for review, see [2]). Additionally, polynucleotides and, more recently, oligonucleotides have been shown to induce cytokine release in a number of species [3]. Thus, oligonucleotides may prove to be useful tools to dissect the roles of various cytokines, useful therapeutic agents for cytokine-mediated diseases and may be interesting adjuvants as a result of their cytokine-releasing properties.

In this chapter, the basic properties of antisense oligonucleotides, the status of the technology and of antisense drugs designed to inhibit cytokines and the proinflammatory effects of oligonucleotides will be reviewed.

Overview of antisense technology

In the past several years, substantial interest in the use of oligonucleotide analogues as drugs has developed. Although oligonucleotide-based drugs can theoretically be designed to inhibit transcription either by interacting with chromatin *via* DNA strand invasion [4, 5] or triple-strand binding [6] or by binding to required transcription factors [7], the greatest excitement and the most substantial progress has been in the creation of oligonucleotide analogues designed to work *via* an antisense mechanism [7–9]. The theoretical specificity of antisense drugs makes the concept highly seductive from a therapeutic perspective and as a tool to help dissect biological processes.

Novel Cytokine Inhibitors, edited by Gerry A. Higgs and Brian Henderson
© 2000 Birkhäuser Verlag Basel/Switzerland

Molecular mechanisms of action

Figure 1 shows a simplified view of RNA intermediary metabolism. Conceptually, virtually any step in the process of synthesis, maturation or degradation of mRNA is amenable to antisense inhibition. In fact, antisense drugs that interfere with 5' cap formation, splicing and translation have been reported (for review, see [10]). Additionally, antisense inhibitors that cause degradation of target RNA by serving as substrates for double-strand RNases have been reported [11]. However, the mechanism of action about which most is known and the mechanism by which all antisense drugs in development are thought to work is RNase H [12]. The RNaseHs are a group of enzymes that cleave RNA in a DNA-RNA duplex (for review, see [13]). Very recently, a human RNase H has been cloned and expressed [14].

Proof of mechanism of action

Antisense technology is less than 10 years old. It is thus clear that there are many questions left to be answered than for which we have answers. One of the most crucial issues is to determine the types of experiments required to provide sufficient data to conclude that effects observed are likely to be due to an antisense mechanism.

Until more is understood about how antisense drugs work, it is essential to positively demonstrate effects consistent with an antisense mechanism. For RNase H-activating oligonucleotides, Northern blot analysis showing selective loss of the target RNA is the best choice, and many laboratories are publishing reports of such activities *in vitro* and *in vivo* [15–18]. Ideally, a demonstration that closely related isotypes are unaffected should be included. In brief, then, for proof of mechanism, the following steps are recommended:

- Perform careful dose-response curves *in vitro*, using several cell lines and methods of *in vitro* delivery.
- Correlate the rank order potency *in vivo* with that observed *in vitro* after thorough dose-response curves are generated *in vivo*.
- Perform careful „gene walks" for all RNA species and oligonucleotide chemical classes.
- Perform careful time courses before drawing conclusions about potency.
- Directly demonstrate the proposed mechanism of action by measuring the target RNA and/or protein.
- Evaluate specificity and therapeutic indices *via* studies on closely related isotypes and with appropriate toxicological studies.
- Perform sufficient pharmacokinetics to define rational dosing schedules for pharmacological studies.

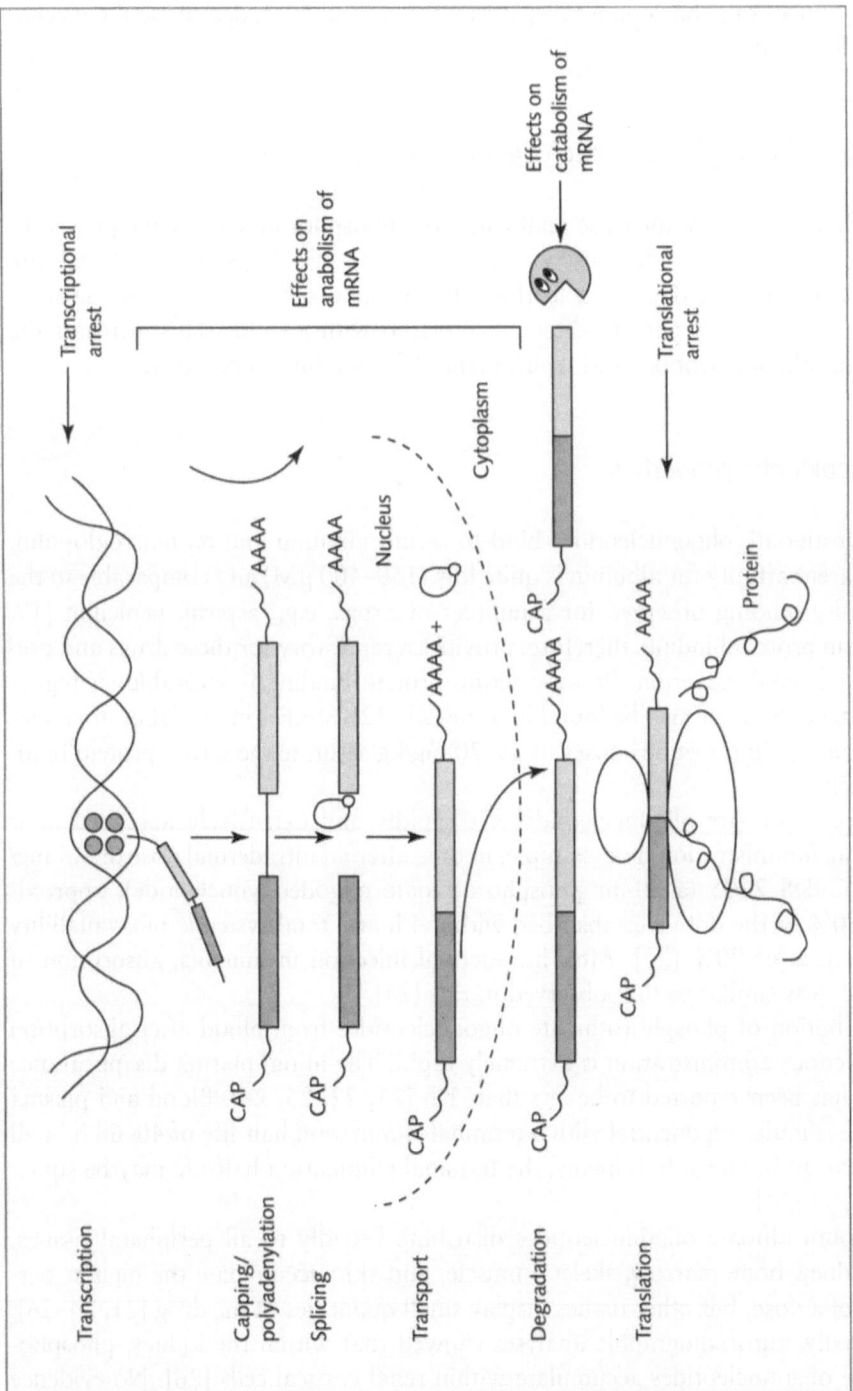

Figure 1:
RNA processing.

- When control oligonucleotides display surprising activities, determine the mechanisms involved.

Properties of phosphorothioate oligonucleotides

Of the first generation antisense analogues, the phosphorothioates have proven to have the best properties and consequently all the antisense drugs in clinical trials are phosphorothioate oligodeoxynucleotides. In phosphorothioates, one of the non-bridging oxygens of a phosphodiester is replaced with a sulfur. This increases the stability of oligonucleotides and their potential for binding to proteins.

Pharmacokinetic properties

Phosphorothioate oligonucleotides bind to serum albumin and α_2 macroglobulin. The apparent affinity for albumin is quite low (150–400 µM) and comparable to the low-affinity binding observed for a number of drugs, e.g., aspirin, penicillin [19, 20]. Serum protein binding, therefore, provides a repository for these drugs and prevents rapid renal excretion. Because serum protein binding is saturable, at higher doses, intact oligomer may be found in urine [21, 22]. Studies in our laboratory suggest that in rats intravenous doses of 15–20 mg/kg saturate the serum protein binding capacity.

Phosphorothioate oligonucleotides are rapidly and extensively absorbed after parenteral administration. For example, in rats, after an intradermal dose of 3.6 mg/kg of ^{14}C-ISIS 2105 (a 20-mr phosphorothioate oligodeoxynucleotide), approximately 70% of the dose was absorbed within 4 h and total systemic bioavailability was in excess of 90% [23]. After intradermal injection in humans, absorption of ISIS 2105 was similar to that observed in rats [24].

Distribution of phosphorothioate oligonucleotides from blood after absorption or intravenous administration is extremely rapid. The initial plasma disappearance half-life has been reported to be less than 1 h [21, 22, 25, 26]. Blood and plasma clearance is multi-exponential with a terminal elimination half-life of 40–60 h in all species except humans. In humans, the terminal elimination half-life may be somewhat longer [24].

Phosphorothioate oligonucleotides distribute broadly to all peripheral tissues. Liver, kidney, bone marrow, skeletal muscle, and skin accumulate the highest percentage of a dose, but other tissues display small quantities of the drug [21, 24–26]. Additionally, autoradiographic analyses showed that within the kidney, phosphorothioate oligonucleotides accumulate within renal cortical cells [26]. No evidence of significant penetration of the blood brain barrier has been reported. The rates of incorporation and clearance from tissues vary as a function of the organ studied, with the liver accumulating the drug most rapidly (~20% of a dose within 1–2 h) and other tissues accumulating the drug more slowly. Similarly, elimination of the drug is more rapid from the liver than any other tissue, e.g., terminal half-life from the liver is 62 h; from renal medula, 156 h.

Very recently the distribution within the liver after an intravenous dose in rats has been reported [27]. The distribution within cells in the liver was time- and dose-dependent. At low doses, most of the drug was distributed to Kupffer and endothelial cells. At higher doses, these cells became saturated, and a higher fraction of the drug became distributed into the hepatocyte fraction.

At relatively low doses, clearance of phosphorothioate oligonucleotides is due primarily to metabolism [23, 25, 26, 28]. Metabolism is mediated by exo- and endonucleases that result in shorter oligonucleotides and, ultimately, in nucleosides that are degraded by normal metabolic pathways. Although no direct evidence of base excision or modification has been reported, these are theoretical possibilities that may occur. In one study, a larger molecular weight radioactive material was observed in urine but not fully characterized [21]. Clearly, the potential for conjugation reactions and extension of oligonucleotides *via* these drugs serving as primers from polymerases must be explored in more detail.

In summary, pharmacokinetic studies of several phosphorothioate oligonucleotides demonstrate that they are well absorbed from parenteral sites, are distributed broadly to all peripheral tissues, do not cross the blood brain barrier and are eliminated primarily by metabolism. In short, once a day or every other day, systemic dosing should be feasible. Although the similarities between oligonucleotides of different sequences are far greater than the differences, additional studies are required before determining whether there are subtle effects of sequence on the pharmacokinetic profile of this class of drugs.

Pharmacological effects

Numerous studies have been reported that demonstrate a variety of effects of antisense phosphorothioate oligonucleotides in various animal models (for review, see [29]). In general, the sophistication of experimentation and the types of controls employed have improved as experience has been gained, and several recent studies provide sufficient data to conclude that the effects observed are likely to be due to an antisense mechanism. For example, Dean and McKay [16] provide direct proof of systemic antisense activity and the potential specificity of such drugs. In this study, antisense oligonucleotides designed to inhibit murine protein kinase C-α (PKC-α) were administered in various doses once a day intraperitoneally. Various tissues were obtained, and the effects of the drugs on PKC-α RNA levels were evaluated by Northern blots. Selective loss of PKC-α RNA in liver was reported with an ID_{50} of 10–20 mg/kg. No evidence of tolerance was observed after as many as seven doses, and the duration of action was at least 24 h. Probing with cDNAs for other isotypes of PKC showed that there were no effects on even very closely related isotypes. Several other studies have demonstrated a sequence-specific pharmacological effect by the antisense oligonucleotide but not control oligonucleotides.

Toxicological effects

The toxicities of numerous phosphorothioate oligodeoxynucleotides have been evaluated in several animal species and the toxicities have proven to be quite consistent with only modest effects noted due to changes in sequences (for review, see [30–32]). In the rodent, the most prominent toxicity appears to be related to cytokine release. Animals treated chronically develop multi-focal, multi-organ inflammation, the most prominent early indication of which is splenomegaly. Although some sequences are more potent than others, this is clearly a class effect of the phosphorothioates. The pro-inflammatory effects of phosphorothioates are far less prominent in species other than rodents. For example, in humans treated chronically with phosphorothioates, inflammation has not proven to be a significant side-effect, except in the eye after intravitreal dosing [33].

In monkeys, the dose-limiting toxicity is hypotension associated with complement activation. Complement activation quite consistently occurs at plasma concentrations of approximately 70 mg/ml but hypotension in the monkeys is very sporadic [30, 31]. Again, this toxicity has not been observed in humans. In humans, several thousand patients have now been treated with several phosphorothioate oligonucleotides, and more than two hundred patients have been treated for six months or longer. Thus, there is enough experience to reach initial conclusions. After systemic administration at doses as high as 6–7 mg/kg every other day, few adverse events have been noted. The most frequent side-effects are related to clotting resulting in plasma concentration-dependent increase in activated partial thromboplastin. This has not been dose-limiting, but intravenous bolus administration has not been attempted to avoid peak plasma concentrations that might be problematic. Additionally, in some patients treated with doses in excess of 4 mg/kg, transient thrombocytopenia has been observed. This does not appear to be due to effects in the bone marrow. Rather, it may be related to margination effects (for review, see [12]).

Clinical activities

Although preliminary evidence of activity has been reported for several antisense anticancer drugs [34–36], and reductions in target RNA levels have been observed [36], only two antisense drugs have progressed sufficiently in clinical trials to display statistically significant effects.

We have recently filed a new drug application for fomivirsen. Fomivirsen is an antisense inhibitor of cytomegalovirus (CMV) replication that has been shown to be highly safe and effective in the treatment of CMV-induced retinitis patients [33, 37]. The drug is administered intravitreally every two to four weeks. It results in prolonged control of CMV retinitis that is highly statistically significantly different from other treatments, with no systemic toxicities. Mild to moderate inflammation in the eye was the only side-effect of note.

ISIS 2302 is an antisense inhibitor of intercellular adhesion molecule 1 (ICAM-1), currently in clinical trials evaluating several inflammatory diseases. A randomized double-blind placebo controlled trial resulted in substantial improvement in Crohn's disease activity index while producing a dramatic reduction in steroid doses required (p = 0.0001) that lasted for six months after one month of treatment. No adverse events were reported in this study [38–40].

Medicinal chemistry

Early on it was anticipated that there would be significant limitations on the use of natural DNA as antisense therapeutics; as a result, a significant number of modifications of the oligonucleotide structure have been made [41–44]. A thorough review of the medicinal chemistry of oligonucleotides is beyond the scope of this article. However, it is worth highlighting some of the more promising modifications. Figure 2 shows a dimeric structure with examples of the positions in which modifications have been synthesized, incorporated into oligonucleotides, and tested for effects on hybridization and nuclease resistance, and in some cases pharmacological activity. Many modifications have been performed on the nucleobase. Among the more promising are substitutions at the C5 position of pyrimidines, resulting in increased hybridization affinity and maintaining recognition of the heteroduplex by RNase H. In particular, 5-propynyl C and T substitutions increase the Tm of an oligonucleotide by as much as 2°C per modification and still support RNase H activity [45, 46]. The more pyrimidine substitutions that are made in a given oligonucleotide, the more benefit will be derived. The 5-propynyl substitutions do not provide any significant nuclease resistance; therefore, the internucleosidic linkage still must be protected by modifications such as phosphorothioate substitutions.

A number of modifications on the 2'-position of the sugar have resulted in increased affinity for RNA and in some cases increased nuclease resistance [47–51]. However, oligonucleotides modified with current 2'-substitutions such as O-methyl, fluoro, O-propyl, O-allyl, and methoxyethoxy do not support RNase H activity. Therefore, either non-RNase H mechanisms must be exploited to fully capitalize on the benefits of such oligonucleotides, or alternatively, a chimeric or gap strategy must be used to retain some RNase H activity [52–54].

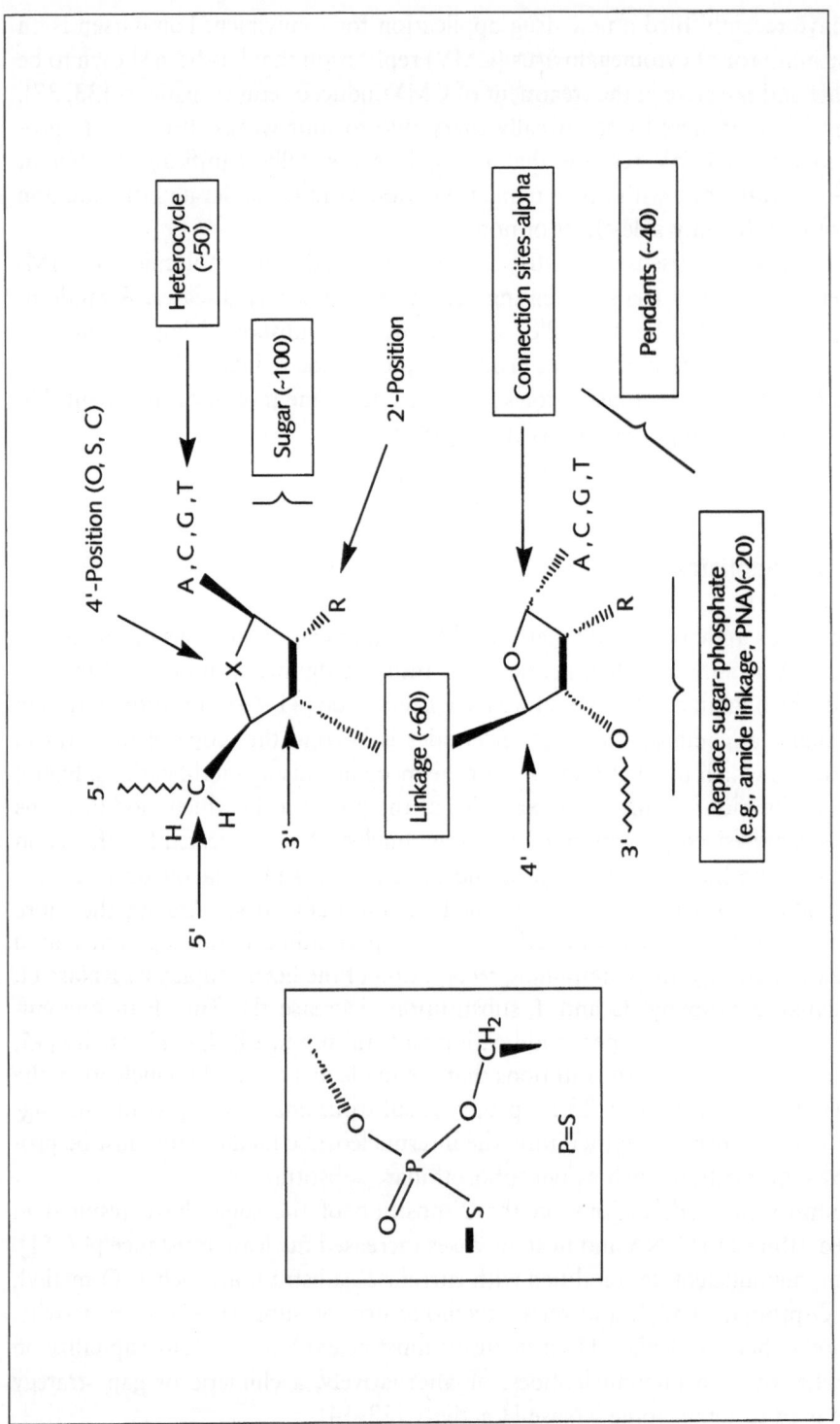

Figure 2
Positions and types of oligonucleotide modifications.

Modifications to the phosphate backbone have been made with the aim of increasing nuclease resistance, increasing affinity, and possibly enhancing the pharmacokinetic properties of the oligonucleotide. A number of modifications on either the phosphate backbone or substitution of the phosphate groups have been made and are providing promising biophysical and biological activity. Methylphosphonate oligonucleotides eliminate the charge on the backbone but introduces a chiral center. Although much has been written about the benefits of methylphosphonate substitutions [55–57], they have not achieved widespread use. This may be due in part to not supporting RNase H activity and the low affinity of chirally impure oligonucleotides. Recently, prepared chirally pure dimers of methylphosphonates have demonstrated marked increased affinity when multiple dimers are incorporated into oligonucleotides [58]. A number of modifications have been made in which the phosphate linkage has been replaced (dephosphono linkages) with achiral linkages. In particular, the methyleneimino, amide-3 and thioformacetal linkages [44, 59–61] are very promising. In each case the linkages support hybridization and may even increase affinity and provide nuclease resistance; they are also compatible with existing phosphoromidite chemistries. Currently, these substitutions are prepared as dimers and incorporated into oligonucleotides as such, with a phosphate linkage between each dimer. Like the 2' modifications, replacement of the phosphate linkage results in loss of RNase H activity; therefore, similar strategies will need to be employed to take full advantage of them.

More radical substitutions have been made in which the sugar and phosphate residues were replaced with an alkylamide linkage [5]. Surprisingly, this modification, peptide nucleic acid (PNA), increased affinity for both RNA and DNA and preserved specificity, and is not susceptible to either nuclease or peptidase cleavage [62–64]. Similarly, morpholino nucleoside oligomers with a carbamate internucleosidic linkage have been prepared that preserve hybridization [65]. Finally, various functional groups have been covalently attached to oligonucleotides to change their physical properties, provide ligands for interacting with specific receptors, and provide additional nuclease resistance [66–69].

Although each of these modifications have improved the properties of the current first-generation phosphorothioate oligodeoxynucleotides, it is unlikely that any single chemical modification will solve all the issues. Therefore, a combination of approaches will probably have to be undertaken to provide the optimum benefit chemical modifications have to offer.

Conclusions

Based on progress in studying phosphorothioate oligodeoxynucleotides, answers to a number of questions are now available. Clearly, antisense oligonucleotide analogues can be created with pharmacokinetic properties that are acceptable for parenteral administration. Equally clear, well-controlled studies have demonstrated potent systemic pharmacological activities in animals that are likely due to the antisense mechanism. Although phosphorothioate oligodeoxynucleotides have clear toxicological limits, studies in animals and humans suggest that the therapeutic index of phosphorothioate oligodeoxynucleotides may be acceptable for a number of indications. Moreover, progress in scale-up and cost reduction has been substantial enough to provide confidence that the cost of therapy with these drugs should not be prohibitive.

Progress in the medicinal chemistry of oligonucleotides has been equally gratifying. Again, much remains to be learned, but there is no question that antisense drugs with improved properties relative to phosphorothioate oligodeoxynucleotides have been created. Tests in progress in animals should provide significant data to support clinical trials with new chemical classes in the near future.

Long-term therapeutic potential

The long-term therapeutic potential of antisense oligonucleotides is not defined yet. And it is still possible that with a more sophisticated understanding of this technology, researchers will conclude that the technology has an as yet unidentified fatal flaw. Nevertheless, the progress to date provides justification for continued exploration and optimization. It will be several years before the therapeutic value of phosphorothioate oligodeoxynucleotides is understood and, it is hoped, many years before advances in the medicinal chemistry of oligonucleotides no longer result in new classes of antisense drugs with improved therapeutic properties and utilities.

Inhibition of cytokine action with oligonucleotides

Cytokines

Although antisense oligonucleotides have been used to inhibit the expression of a large number of cytokines, progress in connecting antisense compounds to products in clinical development has been less than in other areas. In many cases the oligonucleotides were used as a tool to demonstrate that the cytokine serves as an autocrine growth factor for the targeted cell population. For example, antisense oligonucleotides to interleukin-1α (IL-1α) and IL-4 were used to demonstrate that the respective interleukins were autocrine growth factors for Th type-2 cells [70, 71] while IL-2 antisense oligonucleotides were used to demonstrate that IL-2 is an autocrine growth factor for Th type-1 cells [72]. The growth-inhibitory effects of the oligonucleotides were reversed by the addition of the respective cytokines back to the culture. Similarly, IL-6 was shown to be an autocrine growth factor for Kaposi's sarcoma-derived cells and human myeloma cell lines, IL-8 was shown to be an autocrine growth factor for human melanoma cells, and colony stimulating factor-1 (CSF-1) was shown to be an autocrine growth factor for murine monocytes [72–75]. In each case, phosphodiester oligodeoxynucleotides were used to inhibit the expression of the target cytokine.

An antisense, but not sense, oligonucleotide to IL-1β inhibited lymphokine-activated killer cell induction [76]. The effect of the antisense oligonucleotide was attenuated by inclusion of IL-1β protein in the culture medium. Oligonucleotides to IL-1α have been used to prevent senescence in human umbilical vein endothelial cells and block the growth-inhibitory effects of poly (rI):poly (rC) in endothelial cells [77, 78]. These studies led to the conclusion that IL-1α expression inhibited endothelial cell growth, ultimately leading to cellular senescence.

Benbernou et al. used antisense oligonucleotides designed to hybridize to murine IL-4 to inhibit IL-4-mediated IgE synthesis in *Nippostrongylus brasilienis*-infected rat spleen cells [79, 80]. Two of the three antisense oligonucleotides used inhibited IgE and IgG2a synthesis. Comparison of the sequences between mouse and rat IL-4 mRNA in the regions where the oligonucleotides were designed to hybridize, revealed several sequence differences between rat and mouse IL-4 mRNA. The most active oligonucleotide contained mismatches in 3 of 15 bases, while an oligonucleotide which did not exhibit activity had 3 of 20 bases which should not hybridize to rat IL-4 mRNA. Although the effects of the active oligonucleotides were reversed by the addition of IL-4 to the cells, the authors were unable to demonstrate an effect of the oligonucleotides on IL-4 production due to the low levels produced in the cells. Therefore, mechanistically it was not proven that the oligonucleotides were in fact acting by inhibiting IL-4 synthesis.

Cytokine receptors

Another approach to blocking the activity of cytokine function would be to design antisense oligonucleotides which target cytokine receptors. A potential advantage of targeting cytokine receptors is that in many cases local application of the oligonucleotide would be feasible. Burch and Mahan used phosphodiester and phosphorothioate oligonucleotides to inhibit the expression of murine and human type 1 interleukin receptors [81]. They demonstrated a decrease in IL-1-binding sites by 55% and a corresponding reduction in IL-1-mediated prostaglandin E2 synthesis. The effects of the oligonucleotides were enhanced by encapsulating them in liposome formulations. These studies suggest that, at least in the case of fibroblast type-1 interleukin-1 receptors, the reduction in receptor number correlates with biological effects. Perhaps most noteworthy was the fact that subcutaneous injection of the oligonucleotides in mice blocked the recruitment of neutrophils to skin following injection of IL-1, demonstrating *in vivo* activity for the oligonucleotides.

Cytokine-induced products

In addition to directly inhibiting the synthesis of cytokines, antisense oligonucleotides have been used to partially inhibit the effect of cytokines by targeting gene products which are induced by the cytokines, or targeting a protein in the signal transduction pathway for the activation of a target cell. We have used phosphorothioate oligonucleotides to block the induction of ICAM-1, vascular cell adhesion molecule-1 (VCAM-1) and E-selectin by tumor necrosis factor, IL-1, IL-4, and interferon-γ (IFNγ) [15, 82]. In each case, a series of phosphorothioate oligonucleotides of 18 to 21 bases in length were designed to hybridize to different regions on the respective mRNA. The most active oligonucleotides for each target were found to hybridize to sequences in the 3'-untranslated regions of the mRNA. The oligonucleotides were shown to selectively inhibit the targeted protein and block adherence of HL-60 cells to cytokine-activated endothelial cells. Oligonucleotides were also described which selectively inhibited synthesis of granulocyte/monocyte-colony stimulating factor (GM-CSF) and granulocyte-colony stimulating factor (G-CSF) in IL-1-activated endothelial cells [83]. Thomae et al. inhibited the synthesis of an inducible form of nitric oxide synthase and nitric oxide production with antisense oligonucleotides in vascular smooth muscle cells [84]. Oligonucleotides which target protein kinase C-α mRNA were used to inhibit IL-1α induction of cyclooxygenase expression in human endothelial cells, suggesting that IL-1 utilizes this particular protein kinase C isoenzyme as part of the signal transduction pathway [85]. In addition, Dean et al. used phosphorothioate oligodeoxynucleotides designed to hybridize to human protein kinase C-α to block phorbol ester induction of ICAM-1 mRNA in human carcinoma cells [86].

Transcription factors

The NF-κB transcription factor complex is a pleiotropic activator that is involved in the induction of a number of cytokine products, cytokine receptors, and cytokine-induced gene products. The active complex is composed of two subunits designated p50 and p65, both of which exhibit sequence similarity to the oncogene *rel*. The p50 and p5 subunits can form both homodimers and heterodimers. Narayan et al. have identified and characterized phosphorothioate antisense oligonucleotides to both the p50 and p65 subunits [87, 88]. Although the researchers did not examine the effects of the oligonucleotides on a cytokine-regulated process, they did demonstrate that antisense oligonucleotides to p65 inhibited adhesion of a wide variety of cells to the extracellular matrix and blocked CD11b induction on HL-60 cells [87, 88]. Similar findings were observed with double-stranded phosphorothioate oligonucleotides containing three repeats of the NF-κB binding site consensus sequence [89]. In addition, the NF-κB transcription factor decoy blocked tumor necrosis factor-induction of ICAM-1 mRNA [89].

Respirable antisense inhibitors of cytokines

Recently the ability to aerosolize and deliver phosphorothioate oligonucleotides to the lung has been reported [90]. Antisense inhibitors for adenosine A1 and Bradykinin B2 receptors were shown to reduce the levels of their targets in the lungs of rabbits. The A1 receptor antagonists inhibited inflammation and bronchoconstriction induced by allergens. The authors of this paper have further reported that they have tested inhibitors of several cytokines, 5-lipoxygenase and leukotriene C-4 synthase as well as other targets [91].

Conclusions

Although antisense technology is still in its infancy and there are still many more questions than answers, progress has been gratifying. Data from clinical trials and studies in animals provide support for continuing optimism that the technology can yield drugs of significant value. There are a number of controversial inhibitors of cytokines and their receptors that have proven to be valuable tools. In the future, several of these are likely to be evaluated as drugs.

References

1   Henderson B, Bodmer MW (eds) (1996) *Therapeutic modulation of cytokines*. CRC Press, Inc, Boca Raton, FL
2   Bennett CF, Crooke ST (1996) Oligonucleotide-based inhibitors of cytokine expression and function. In: B Henderson, M Bodmer (eds): *Therapeutic modulation of cytokines*. CRC Press, Inc, Boca Raton, FL, 171–193
3   Krieg AM (1998) Immune stimulation by oligonucleotides. In: ST Crooke (ed): *Handbook of experimental pharmacology: Antisense research and application*. Springer-Verlag, Berlin, Heidelberg, 243–262
4   Neilsen PE (1991) Sequence-selective recognition of DNA by strand displacement with a thymine-substituted polyamide. *Science* 254: 1497–1500
5   Mollegaard NE, Buchardt O, Egholm M, Nielsen PE (1994) PNA-DNA strand displacement loops as artificial transcription promoters. *Proc Natl Acad Sci USA* 91: 3892–3895
6   Sun JS, Helene C (1993) Oligonucleotide-directed triple-helix formation. *Curr Opin Struct Biol* 3(3): 345–356
7   Bielinska A, Shivdasani RA, Zhang L, Nabel GJ (1990) Regulation of gene expression with double-stranded phosphorothioate oligonucleotides. *Science* 250: 997–1000
8   Stein CA, Cheng Y-C (1993) Antisense oligonucleotides as therapeutic agents-Is the bullet really magical? *Science* 261: 1004–1012

9   Wagner RW (1994) Gene inhibition using antisense oligodeoxynucleotides. *Nature* 372: 333–335

10   Crooke ST (1997) Advances in understanding the pharmacological properties of antisense oligonucleotides. In: B Weiss (ed): *Antisense oligodeoxynucleotides and antisense RNA, novel pharmacological and therapeutic agents*. CRC Press, Boca Raton, 17–25

11   Wu H, Macleod AR, Lima WF, Crooke ST (1998) Identification and partial purification of human double strand RNase activity A novel terminating mechanism for oligoribonucleotide antisense drugs. *J Biol Chem* 273 (5): 2532–2542

12   Crooke ST (1996) Advances in understanding the pharmacological properties of antisense oligonucleotides. *Advances in Pharmacology* 40: 1–49

13   Crouch RJ, Dirksen M-L (1985) Ribonucleases H. In: SM Linn, RJ Roberts (eds): *Nucleases*. Cold Spring Harbor Laboratory Press, Cold Spring Harbor, NY, 211–241

14   Wu H, Lima WF, Crooke ST (1998) Molecular cloning and expression of cDNA for human RNase H. *Antisense Nucleic Acid Drug Dev* 8(1): 53–61

15   Chiang MY, Chan H, Zounes MA, Freier SM, Lima WF, Bennett CF (1991) Antisense oligonucleotides inhibit intercellular adhesion molecule 1 expression by two distinct mechanisms. *J Biol Chem* 266: 18162–18171

16   Dean NM, McKay R (1994) Inhibition of protein kinase C-alpha expression in mice after systemic administration of phosphorothioate antisense oligodeoxynucleotides. *Proc Natl Acad Sci USA* 91: 11762–11766

17   Skorski T, Nieborowska-Skorska M, Nicolaides NC, Szczylik C, Iversen P, Iozzo RV, Zon G, Calabretta B (1994) Suppression of Philadelphia leukemia cell growth in mice by BCR-ABL antisense oligodeoxynucleotide. *Proc Natl Acad Sci USA* 91: 4504–4508

18   Hijiya N, Zhang J, Ratajczak MZ, Kant JA, DeRiel K, Herlyn M, Zon G, Gewirtz AM (1994) Biologic and therapeutic significance of MYB expression in human melanoma. *Proc Natl Acad Sci USA* 91: 4499–4503

19   Greig MJ, Gaus H, Cummins LL, Sasmor H, Griffey RH (1995) Measurement of macromolecular binding using electrospray mass spectrometry Determination of dissociation constants for oligonucleotide: Serum albumin complexes. *J Am Chem Soc* 117 (43): 10765–10766

20   Joos RW, Hall WH (1969) Determination of binding constants of serum albumin for penicillin. *J Pharmacol Exp Ther* 166: 113

21   Agrawal S, Temsamani J, Tang JY (1991) Pharmacokinetics, biodistribution, and stability of oligodeoxynucleotide phosphorothioates in mice. *Proc Natl Acad Sci USA* 88: 7595–7599

22   Srinivasan SK, Iversen P (1995) Review of *in vivo* pharmacokinetics and toxicology of phosphorothioate oligonucleotides. *J Clin Lab Anal* 9: 129–137

23   Cossum PA, Truong L, Owens SR, Markham PM, Shea JP, Crooke ST (1994) Pharmacokinetics of a ^{14}C-labeled phosphorothioate oligonucleotide ISIS 2105, after intradermal administration to rats. *J Pharm Exp Ther* 269: 89–94

24   Crooke ST, Grillone LR, Tendolkar A, Garrett A, Fratkin MJ, Leeds J, Barr WH (1994)

A pharmacokinetic evaluation of 14C-labeled afovirsen sodium in patients with genital warts. *Clin Pharm Therap* 56: 641–646

25   Cossum PA, Sasmor H, Dellinger D, Truong L, Cummins L, Owens SR, Markham PM, Shea JP, Crooke S (1993) Disposition of the ^{14}C-labeled phosphorothioate oligonucleotide ISIS 2105 after intravenous administration to rats. *J Pharmacol Exp Ther* 267: 1181–1190

26   Sands H, Gorey-Feret LJ, Cocuzza AJ, Hobbs FW, Chidester D, Trainor GL (1994) Biodistribution and metabolism of internally ^{3}H-labeled oligonucleotides. I. Comparison of a phosphodiester and a phosphorothioate. *Mol Pharmacol* 45: 932–943

27   Graham MJ, Crooke ST, Monteith DK, Cooper SR, Lemonidis-Farrar KM, Stecker KK, Martin MJ, Crooke RM (1998) *In vivo* distribution and metabolism of a phosphorothioate oligonucleotide within rat liver after intravenous administration. *J Pharmacol Exp Ther* 286: 447–458

28   Sands H, Gorey-Feret LJ, Ho SP, Bao Y, Cocuzza AJ, Chidester D, Hobbs FW (1995) Biodistribution and metabolism of internally ^{3}H-labeled oligonucleotides II 3',5'-blocked oligonucleotides. *Mol Pharmacol* 47: 636–646

29   Crooke ST, Bennett CF (1996) Progress in antisense oligonucleotide therapeutics. *Ann Rev Pharmacol Toxicol* 36: 107–129

30   Henry SP, Taylor J, Midgley L, Levin AA, Kornbrust DJ (1997) Evaluation of the toxicity profile of ISIS 2302, a phosphorothioate oligonucleotide in a 4-week study in CD-1 mice. *Antisense Nucleic Acid Drug Dev* 7(5): 473–481

31   Levin AA, Monteith DK, Leeds JM, Nicklin PL, Geary RS, Butler M, Templin MV, Henry SP (1998) Toxicity of oligodeoxynucleotide therapeutic agents. In: *Handbook of experimental pharmacology: Antisense research and application.* Springer-Verlag, Berlin, Heidelberg,

32   Monteith DK, Henry SP, Howard RB, Flournoy S, Levin AA, Bennett F, Crooke ST (1997) Immune stimulation – a class effect of phosphorothioate oligodeoxynucleotides in rodents. *Anti-cancer Drug Design* 12(5): 421–432

33   Muccioli C, Goldstein DA, Johnson DW, Perez JE, Mora-Duarte JF, Sheppard JD, Mansour SE, Chan CK, Palestine AG, Grillone LR et al (1998) Fomivirsen safety and efficacy in the treatment of CMV retinitis: A Phase 3, controlled, multicenter study comparing immediate versus delayed treatment. In: *5th Conference on retroviruses and opportunistic infections,* Sheraton Chicago Hotel, Chicago, IL

34   Gewirtz AM (1997) Developing oligonucleotide therapeutics for human leukemia. *Anti-Cancer Drug Design* 12(5): 341–358

35   Sikic GI, Yuen AR, Halsey J, Fisher GA, Pribble JP, Smith RM, Geary R, Dorr A (1997) A Phase I trial of an antisense oligonucleotide targeted to protein kinase C-α (ISIS 3521) delivered by twenty-one day continuous intravenous infusion. In: *33rd Annual Meeting of the American Society of Clinical Oncology,* Denver, CO

36   O'Dwyer PJ, Stevenson JP, Gallagher M, Mitchell E, Friedland D, Rose L, Cassella A, Holmlund J, Dean N, Dorr A et al (1998) Phase I/phamracokinetic/pharmacodynamic

trial of raf-1 antisense ODN (ISIS 5132, CGP 69846A). In: *34th Annual Meeting of the American Society of Clinical Oncology*, Los Angeles, CA

37 Muccioli C, Goldstein DA, Perez JE, Mora-Duarte JF, Sheppard JD, Mansour SE, Chan CK, Palestine AG, Holland GN, Pelkey WL et al (1998) Safety and efficacy of fomivirsen in the treatment of CMV retinitis: a phase 3, controlled, multicenter study comparing immediate versus delayed treatment. In: *8th international Congress on Infectious Diseases*, Boston, MA

38 Glover JM, Leeds JM, Mant TGK, Amin D, Kisner D, Zuckerman J, Geary RS, Levin A, Shanahan JWR (1997) Phase 1 safety and pharmacokinetic profile of an intercellular adhesion molecule-1 antisense oligodeoxynucleotide (ISIS 2302). *J Pharmacol Exp Ther* 282: 1173-1180

39 Yacyshyn BR, Bowen-Yacyshyn MB, Jewell L, Rothlein RA, Mainolfi E, Tami JA, Bennett CF, Kisner DL, Shanahan WR (1998) A placebo-controlled trial of ISIS 2302 (ICAM-1 antisense oligonucleotide) in the treatment of steroid-dependent Crohn's disease. *Gastroenterology* 114: 1133–1142

40 Shanahan WR (1998) Properties of ISIS 2302, an inhibitor of intercellular adhesion molecule-1, in humans. In: ST Crooke (ed): *Handbook of experimental pharmacology: Antisense research and application*. Springer-Verlag, Berlin, Heidelberg, 499–524

41 Milligan JF, Matteucci MD, Martin JC (1993) Current concepts in antisense drug design. *J Med Chem* 36: 1923–1937

42 Sanghvi YS, Cook PD (1993) Towards second-generation synthetic backbones for antisense oligonucleosides. In: CK Chu, DC Baker (eds): *Nucleosides, nucleotides as antiviral and antitumor agents*. Plenum Press, New York, 309–322

43 Beaucage SL, Iyer RP (1993) The synthesis of modified oligonucleotides by the phosphoramidite approach and their applications. *Tetrahedron* 49: 6123–6194

44 Sanghvi YS, Cook PD (1994) Carbohydrates: Synthetic methods and applications in antisense therapeutics: An overview. In: YS Sanghvi, PD Cook (eds): *Carbohydrate modifications in antisense research*. ACS Symposium Series No 580, American Chemical Society: Washington, DC 1–22

45 Sagi J, Szemzo A, Ebinger K, Szabolocs A, Sagi G, Ruff E, Otvos L (1993) Base-modified oligodeoxynucleotides I Effect of 5-alkyl, 5-(1-alkenyl) and 5-(1-alkynyl) substitution of the pyrmidines on duplex stability and hydrophobicity. *Tetrahedron Letters* 34: 2191–2194

46 Wagner RW, Matteucci MD, Lewis JG, Gutierrez AJ, Moulds C, Froehler BC (1993) Antisense gene inhibition by oligonucleotides containing C-5 propyne pyrimidines. *Science* 260 (5113): 1510–1513

47 Iribarren AM, Sproat BS, Neuner P, Sulston I, Ryder U, Lamond AI (1990) 2'-O-alkyl oligoribonucleotides as antisense probes. *Proc Natl Acad Sci USA* 87(19): 7747–7751

48 Kawasaki AM, Casper MD, Freier SM, Lesnik EA, Zounes MC, Cummins LL, Gonzalez C, Cook PD (1993) Uniformly modified 2'-deoxy-2'-fluoro phosphorothioate oligonucleotides as nuclease-resistant antisense compounds with high affinity and specificity for RNA targets. *J Med Chem* 36: 831–841

49 Lesnik EA, Guinosso CJ, Kawasaki AM, Sasmor H, Zounes M, Cummins LL, Ecker DJ, Cook PD, Freier SM (1993) Oligodeoxynucleotides containing 2'-O-modified adenosine: Synthesis and effects on stability of DNA: RNA duplexes. *Biochemistry* 32: 7832–7838

50 Martin P (1995) New access to 2'-O-alkylated ribonucleosides and properties of 2'-O-alkylated oligoribonucleotides. *Helv Chim Acta* 78: 486–504

51 Cummins LL, Owens SR, Risen LM, Lesnik EA, Freier SM, McGee D, Guinosso CJ, Cook PD (1995) Characterization of fully 2'-modified oligoribonucleotide hetero- and homoduplex hybridization and nuclease sensitivity. *Nucleic Acids Res* 23 (11): 2019–2024

52 Monia BP, Lesnik EA, Gonzalez C, Lima WF, McGee D, Guinosso CJ, Kawasaki AM, Cook PD, Freier SM (1993) Evaluation of 2' modified oligonucleotides containing deoxy gaps as antisense inhibitors of gene expression. *J Biol Chem* 268: 14514–14522

53 Agrawal S, Mayrand SH, Zamecnik PC, Pederson T (1990) Site-specific excision from RNA by RNase H and mixed-phosphate- backbone oligodeoxynucleotides. *Proc Natl Acad Sci USA* 87: 1401–1405

54 Lamond AI, Barabino S, Blencowe BJ, Sproat B, Ryder U (1990) Studying pre-mRNA splicing using antisense 2-OMe RNA oligonucleotides. *Mol Biol Rep* 14: 201

55 Miller PS, McParland KB, Jayaraman K, Ts'o P (1981) Biochemical and biological effects of nonionic nucleic acid methylphosphonates. *Biochemistry* 20: 1874–1880

56 Miller PS, Ts'o P (1987) A new approach to chemotherapy based on molecular biology and nucleic acid chemistry: Matagen (masking tape for gene expression). *Anti-Cancer Drug Design* 2: 117–128

57 Miller PS (1991) Oligonucleoside methylphosphonates as antisense reagents. *BioTechnology* 9: 358–362

58 Arnold LJ (1995) The development of oligonucleotides containing chirally pure methylphosphonates as therapeutic agents. *Antisense Res Dev* 5: 108

59 Vasseur JJ, Debart F, Sanghvi YS, Cook PD (1992) Oligonucleosides: synthesis of a novel methylhydroxylamine-linked nucleoside dimer and its incorporation into antisense sequences. *J Am Chem Soc* 114 (10): 4006–4007

60 De Mesmaeker A, Lebreton J, Waldner A, Fritsch V, Wolf RM, Freier SM (1993) Amides as substitute for the phosphodiester linkage in antisense oligonucleotides. *Synlett* 10: 733–736

61 Jones RJ, Lin K-Y, Milligan JF, Wadwani S, Matteucci MD (1993) Synthesis and binding properties of pyrimidine oligodeoxynucleoside analogs containing neutral phosphodiester replacements The formacetal and 3'-thioformacetal internucleoside linkages. *J Org Chem* 58: 2983–2991

62 Crooke ST (1995) *Therapeutic applications of oligonucleotides*. RG Landes Company, Austin, 138

63 Buchardt O, Egholm M, Berg RH, Nielsen PE (1993) Peptide nucleic acids (PNA) and their potential applications in medicine and biotechnology. *Trends BioTechnology* 11: 384–386

64 Egholm M, Buchardt O, Christensen L, Behrens C, Freier SM, Driver DA, Berg RH, Kim SK, Norden B, Nielsen PE (1993) PNA hybridizes to complementary oligonucleotides obeying the Watson-Crick hydrogen-bonding rules. *Nature* 365: 566–568

65 Stirchak EP, Summerton JE, Weller DD (1989) Uncharged stereoregular nucleic acid analogs: 2 Morpholino nucleoside oligomers with carbamate internucleotide linkages. *Nucleic Acids Research* 17: 6129–6141

66 Letsinger RL, Zhang GR, Sun DK, Ikeuchi T, Sarin PS (1989) Cholesteryl-conjugated oligonucleotides: synthesis, properties, and activity as inhibitors of replication of human immunodeficiency virus in cell culture. *Proc Natl Acad Sci USA* 86: 6553–6556

67 Torrence PF, Maitra RK, Lesiak K, Khamnei S, Zhou A, Silverman RH (1993) Targeting RNA for degradation with a (2'-5')oligoadenylate-antisense chimera. *Proc Natl Acad Sci USA* 90: 1300–1304

68 Manoharan M, Johnson LK, Bennett CF, Vickers TA, Ecker DJ, Cowsert LM, Freier SM, Cook PD (1994) Cholic acid-oligonucleotide conjugates for antisense applications. *Bioorganic & Medicinal Chemistry Letters* 4: 1053–1060

69 Manoharan M, Tivel KL, Andrade LK, Mohan V, Condon TP, Bennett CF, Cook PD (1995) Oligonucleotide conjugates: Alteration of the pharmacokinetic properties of antisense agents. *Nucleosides Nucleotides* 14 (3–5): 969–973

70 Harel-Bellan A, Durum S, Muegge K, Abbas AK, Farrar WL (1988) Specific inhibition of lymphokine biosynthesis and autocrine growth using antisense oligonucleotides in TH1 and Th2 helper T cell clones. *J Exp Med* 168: 2309–2318

71 Zubiaga AM, Munoz E, Huber BT (1991) Production of IL-1 alpha by activated Th type 2 cells Its role as an autocrine growth factor. *J Immunol* 146: 3849–3856

72 Miles SA, Rezai AR, Salazar-Gonz lez JF, Meyden MV, Stevens RH, Logan DM, Mitsuyasu RT, Taga T, Hirano T, Kishimoto T et al (1990) AIDS Kaposi sarcoma-derived cells produce and respond to interleukin 6. *Proc Natl Acad Sci USA* 87: 4068–4072

73 Levy Y, Tsapis A, Brouet J-C (1991) Interleukin-6 antisense oligonucleotides inhibit the growth of human myeloma cell lines. *J Clin Invest* 88: 696–699

74 Schadendorf D, Moller A, Algermissen B, Worm M, Sticherling M, Czarnetzki BM (1993) IL-8 produced by human malignant melanoma cells *in vitro* is an essential autocrine growth factor. *J Immunol* 151: 2667–2675

75 Birchenall-Roberts MC, Ferrer C, Ferris D, Falk LA, Kasper J, Gretchen W, Ruscetti FW (1990) Inhibition of murine monocyte proliferation by a colony-stimulating factor-1 antisense oligodeoxynucleotide Evidence for autocrine regulation. *J Immunol* 145: 3290–3296

76 Fujiwara T, Grimm EA (1992) Specific inhibition of interleukin 1β gene expression by an antisense oligonucleotide: Obligatory role of interleukin 1 in the generation of lymphokine-activated killer cells. *Cancer Res* 52: 4954–4959

77 Maier JAM, Voulalas P, Roeder D, Maciag T (1990) Extension of the life-span of human endothelial cells by an interleukin-1α antisense oligomer. *Science* 249: 1570–1574

78 Garfinkel S, Haines DS, Brown S, Wessendorf J, Gillespie DH, Maciag T (1992) Inter-

leukin-1α mediates an alternative pathway for the antiproliferative action of poly(I-C) on human endothelial cells. *J Biol Chem* 267: 24375–24378

79   Benbernou N, Matsiota-Bernard P, Guenounou M (1993) Effect of cytokine-specific antisense oligonucleotides on the immunoglobulin production by rat spleen cells *in vitro. Biochimie* 75: 55–61

80   Benbernou N, Matsiota-Bernard P, Guenounou M (1993) Antisense oligonucleotides to interleukin-4 regulate IgE and IgG2a production by spleen cells from Nippostrongylus brasiliensis-infected rats. *Eur J Immunol* 23: 659–663

81   Burch RM, Mahan LC (1991) Oligonucleotides antisense to the interleukin 1 receptor mRNA block the effects of interleukin 1 in cultured murine and human fibroblasts and in mice. *J Clin Invest* 88: 1190–1196

82   Bennett CF, Condon T, Grimm S, Chan H, Chiang M-Y (1994) Inhibition of endothelial cell-leukocyte adhesion molecule expression with antisense oligonucleotides. *J Immunol* 152: 3530–3540

83   Segal GM, Smith TD, Heinrich MC, Ey FS, Bagby GC (1992) Specific repression of granulocyte-macrophage and granulocyte colony-stimulating factor gene expression in interleukin-1-stimulated endothelial cells with antisense oligodeoxynucleotides. *Blood* 80: 609–616

84   Thomae KR, Geller DA, Billiar TR, Davies P, Pitt BR, Simmons RL, Nakayama DK (1993) Antisense oligodeoxynucleotide to inducible nitric oxide synthase inhibits nitric oxide synthesis in rat pulmonary artery smooth muscle cells in culture. *Surgery* 114: 272–277

85   Maier JAM, Ragnotti G (1993) An oligomer targeted against protein kinase C prevents interleukin-1 induction of cyclooxygenase expression in human endothelial cells. *Exp Cell Res* 205: 52–58

86   Dean NM, McKay R, Condon TP, Bennett CF (1994) Inhibition of protein kinase C-α expression in human A549 cells by antisense oligonucleotides inhibits induction of intercellular adhesion molecule 1 (ICAM-1) mRNA by phorbol esters. *J Biol Chem* 269: 16416–16424

87   Narayan R, Higgins KA, Perez JR, Coleman TA, Rosen CA (1993) Evidence for differential functions of the p50 and p65 subunits of NF-κB with a cell adhesion model. *Mol Cell Biol* 13: 3802–3810

88   Sokoloski JA, Sartorelli AC, Rosen CA, Narayan R (1993) Antisense oligonucleotides to the p65 subunit of NF-kappaB block CD11b expression and alter adhesion properties of differentiated HL-60 granulocytes. *Blood* 82: 625–632

89   Griffin LC, Tidmarsh GF, Bock LC, Toole JJ, Leung LLK (1993) *In vivo* anticoagulant properties of a novel nucleotide-based thrombin inhibitor and demonstration of regional anticoagulation in extracorporeal circuits. *Blood* 81: 3271–3276

90   Nyce JW, Metzger WJ (1997) DNA antisense therapy for asthma in an animal model. *Nature* 385 (6618): 721–725

91   Nyce JW (1997) Respirable antisense oligonucleotides as novel therapeutic agents for asthma and other pulmonary diseases. *Exp Opin Invest Drugs* 6 (9): 1149–1156

Inhibiting cytokine-processing enzymes

Peter I. Croucher, Ingunn Holen and Philip G. Hargreaves

Division of Biochemical and Musculoskeletal Medicine, University of Sheffield Medical School, Beech Hill Road, Sheffield S10 2RX, UK

Introduction

Cytokines and local growth factors are important soluble mediators that play a fundamental role in regulating the growth and differentiated function of many cell types. They mediate their effects by binding to membrane-bound receptors which initiates a complex sequence of signaling events that leads to a cellular response. The magnitude of this effect reflects the complex balance that exists between the different components of the cytokine system. These include the cytokine itself, the membrane-bound receptor, and the presence or absence of soluble receptor and/or receptor antagonists.

The individual components of this system are subject to tight controls and can be regulated at a number of different levels. One of the most important regulatory mechanisms is limited proteolytic processing which may be involved in controlling the activity of a cytokine in a number of ways (Fig. 1). First, certain cytokines are produced as inactive precursors that require proteolytic cleavage of a pro-region to yield the functional cytokine. The best characterised example of such a system is that of interleukin-1β (IL-1β), although other factors may also be regulated in a similar manner. Secondly, proteolytic activity may be utilised to regulate the spatial distribution of cytokines and growth factors. A number of factors, including tumour necrosis factor α (TNFα), transforming growth factor α (TGFα) and macrophage colony stimulating factor (M-CSF), are synthesised as membrane-bound proteins which remain biologically active. Such cytokines can function as juxtacrine factors by interacting directly with specific receptors on adjacent cells [1]. However, proteolytic cleavage at the cell surface can transform membrane-bound cytokines into soluble factors, capable of regulating the function of distant cells. Thirdly, proteinases are able to promote the release of cytokine and growth factor receptors which indirectly serve to regulate cytokine responses. Soluble cytokine receptors generally act as antagonists by sequestering the appropriate ligand and preventing

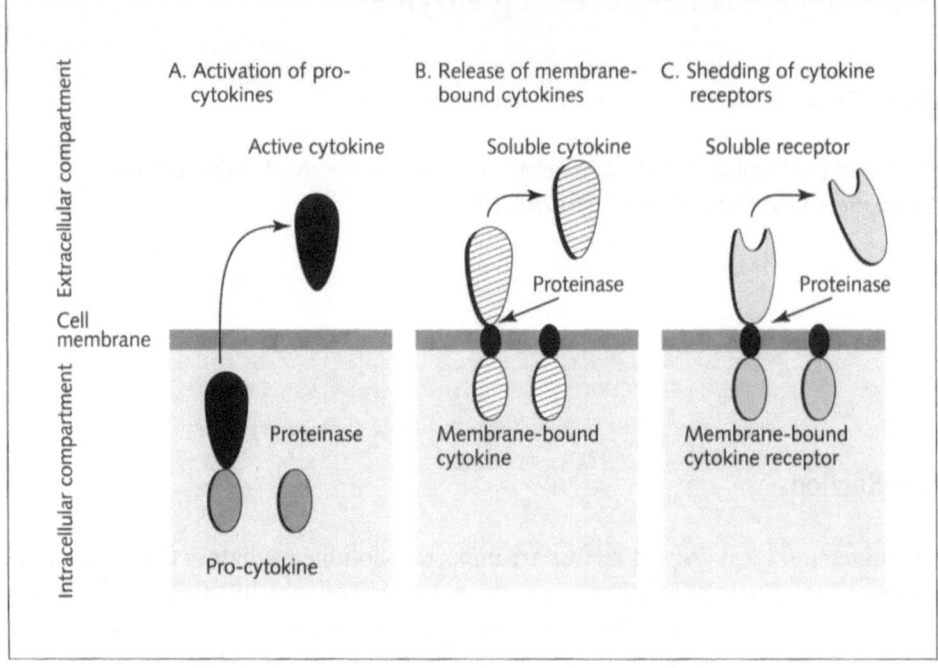

Figure 1

Diagrammatic representation of three mechanisms by which proteinases can regulate the activity of cytokines and their receptors.

association with membrane-bound receptors. However, they may also serve to increase the stability of a cytokine and prolong their half-life in solution [2].

Studies over recent years have shown that proteolytic processing is involved in regulating the function of many important cytokines and their receptors (Tab. 1). These observations have led to increased interest in identification of the proteinases involved, since they may prove to be important therapeutic targets. Although the proteinases responsible for processing the majority of cytokines and their receptors remain unknown, studies have shown that different classes of proteinases appear to be involved at the different stages of regulation. Thus, intracellular processing of pro-cytokines appears to be mediated by specific cysteine proteinases, whereas the release of membrane-bound cytokines and their receptors appears to be mediated by members of the metalloproteinase family. The focus of this chapter will be to describe examples of these proteolytic processing events in more detail and to discuss how proteinase inhibitors can be utilised to modulate their activity.

Table 1 - Cytokines and their receptors which are known to be processed by proteinases[a]

	Alternate mRNA splicing[b]	Proteolytic processing	Proteinase class[c]	References
Activation of cytokine precursors				
IL-1β		+	Cysteine	[3, 4]
IGIF (IL-18)		+	Cysteine	[5]
Release of membrane-bound cytokines				
TNFα		+	Metallo	[6–8]
TGFα		+	Metallo/serine[d]	[9, 10]
M-CSF (CSF-1)		+	Metallo	[10, 78]
Kit ligand-1 (KL-1)		+	Metallo/serine	[9–11]
Kit ligand-2 (KL-2)		+	Metallo/serine	[9–11]
flt3 ligand	+	+	?	[12, 13]
Release of cytokine receptors				
IL-1 receptor (type II)		+	Metallo	[11]
IL-2 receptor		+	Metallo	[11, 14]
IL-4 receptor	+			[15]
IL-5 receptor	+			[16, 17]
IL-6 receptor (gp80)	+	+	Metallo	[18–20]
IL-7 receptor	+			[21]
IL-9 receptor	+			[22]
TNF receptor p55		+	Metallo	[19]
TNF receptor p75		+	Metallo	[23]
M-CSF receptor		+	?	[24]
G-CSF receptor	+			[25]
GM-CSF receptor	+			[26]
PDGF receptor		+	?	[27]
LIF receptor	+			[28, 29]

[a]This is not a definitive list but comprises those that have been most widely studied.
[b]Examples of soluble cytokines and their receptors that can be generated by alternate mRNA splicing which results in a deletion of the region encoding the transmembrane domain.
[c]The class of proteinase generally thought to be responsible for mediating proteolysis.
[d]Although early studies have implicated serine proteinases in processing a number of these cytokines, more recent studies suggest that it is a metalloproteinase that is responsible for release from the cell surface. Serine proteinases may be involved in activating metalloproteinases or mediate trafficking to the cell surface [30].

Regulation of pro-cytokine processing

Proteinases have been shown to activate cytokines by mediating intracellular cleavage of pro-cytokines to yield the mature form. The best characterised example of such processing is the activation of IL-1β, although interleukin-18 (IL-18) has also been reported to be activated in a similar manner.

The interleukin-1 family consists of three members, the agonists IL-1α and IL-1β, and the IL-1 receptor antagonist (IL-1RA) [31]. Although IL-1α and IL-1β are products of separate genes they show approximately 25% identity at the amino acid level and mediate their biological effects through binding to the same receptors [32]. IL-1α and IL-1β are produced as 33 and 31 kDa precursors, respectively, which undergo intracellular proteolysis to generate the mature form. IL-1α has been shown to be cleaved by calpain in a calcium-dependent process [33], although the mature protein remains largely intracellular. In contrast, IL-1β is cleaved by the IL-1β-converting enzyme (ICE) to produce the 17 kDa active form which is subsequently released from the cell. The ICE cleavage site in pro-IL-1β is at Asp^{116}-Ala^{117}, which is a unique prohormone processing site [4]. More recently, an activity that promotes the synthesis of interferon γ (IFNγ) has been described and shown to be produced by activated Kupfer cells and macrophages [34]. This protein has been cloned and found to encode an 18 kDa cytokine known as IFNγ-inducing factor (IGIF) or IL-18 [34]. Recent studies by Gu et al. [5] have shown that IGIF is synthesised as an inactive precursor that is also cleaved by ICE, at an Asp residue, to yield the active protein.

IL-1β converting enzyme

ICE was first isolated from monocytic cells by Black et al. in 1988 [35]. Subsequently, two independent groups cloned and sequenced ICE, identifying it as a member of a novel family of cysteine proteinases [3, 4]. The substrate specificity of ICE has been determined and includes an absolute requirement for Asp in the P_1 position, a strong preference for a large hydrophobic amino acid in P_4, and a stringent requirement for at least 4 amino acids N-terminal to the cleavage site [4, 36]. Recently, this group of proteinases has been renamed as the caspase (for cysteinyl aspartate-specific proteinase) family. Ten members of this family have now been described with ICE corresponding to caspase-1 [37].

Caspase-1 is produced as an inactive 45 kDa cytosolic pro-enzyme. Limited proteolytic cleavage results in the release of 20 kDa and 10 kDa fragments which form a heterodimer, with both contributing to the formation of the catalytic site [38]. The crystal structure of this complex demonstrates that two heterodimers combine, with the resulting tetramer constituting the active form of the proteinase [38]. An Asp-flanked linker segment and the N-terminal pro-domain are believed to be involved in activation and can be released upon cleavage. Although the exact mechanism of caspase-1 activation is unknown, it appears to occur in response to stimuli that upregulate the cellular production of IL-1β [39]. Since activation occurs through cleavage of an internal conserved Asp residue, it has been suggested that autoactivation could take place as part of a caspase cascade [4, 40]. Although few cellular substrates for caspases have been identified, caspase-1 appears to specifically cleave the precursors of IL-1β and IGIF [5, 40]. Caspase-1 and its homologues have also been implicated in protein degradation during apoptosis [41].

Inhibiting caspase-1

Since IL-1β plays such an important role in inflammatory disorders, the proteinase responsible for activating pro-IL-1β is thought to be a potential therapeutic target. Thus, considerable effort has gone into the identification and design of specific inhibitors of caspase-1.

Two naturally occurring caspase-1 inhibitors have been identified, the cowpox virus protein CrmA (Cytokine response modifier A) and the baculovirus protein p35 [42, 43]. CrmA is a 38 kDa protein that binds to caspase-1 during cowpox infection, thereby preventing the infected cells from producing mature IL-1β. CrmA is a potent inhibitor of caspase-1 (K_i = 0.01 nM), but can also inhibit caspase 8 (K_i = 1 nM) and caspases-3 and -6 (K_i in the μM range). p35 acts in similar manner to CrmA by binding caspase-1 and undergoing cleavage [43]. The cleaved products dissociate slowly from caspase-1, preventing the association between caspase-1 and pro-IL-1β. However, it is important to note that no mammalian p35 homologues have been identified and the mammalian CrmA homologues that have been described do not inhibit caspases [40].

Based upon a detailed knowledge of the crystal structure and substrate specificity, a number of synthetic inhibitors of caspase-1 have been described. The general structure of a caspase-1 inhibitor is peptide–CO–CH$_2$–X, where X can be a halide ion (as in chloromethylketones and fluoromethylketones), –N$_2$ (as in diazomethylketones), –OCOR (as in (acyloxy)methylketones) or –OR (as in (phosphinyloxy) methylketones) (Fig. 2). For a recent comprehensive review of these inhibitors see Livingston [44]. The most potent inhibitors are those that contain a peptide sequence that resembles the endogenous substrate (Fig. 2). The tetrapeptide aldehyde Ac-Tyr-Val-Ala-Asp-CHO (L-709,049), which includes a sequence similar to the cleavage site in pro-IL-1β, is reported to be a highly potent caspase-1 inhibitor (K_i = 0.76 nM), but a poor inhibitor of caspase-3 (K_i = 10 μM) [4].

Caspase-1 inhibitors have been examined for their ability to reduce the level of IL-1β in a number of *in vitro* systems. Miller et al. [45] reported that preincubation of human peripheral blood samples with L-709,049 resulted in a dose-dependent reduction in *S. aureus*-induced IL-1β production. The same compound reduced the

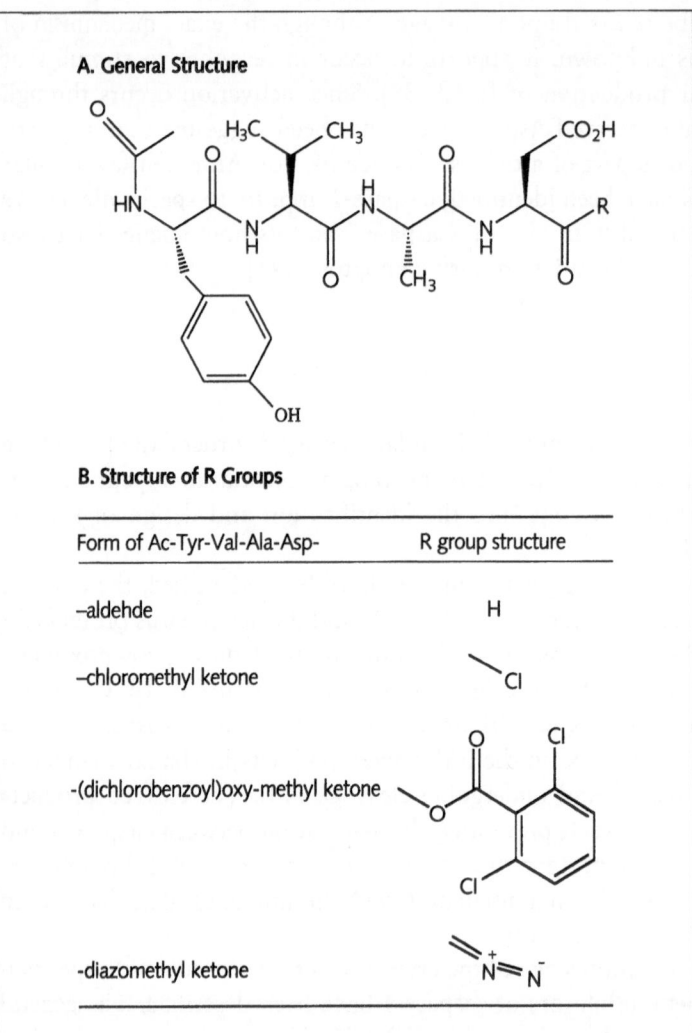

A. General Structure

B. Structure of R Groups

Form of Ac-Tyr-Val-Ala-Asp-	R group structure
–aldehde	H
–chloromethyl ketone	Cl
-(dichlorobenzoyl)oxy-methyl ketone	
-diazomethyl ketone	

Figure 2
An example of the structure of a caspase-1 inhibitor.

level of IL-1β release from LPS-stimulated murine blood samples [46]. Furthermore, Miller et al. [47] have demonstrated that Z-Val-Ala-Asp-CH$_2$O(CO)[2,6-(Cl$_2$)]Ph (WIN 67694) can prevent LPS-stimulated IL-1β release from murine peritoneal macrophages. Several of the caspase-1 inhibitors have also been tested for their ability to inhibit the activity of IL-1β *in vivo* [44]. The inhibitor carbobenzyloxy-Val-Ala-Asp(OEt)-CH$_2$O-dichlorobenzoate (VE-13,045) (50 mg/kg/day) has been shown to reduce serum levels of IL-1β by 50–70% following a challenge with LPS in mice [48]. In addition, VE-13,045 has been shown to reduce paw swelling and proteoglycan degradation in a murine model of collagen type II-induced arthritis [48]. VE-13,045 was suggested to be more effective than either indomethacin or methylprednisolone in these studies. In a murine model of endotoxic shock, intraperitoneal (i.p.) administration of the inhibitor L709,049 (50mg/kg) reduced the LPS-induced increase in concentration of IL-1β in plasma and peritoneal fluid, while production of IL-1α and IL-6 remained unaffected [46]. Furthermore, in an alternative murine model of inflammation, administration of the inhibitor WIN 67694 (i.p. 10–100 mg/kg) reduced the level of IL-1β in a dose-dependent manner, again without affecting the level of IL-1α, IL-6 or TNFα [47].

Although inhibitors of caspase-1 appear to be able to specifically block processing of IL-1β and can reduce circulating levels of this cytokine *in vivo*, several potential problems need to be addressed before these compounds can be utilised therapeutically. Since the active site of members of the caspase family has been shown to be completely conserved, small proteinase inhibitors may inhibit more than one family member. Consequently, the specificity of these inhibitors for caspase-1 over other members of the caspase family require to be determined. When caspase-1 is completely blocked or absent, the release of pro-IL-1β from activated cells is unaffected. Although pro-IL-1β may be bound by the soluble IL-1 receptor in the extracellular environment, concerns have been raised that "bystander proteases", such as neutrophil elastase or cathepsin G, could generate active IL-1β from this pool [46]. However, studies have shown that following treatment with L-709,049 the amount of pro-IL-1β in the peritoneal fluid of mice was unchanged [46]. Furthermore, although caspase-1 knockout mice continue to synthesise pro-IL-1β, they do not produce active IL-1β, suggesting that other proteinases do not process pro-IL-1β *in vivo* [49].

Processing of membrane-bound cytokines

A number of different cytokines are processed from membrane-bound proteins to active soluble forms by the action of a proteinase (Tab. 1). Our understanding of the mechanism of release of these cytokines is not as advanced as that of the intracellular activation of pro-IL-1β. However, over the last year, the identity of the proteinase responsible for one such processing event has become known. Thus, the best-characterised example of the release of a membrane-bound cytokine is that of TNFα.

TNFα is a potent pro-inflammatory cytokine produced mainly by activated monocytes and macrophages [50]. TNFα is produced as a 26 kDa type II transmembrane protein, which is subsequently processed to a 17 kDa secreted form by proteolysis at Ala^{76}-Val^{77} [51–53]. The activity responsible for the processing of TNFα was suggested to be a matrix metalloproteinase (MMP) since cleavage of a

Table 2 - Inhibition of TNFα processing by hydroxamate-based metalloproteinase inhibitors.

Cells	Inhibitor	Stimulus	References
Human monocytic cells (THP-1) Human T cells	TAPI[a]	LPS	[6]
Human leukocytes	BB-2284 BB-2116 BB-2275	LPS	[7]
Human monocytes Human T cells (Jurkat) Human B cells (EHM) Human monocytic cells (Mono Mac 6 and THP-1)	GI 129471	PMA or LPS	[8]
Human T cells	TAPI	PMA + ionomycin PMA + anti-CD3	[23]
Simian kidney cell line (COS-7[b]) Human monocytic cells (THP-1) Human monocytes	TAPI	PMA or LPS	[11, 19]
RA synovial membrane cells	BB-2275	none	[54]
Monocytic cells (THP-1 and U937) Human monocytes	RU36156[c]	PMA or LPS	[10]

[a]*TNFα protease inhibitor*
[b]*COS-7 cells transfected with TNFα cDNA*
[c]*identical compound to BB-94 (see Tab. 3)*

recombinant TNFα precursor could be achieved by certain members of the MMP family and suppressed by hydroxamate-based metalloproteinase inhibitors (Tab. 2) [7]. However, other reports demonstrated that a number of hydroxamate-based compounds could inhibit processing of TNFα (Tab. 2) (Fig. 3), whereas naturally occurring inhibitors of matrix metalloproteinases, tissue inhibitor of metalloproteinase-1 and -2 (TIMP-1 and -2), were unable to block this event [6, 8, 55]. These data led to the conclusion that TNFα was processed by a non-matrix type metalloproteinase.

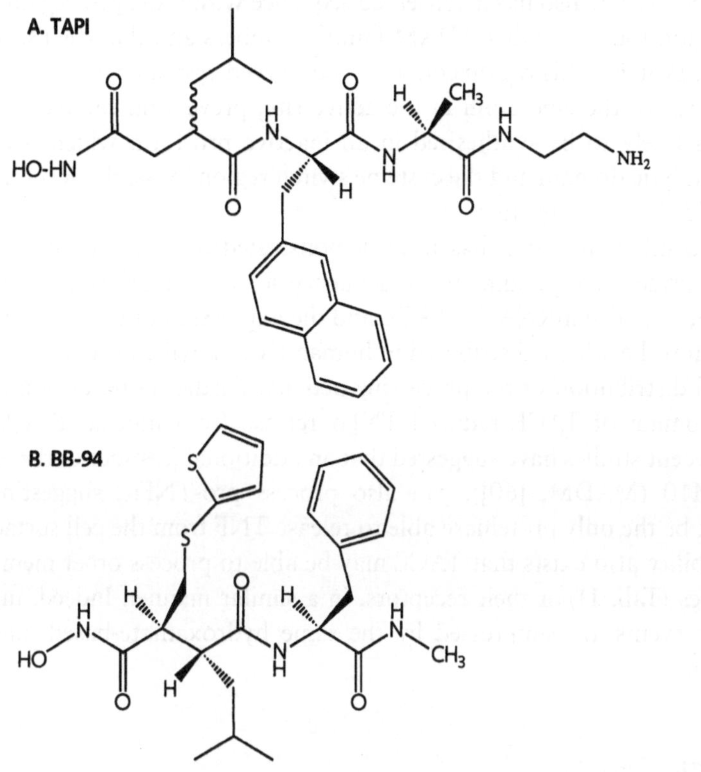

Figure 3
The structure of two hydroxamate-based metalloproteinase inhibitors.

TNFα-converting enzyme (TACE)

The enzyme responsible for processing TNFα, the TNFα converting enzyme or TACE, was subsequently cloned [56, 57]. The amino acid sequence of TACE revealed the presence of a number of characteristic domains including a signal sequence, a pro-region, a metalloproteinase domain (preceded by a potential furin cleavage site), a disintegrin-like region, a cysteine-rich domain, an EGF-like repeat, a transmembrane region and a cytoplasmic tail. The presence of this conserved structure and significant sequence identity defined this as a member of the mammalian adamalysin , or ADAM (for a disintegrin and a metalloproteinase [58]), family. TACE has now been included in the ADAM nomenclature as ADAM 17. The structure of TACE differs from other ADAM family members in that a crambin-like domain follows the EGF-like domain. However, the crystal structure of the catalytic domain of TACE has been recently reported and shown to resemble other adamalysins [59]. TACE also has a conserved sequence within the pro-region that is similar to a region found in other ADAM family members and the MMPs. Known as the "cysteine switch", this region contains a conserved cysteine residue (Cys^{184} in TACE) that binds to the zinc atom in the active site, preventing catalytic activity. Thus, TACE is likely to be synthesised in an inactive pro-form which is cleaved between the catalytic domain and the cysteine switch region, possibly by a furin-like enzyme, to yield the active proteinase.

In vitro, recombinant TACE has been demonstrated to cleave both precursor TNFα and a peptide corresponding to the sequence surrounding the cleavage site at Ala^{76}-Val^{77} [56, 57]. The mRNA for TACE and the expression of the TACE protein has been shown to be widely distributed in human tissues, which is consistent with the widespread distribution of the processing activity. Furthermore, deletion of the zinc-binding domain of TACE reduced TNFα release from murine T cells [56]. Interestingly, recent studies have suggested that an additional member of the ADAM family, ADAM10 (MADM, [60]), can also process pro-TNFα, suggesting that TACE may not be the only proteinase able to release TNF from the cell surface [61, 62]. The possibility also exists that TACE may be able to process other membrane-bound cytokines (Tab. 1), or their receptors, in a similar manner. Indeed, many of these shedding events are suppressed by the same hydroxamate-based inhibitors (Tabs. 1 and 3).

Inhibiting TACE

A number of studies have shown that hydroxamate-based metalloproteinase inhibitors can block processing of TNFα (Tab. 2). Mohler et al. [6] showed that a metalloproteinase inhibitor (TAPI) specifically blocked TNFα processing in a monocytic cell line (THP-1) and human peripheral blood T cells. This inhibitor was also able to reduce the serum concentration of TNFα in mice challenged with LPS, an effect that protected these animals from a lethal dose combination of LPS and D-galactosamine. McGeehan et al. [8] have demonstrated that the inhibitor GI 129471 could block TNFα secretion from human peripheral blood monocytes and monocytic cells but was without effect upon levels of other soluble cytokines (IL-1β, IL-2 or IL-6). In addition, GI 129471 inhibited the LPS-induced increase in serum concentrations of TNFα in mice. Gearing et al. [7] made similar observations using three different hydroxamate-based compounds. Thus, hydroxamate-based metalloproteinase inhibitors appear to be effective at preventing processing of TNFα both *in vitro* and *in vivo*. However, it should be noted that many of these compounds are likely to inhibit other membrane-processing events (see next section, [63]) and members of the MMP family. For example, Williams et al. [54] have shown that in cultures of rheumatoid synovial cells, one such inhibitor (Tab. 2)

Table 3 - Hydroxamate-based metalloproteinase inhibitors known to block shedding of specific cytokine receptors

Cytokine receptor	Inhibitor	Reference
IL-1 receptor	TAPI	[11]
IL-2 receptor	TAPI	[11]
IL-6 receptor	TAPI	[19]
	TAPI-2	[30]
	RU36156[a]	[10]
	BB94	[20]
TNF receptor p55	TAPI	[19]
	BB-2275	[54]
	RU36156	[10]
TNF receptor p75	TAPI	[23]
	BB-2275	[54]
	RU36156	[10]

[a]*identical compound to BB-94*

decreased the level of soluble TNF receptors and inhibited MMP activity in addition to reducing the concentration of soluble TNFα. Thus, the consequences of using such broad-spectrum inhibitors *in vivo* need to be determined. However, as TACE has now been identified, and the crystal structure determined, it is likely that this will lead to a more rational design of inhibitors [59]. Indeed, whilst the structure bears close resemblance to that of other ADAM proteins, several unique features are present in TACE which may make it possible to design more selective inhibitors of this proteinase.

Shedding of cytokine receptors

The final point at which proteinases play an important role in regulating the response of cells or tissues to cytokines is by the generation of soluble cytokine receptors (see chapter by Meager, this volume). Soluble receptors can be produced by either alternate mRNA splicing or by limited proteolytic cleavage of the membrane-bound receptor (Tab. 1). Soluble receptors that are generated by proteolytic activity include those that are important in mediating inflammatory responses, such as the receptors for TNFα, IL-1 and IL-6. Although our understanding of the proteinases responsible, and the mechanisms by which this process occurs, is less clear than in the previous two sections, studies have shown that proteinase inhibitors can block these events. The two best-characterised examples are the releases of the receptors for TNF and IL-6.

Tumour necrosis factor receptor shedding

TNFα causes its biological effect by binding with the TNF receptor, of which two forms have been identified, a 55 kDa protein (p55, TNFR-I) and a 75 kDa protein (p75, TNFR-II). Early studies demonstrated the presence of TNF-binding proteins in urine and that these proteins cross-reacted immunologically with cell-surface TNF receptors [64]. Subsequent studies revealed soluble TNF-binding proteins to be identical to the membrane-bound receptor, and it became clear that they were likely to be derived by proteolysis [65, 66]. The proteinase responsible for shedding of both TNFR-I and -II has yet to be identified. Studies have demonstrated that shedding of both receptors could be promoted by treatment with phorbol 12-myristate 13-acetate (PMA) and inhibited by certain serine proteinase inhibitors [67]. In addition, Porteu et al. [68] showed that neutrophil elastase could release a fragment of the TNFR-II capable of binding TNFα. However, extensive mutagenesis studies have shown that deletion of the putative elastase cleavage sites fails to reduce PMA-induced shedding [69]. Arribas et al. [30] have also shown that serine proteinase inhibitors block the maturation and movement of membrane-bound proteins to the cell surface rather than suppressing release from the cell membrane. A number of recent studies have shown that release of TNFR-I and -II can be blocked by hydroxamate-based inhibitors (Tab. 3) [10, 19, 23, 54]. Interestingly, Smith et al. [70] demonstrated that transfection of DLD colon carcinoma cells with a tissue inhibitor of metalloproteinase-3 (TIMP-3) cDNA prevented release of TNFR-I, providing further evidence that a metalloproteinase may be responsible for releasing the TNFRs.

Interleukin-6 receptor shedding

Interleukin-6 elicits its effects *via* the IL-6 receptor (IL-6R) complex which consists of two subunits. IL-6 initially binds to the IL-6R (also known as gp80) and the resulting complex associates with a second, signal-transducing subunit known as gp130. This leads to homodimerisation of gp130 and activation of signalling pathways within the cell. A soluble form of the IL-6R (sIL-6R) has been identified [71] that acts as an agonist by binding IL-6, associating with gp130, and promoting signal transduction. Initial studies reported that sIL-6R could be generated by alternate mRNA splicing [18]. However, Mullberg et al. [72, 73] showed that sIL-6R could also be generated by shedding from the cell surface, and that this process could be induced by PMA, but did not require new protein synthesis. Cleavage was reported to occur at Gln^{357}-Asp^{358}, a single residue proximal to the predicted transmembrane region [74]. Attempts to identify the proteinase responsible for this activity have proved unsuccessful. Inhibitors of serine and cysteine proteinases and several metalloproteinase inhibitors, including TIMP-1 and TIMP-2, have no effect on this activity [19, 20]. However, hydroxamate-based inhibitors (Tab. 3) [10, 19, 20, 30] can inhibit shedding of the IL-6R. Furthermore, Hargreaves et al. [20] have shown that TIMP-3 can also prevent release of IL-6R from human myeloma cells.

Narazaki et al. [75] have reported the presence of a soluble form of gp130 (sgp130) in serum. Sgp130 is believed to function as an antagonist to IL-6 by binding the IL-6/sIL-6R complex and preventing signalling through membrane-bound gp130. Studies have suggested that sgp130 can also be generated by both alternate mRNA splicing and proteolytic processing [76, 77]. Furthermore, our own data suggest that this process can be promoted by PMA and inhibited by hydroxamate-based inhibitors (P.G. Hargreaves, N.L. Dury and P.I. Croucher, unpublished observations). These observations suggest that both IL-6R and gp130 can be released from the cell in a similar manner.

Inhibiting cytokine receptor shedding

Although the mechanism of release of membrane-bound cytokine receptors is poorly understood, it is clear that this activity can be inhibited by hydroxamate-based metalloproteinase inhibitors (Tab. 3). This applies not only to the TNFα and IL-6 receptors, but to all of the shed cytokine receptors examined to date. Since the inhibition profile is identical to that for TNFα processing, there is a strong possibility that members of the ADAM family of proteinases are also responsible for releasing membrane-bound cytokine receptors. However, as there is no consensus cleavage site within these membrane-bound proteins, it is unknown whether the individual processing events are mediated by specific proteinases or whether a single proteinase, with more relaxed specificity, could be responsible for more than one processing event. Indeed, it is possible that TACE itself may be able to release cytokine receptors from the cell surface. Thus, until the identities of the proteinases responsible for releasing cytokine receptors from the cell surface become clear, it is unlikely that specific inhibitors of these events will be developed. However, the broad-spectrum inhibitors that are currently available will continue to prove of value in investigating the mechanisms of release of these receptors.

Conclusions

Proteinases clearly play an important role in regulating the activity of cytokines and modulating the cellular response to these factors. They are involved at several different levels, including the activation of pro-cytokines, the generation of soluble cytokines and the release of soluble cytokine receptors. However, our understanding of these complex processes is incomplete, with our level of knowledge represented by the examples described in this chapter. Thus, the intracellular processing of IL-1β by caspase-1 constitutes the best-characterised system, with caspase-1 inhibitors being shown to be effective both *in vitro* and in murine models of inflammation. The experience from these systems suggests that targeting this proteinase may have important therapeutic potential, since inhibiting caspase-1 does not appear to affect other systems. Our understanding of the mechanisms responsible for releasing membrane-bound cytokines is less clear. Although the proteinase responsible for processing TNFα has recently been identified, specific inhibitors of this enzyme (TACE) are not yet available. Hydroxamate-based compounds are effective at blocking the activity of TACE, although they also inhibit the release of other membrane-bound cytokines and cytokine receptors. Consequently, the therapeutic value of inhibiting TACE may only be determined once more selective inhibitors become available. Finally, the proteinase(s) responsible for shedding of cytokine receptors remain unknown, although members of the ADAM family represent strong candidates. Whether inhibitors of receptor shedding will prove to be valuable therapeutic targets will only become clear when these proteinases have been discovered and selective inhibitors identified. However, it is clear that proteinases will continue to be targets for therapeutic intervention in the future.

References

1   Massague J (1990) Transforming growth factor-α. A model for membrane-anchored growth factors. *J Biol Chem* 265: 21393–21396

2   Rose-John S, Heinrich PC (1994) Soluble receptors for cytokines and growth factors: generation and biological function. *Biochem J* 300: 281–291

3   Cerretti DP, Kozlosky CJ, Mosley B, Nelson N, Van Ness K, Greenstreet TA, March CJ, Kronheim SR, Druck T, Cannizzaro LA et al (1992) Molecular cloning of the interleukin 1β converting enzyme. *Science* 256: 97–100

4   Thornberry NA, Bull HG, Calaycay JR, Chapman KT, Howard AD, Kostura MJ, Miller DK, Molineaux SM, Weider JR, Aunins J et al (1992) A novel heterodimeric cyteine protease is required for interleukin-1β processing in monocytes. *Nature* 356: 768–774

5   Gu K, Kuida K, Tsutsui H, Ku G, Hsiao K, Fleming MA, Hayashi N, Higashino K, Okamura H, Nakanishi K et al (1997) Activation of interferon-γ-inducing factor mediated by interleukin-1β converting enzyme. *Science* 275: 206–209

6   Mohler KM, Sleath PR, Fitzner JN, Cerretti DP, Alderson M, Kerwar SS, Torrance DS, Otten-Evans C, Greenstreet T, Weerawarna K et al (1994) Protection against a lethal dose of endotoxin by an inhibitor of tumour necrosis factor processing. *Nature* 370: 218–220

7   Gearing AJH, Beckett P, Christodoulou M, Churchhill M, Clements J, Davidson AH, Drummond AH, Galloway WA, Gilbert R, Gordon JL et al (1994) Processing of tumour necrosis factor-α precursor by metalloproteinases. *Nature* 370: 555–557

8   McGeehan GM, Becherer JD, Champion B, Connolly KM, Conway JG, Furdon P, Karp S, Kidao S, McElroy AB, Nichols J, Pryzwansky KM et al (1994) Regulation of tumour necrosis factor-α processing by a metalloproteinase inhibitor. *Nature* 370: 558–561

9   Pandiella A, Bosenberg MW, Huang EJ, Besmer P, Massague J (1992) Cleavage of membrane-anchored growth factors involves distinct protease activities regulated through common mechanisms. *J Biol Chem* 267: 24028–24033

10   Gallea-Robache S, Morand V, Millet S, Bruneau J-M, Bhatnagar N, Chouaib S, Roman-Roman S (1997) A metalloproteinase inhibitor blocks the shedding of soluble cytokine receptors and processing of transmembrane cytokine precursors in human monocytic cells. *Cytokine* 9: 340–346

11   Mullberg J, Rauch CT, Wolfson MF, Castner B, Fitzner JN, Otten-Evans C, Mohler KM, Cosman D, Black RA (1997) Further evidence for a common mechanism for shedding of cell surface proteins. *FEBS Lett* 401: 235–238

12   Lyman SD, James L, Escobar S, Downey H, de Vries P, Brasel K, Stocking K, Beckmann MP, Copeland NG, Cleveland LS et al (1995) Identification of soluble and membrane-bound isoforms of the murine flt3 ligand generated by alternative splicing of mRNAs. *Oncogene* 10: 149–157

13   Lyman SD, James L, Vanden Bos T, de Vries P, Brasel K, Gliniak B, Hollingsworth LT, Picha KS, McKenna HJ, Splett RR et al (1993) Molecular cloning of a ligand for the flt3/flk-2 tyrosine kinase receptor: a proliferative factor for primative hematopoietic cells. *Cell* 75: 1157–1167

14   Loughnan MS, Sanderson CJ, Nossal GJV (1988) Soluble interleukin 2 receptors are released from the cell surface of normal murine B lymphocytes stimulated with interleukin 5. *Proc Natl Acad Sci USA* 85: 3115–3119

15   Mosley B, Beckmann MP, March CJ, Idzerda RL, Gimpel SD, VandenBos T, Friend D, Alpert A, Anderson D, Jackson J et al (1989) The murine interleukin-4 receptor: molecular cloning and characterization of secreted and membrane bound forms. *Cell* 59: 335–348

16   Tavernier J, Devos R, Cornelis S, Tuypens T, Van der Heyden J, Fiers W, Plaetinck G (1991) A human high affinity interleukin-5 receptor (IL5R) is composed of an IL5-specific α chain and a β chain shared with the receptor for GM-CSF. *Cell* 66: 1175–1184

17   Tavernier J, Tuypens T, Plaetinck G, Verhee A, Fiers W, Devos R (1992) Molecular basis of the membrane-anchored and two soluble isoforms of the human interleukin 5 receptor α subunit. *Proc Natl Acad Sci USA* 89: 7041–7045

18   Lust JA, Donovan KA, Kline MP, Greipp PR, Kyle RA, Maihle NJ (1992) Isolation of an mRNA encoding a soluble form of the human interleukin-6 receptor. *Cytokine* 4: 96–100

19   Mullberg J, Durie FH, Otten-Evans C, Alderson MR, Rose-John S, Cosman D, Black

RA, Mohler KM (1995) A metalloproteinase inhibitor blocks shedding of the IL-6 receptor and the p60 TNF receptor. *J Immunol* 155: 5198–5205

20 Hargreaves PG, Wang F, Antcliff J, Murphy G, Lawry J, Russell RGG, Croucher PI (1998) Human myeloma cells shed the interleukin-6 receptor: Inhibition by tissue inhibitor of metalloproteinase-3 and a hydroxamate-based metalloproteinase inhibitor. *Br J Haematol* 101: 694–707

21 Goodwin RG, Friend D, Ziegler SF, Jerzy R, Falk BA, Gimpel S, Cosman D, Dower SK, March CJ, Namen AE et al (1990) Cloning of the human and murine interleukin-7 receptors: demonstration of a soluble form and homology to a new receptor superfamily. *Cell* 60: 941–951

22 Renauld J-C, Druez C, Kermouni A, Houssiau F, Uyttenhove C, Van Roost E, Van Snick J (1992) Expression cloning of the murine and human interleukin 9 receptor cDNAs. *Proc Natl Acad Sci USA* 89: 5690–5694

23 Crowe PD, Walter BN, Mohler KM, Otten-Evans C, Black RA, Ware CF (1995) A metalloproteinase inhibitor blocks shedding of the 80-kD TNF receptor and TNF processing in T lymphocytes. *J Exp Med* 181: 1205–1210

24 Downing JR, Roussel MF, Sherr CJ (1989) Ligand and protein kinase C downmodulate the colony stimulating factor 1 receptor by independent mechanisms. *Mol Cell Biol* 9: 2890–2896

25 Fukunaga R, Seto Y, Mizushima S, Nagata S (1990) Three different mRNAs encoding human granulocyte colony-stimulating factor receptor. *Proc Natl Acad Sci USA* 87: 8702–8706

26 Raines MA, Liu L, Quan SG, Joe V, DiPersio JF, Golde DW (1991) Identification and molecular cloning of a soluble human granulocyte-macrophage colony-stimulating factor receptor. *Proc Natl Acad Sci USA* 88: 8203–8207

27 Tiesman J, Hart CE (1993) Identification of a soluble receptor for platelet-derived growth factor in cell-conditioned medium and human plasma. *J Biol Chem* 268: 9621–9628

28 Gearing DP, Thut CJ, VandenBos T, Gimpel SD, Delaney PB, King J, Price V, Cosman D, Beckmann MP (1991) Leukemia inhibitory factor receptor is structurally related to the IL-6 signal transducer, gp130. *EMBO J* 10: 2839–2848

29 Layton MJ, Cross BA, Metcalf D, Ward LD, Simpson RJ, Nicola NA (1992) A major binding protein for leukemia inhibitory factor in normal mouse serum: Identification as a soluble form of the cellular receptor. *Proc Natl Acad Sci USA* 89: 8618–8620

30 Arribas J, Coodly L, Vollmer P, Kishimoto TK, Rose-John S, Massague J (1996) Diverse cell surface protein ectodomains are shed by a system sensitive to metalloproteinase inhibitors. *J Biol Chem* 271: 11376–11382

31 Dinarello CA, Wolff SM (1993) The role of interleukin-1 in disease. *N Engl J Med* 328: 106–113

32 Dinarello CA (1991) Interleukin-1 and interleukin-1 antagonism. *Blood* 77: 1627–1652

33 Kobayashi Y, Yamamoto K, Saido T, Kawasaki H, Oppenheim JJ, Matsushima K (1990)

Identification of calcium-activated neutral protease as a processing enzyme of human interleukin 1α. *Proc Natl Acad Sci USA* 87: 5548–5552

34 Okamura H, Tsutsui H, Komatsu T, Yutsudo M, Hakura A, Tanimoto T, Torigoe K, Okura T, Nakuda Y, Hattori K et al (1995) Cloning of a new cytokine that induces INF-γ production by T cells. *Nature* 378: 88–92

35 Black RA, Kronheim SR, Cantrell M, Deeley MC, March CJ, Prickett KS, Wignall J, Conlon PJ, Cosman D, Hopp TP et al (1988) Generation of biologically active interleukin 1β by proteolytic cleavage of the inactive precursor. *J Biol Chem* 263: 9437–9442

36 Talanian RV, Quinlan C, Trautz S, Hackett MC, Mankovich JA, Banach D, Ghayur T, Brady KD, Wong WW (1997) Substrate specificity of caspase family proteases. *J Biol Chem* 272: 9677–9682

37 Villa P, Kaufman SH, Earnshaw WC (1997) Caspases and caspase inhibitors. *Trends Biochem Sci* 22: 388–393

38 Walker NP, Talanian RV, Brady KD, Dang LC, Bump NJ, Ferenz CR, Franklin S, Ghayur T, Hackett MC, Hammill LD et al (1994) Crystal structure of the cysteine protease interleukin-1β converting enzyme: a (p20/p100)2 homodimer. *Cell* 78: 343–352

39 Dinarello CA, Margolis NH (1995) Stopping the cuts. *Curr Biol* 5: 587–590

40 Salvesen GS, Dixit VM (1997) Caspases:intracellular signalling by proteolysis. *Cell* 91: 443–446

41 Nicholson DW, Thornberry NA (1997) Caspases: killer proteases. *Trends Biochem Sci* 22: 299–306

42 Ray CA, Black RA, Kronheim SR, Greenstreet TA, Sleath PR, Salvesen GS, Pickup DJ (1992) Viral inhibition of inflammation: cowpox virus encodes an inhibitor of the interleukin-1β converting enzyme. *Cell* 69: 597–604

43 Bump NJ, Hackett M, Hugunin M, Seshagiri S, Brady K, Chen P, Ferenz C, Franklin S, Ghayur T, Li P et al (1995) Inhibition of ICE family proteases by baculovirus antiapoptotic protein p35. *Science* 269: 1885–1888

44 Livingston DJ (1997) *In vitro* and *in vivo* studies of ICE inhibitors. *J Cell Biochem* 64: 19–26

45 Miller DK, Calaycay JR, Chapman KT, Howard AD, Kostura MJ, Molineaux SM, Thornberry NA (1993) The IL-1β converting enzyme as a therapeutic target. *Ann N Y Acad Sci* 696: 133–148

46 Fletcher DS, Agarwal L, Chapman KT, Chin J, Egger LA, Limjuko G, Luell S, MacIntyre DE, Peterson EP, Thornberry NA et al (1995) A synthetic inhibitor of interleukin-1β converting enzyme prevents endotoxin-induced interleukin-1β production *in vitro* and *in vivo*. *J Interferon Cytokine Res* 15: 243–248

47 Miller BE, Krasney PA, Gauvin DM, Holbrook KB, Koonz DJ, Abruzzese RV, Miller RE, Pagni KA, Dolle RE, Ator MA et al (1995) Inhibition of mature IL-1β production in murine macrophages and a murine model of inflammation by WIN 67694, an inhibitor of IL-1β converting enzyme. *J Immunol* 154: 1331–1338

48 Ku G, Faust T, Lauffer LL, Livingston DJ, Harding MW (1996) Interleukin-1β convert-

ing enzyme inhibition blocks progression of type II collagen-induced arthritis in mice. *Cytokine* 8: 377–386

49   Li P, Allen H, Banjerjee S, Seshadri T (1997) Characterization of mice deficient in interleukin-1β converting enzyme. *J Cell Biochem* 64: 27–32

50   Beutler B, Cerami A (1989) The biology of cachetin/TNF: a primary mediator of the host response. *Annu Rev Immunol* 7: 625–655

51   Aggarawal BB, Kohr WJ, Hass PE, Moffat B, Spencer SA, Henzel W, Bringman TS, Nedwin GE, Goeddel DV, Harkins RN (1985) Human tumor necrosis factor: production, purification and characterization. *J Biol Chem* 260: 2345–2354

52   Wang AM, Creasey AA, Ladner MB, Lin LS, Strickler J, Van Arsdell JN, Yamamoto R, Mark DF (1985) Molecular cloning of the complementary DNA for human tumor necosis factor. *Science* 228: 149–154

53   Kriegler M, Perez C, De Fray K, Albert I, Lu SD (1988) A novel form of TNF/cachetin is a surface cytotoxic transmembrane protein: ramifications for the complex physiology of TNF. *Cell* 53: 45–53

54   Williams LM, Gibbons DL, Gearing A, Maini RN, Feldmann M, Brennan FM (1996) Paradoxical effects of a synthetic metalloproteinase inhibitor that blocks both p55 and p75 TNF receptor shedding and TNFα processing in RA synovial membrane cell culture. *J Clin Invest* 97: 2833–2841

55   Black RA, Durie FH, Otten-Evans C, Miller R, Slack JL, Lynch DH, Castner B, Mohler KM, Gerhart M, Johnson RS et al (1996) Relaxed specificity of matrix metalloproteinases (MMPS) and TIMP insensitivity of tumour necrosis factor-α (TNF-α) production suggest the major TNF-α converting enzyme is not an MMP. *Biochem Biophys Res Commun* 225: 400–405

56   Black RA, Rauch CT, Kozlosky CJ, Peschon JJ, Slack JL, Wolfson MF, Castner BJ, Stocking KL, Reddy P, Srinivasan S et al (1997) A metalloproteinase disintegrin that releases tumour-necrosis factor-α from cells. *Nature* 385: 729–733

57   Moss ML, Jin S-LC, Milla ME, Burkhart W, Carter HL, Chen W-J, Clay WC, Didsbury JR, Hassler D, Hoffman CR et al (1997) Cloning of a disintegrin metalloproteinase that processes precursor tumour-necrosis factor-α. *Nature* 385: 733–736

58   Wolfsberg TG, Primakoff P, Myles DG, White JM (1995) ADAM, a novel family of membrane proteins containing a disintegrin and metalloproteinase domain: Multipotential functions in cell-cell and cell-matrix interactions. *J Cell Biol* 131: 275–278

59   Maskos K, Fernandez-Catalan C, Huber R, Bourenkov GP, Bartunik H, Ellestad GA, Reddy P, Wolfson MF, Rauch CT, Castner BJ et al (1998) Crystal structure of the catalytic domain of human tumour necrosis factor-α-converting enzyme. *Proc Natl Acad Sci USA* 95: 3408–3412

60   Howard L, Lu X, Mitchell S, Griffiths S, Glynn P (1996) Molecular cloning of MADM: a catalytically active mammalian disintegrin-metalloproteinase expressed in various cell types. *Biochem J* 317: 45–50

61   Lunn CA, Fan X, Dalie B, Miller K, Zavodny PJ, Narula SK, Lundell D (1997) Purification of ADAM 10 from bovine spleen as a TNFα convertase. *FEBS Lett* 400: 333–335

62 Rosendahl MS, Ko SC, Long DL, Brewer MT, Rosenzweig B, Hedl E, Anderson L, Pyle SM, Moreland J, Meyers MA et al (1997) Identification and characterization of a pro-tumor necrosis factor-α-processing enzyme from the ADAM family of zinc metallopro-teinases. *J Biol Chem* 272: 24588–24593

63 Hooper HM, Karran EH, Turner AJ (1997) Membrane protein secretases. *Biochem J* 321: 265–279

64 Engelmann H, Aderka D, Rubenstein M, Rotman D, Wallach D (1989) A tumor necro-sis factor-binding protein purified to homogeneity from human urine protects cells from tumor necrosis factor toxicity. *J Biol Chem* 264: 11974–11980

65 Nophar Y, Kemper O, Brakebusch C, Englemann H, Zwang R, Aderka D, Holtmann H, Wallach D (1990) Soluble forms of tumour necrosis factor receptors (TNF-Rs). The cDNA for the typeI TNF-R, cloned using amino acid sequence data of its soluble form, encodes both cell surface and a soluble form of the receptor. *EMBO J* 9: 3269–3278

66 Heller RA, Song K, Onasch MA, Fischer WH, Chang D, Ringold GM (1990) Comple-mentary DNA cloning of a receptor for tumor necrosis factor and demonstration of a shed form of the receptor. *Proc Natl Acad Sci USA* 87: 6151–6155

67 Hwang C, Gatanaga M, Granger GA, Gatanaga T (1993) Mechanism of release of sol-uble forms of tumor necrosis factor/lymphotoxin receptors by phorbol myristate acetate-stimulated human THP-1 cells *in vitro*. *J Immunol* 151: 5631–5638

68 Porteu F, Brockhaus M, Wallach D, Engelmann H, Nathan CF (1991) Human neu-trophil elastase releases a ligand-binding fragment from the 75-kDa tumor necrosis fac-tor (TNF) receptor. *J Biol Chem* 266: 18846–18853

69 Hermann C, Chernajovsky Y (1998) Mutation of proline 211 reduces shedding of the human p75 TNF receptor. *J Immunol* 160: 2478–2487

70 Smith MR, Kung H-f, Durum SK, Colburn NH, Sun Y (1997) TIMP-3 induces cell death by stabilizing TNF-α receptors on the surface of human colon carcinoma cells. *Cytokine* 9: 770–780

71 Novick D, Engelmann H, Wallach D, Rubinstein M (1989) Soluble cytokine receptors are present in normal human urine. *J Exp Med* 170: 1409–1414

72 Mullberg J, Schooltink H, Stoyan T, Heinrich PC, Rose-John S (1992) Protein kinase C acitivity is rate limiting for shedding of the interleukin-6 receptor. *Biochem Biophys Res Commun* 189: 794–800

73 Mullberg J, Schooltink H, Stoyan T, Gunther M, Graeve L, Buse G, Mackiewicz A, Heinrich PC, Rose-John S (1993) The soluble interleukin-6 receptor is generated by shedding. *Eur J Immunol* 23: 473–480

74 Mullberg J, Oberthur W, Lottspeich F, Mehl E, Dittrich E, Graeve L, Heinrich PC, Rose-John S (1994) The soluble human IL-6 receptor. Mutational characterization of the pro-teolytic cleavage site. *J Immunol* 152: 4958–4968

75 Narazaki M, Yasukawa K, Saito T, Ohsugi Y, Fukui H, Koishihara Y, Yancopoulos G, Taga T, Kishimoto T (1993) Soluble forms of the interleukin-6 signal-transducing recep-tor component gp130 in human serum possessing a potential to inhibit signals through membrane-anchored gp130. *Blood* 82: 1120–1126

76   Diamant M, Rieneck K, Mechti N, Zhang X-G, Svenson M, Bendtzen K , Klein B, (1997) Cloning and expression of an alternative spliced mRNA encoding a soluble form of the human interleukin-6 signal transducer gp130. *FEBS Lett* 412: 379–384

77   Mullberg J, Dittrich E, Graeve L, Gerhartz C, Yasukawa K, Taga T, Kishimoto T, Heinrich PC, Rose-John S (1993) Differential shedding of the two subunits of the interleukin-6 receptor. *FEBS Lett* 332: 174–178

78   Stein J, Rettenmier C (1991) Proteolytic processing of a plasma membrane-bound precursor to human macrophage-colony stimulating factor (CSF-1) is accelerated by phorbol ester. *Oncogene* 6: 601–605

Cytokine-neutralizing therapeutic antibodies

Amanda Suitters and Roly Foulkes

Celltech Chiroscience, 216 Bath Road, Slough, Berkshire SL1 4EN, UK

Introduction

Monoclonal antibodies (Mabs) offer the potential as useful therapeutic agents because of their high affinity and selectivity for the target antigen. The standard approach in making Mabs has been to immunise rodents, usually mice, with the desired human protein and generate a mouse anti-human IgG through hybridoma technology. Indeed mouse Mabs are used clinically in acute diseases, such as the use of OKT3 (an anti-CD3 Mab) for the treatment of acute rejection episodes following transplantation. The major drawback for these murine Mabs is the high prevalence of anti-mouse antibody responses in the recipient (the so-called HAMA response). This has two consequences: it results in rapid clearance of the Mab and could cause a large immune response if it (or any other murine Mab) was given a second time. Such Mabs are therefore unsuitable for chronic diseases where repeat therapy is required. Immunogenicity can be reduced using human Mabs produced, for example, by phage display or to humanise or CDR graft a mouse antibody. This reduces the number of mouse residues in the Mab to about 12–15%, the vast majority of which are contained in the CDR regions. Such engineered human antibodies are now having a large impact on the treatment of chronic diseases. Although much of the early pioneering work was done with Mabs in different animal systems, the last few years have seen a plethora of clinical studies performed. These advances have resulted in the registration throughout the world of the first of these drugs for a wide range of diseases.

This chapter summarises the information to date on the rationale, development and clinical use of Mabs in the treatment of a whole range of acute and chronic diseases.

Septic shock

One of the earliest concerted attempts to utilize the wealth of convincing preclinical data to produce a therapeutic Mab was in inhibition of tumour necrosis factor

Novel Cytokine Inhibitors, edited by Gerry A. Higgs and Brian Henderson
© 2000 Birkhäuser Verlag Basel/Switzerland

(TNF) for the treatment of septic shock. TNF has been thought of as one of a number of host-derived mediators which, either singly or in combination, appear able to mediate some of the systemic inflammatory events which follow a septic challenge. There is a large body of evidence from both experimental animal studies and clinical monitoring implicating inappropriate production of TNF as a key factor in the degeneration of patients with septic shock. The magnitude of the contribution of any given mediator to clinical status remains open to debate. However, for the purposes of the development of new therapies, the primary focus has been to select individual mediators whose antagonism may be of clinical benefit. Prominent recent examples of this include interleukin-1 (IL-1), platelet activating factor (PAF), and TNF.

In humans, numerous assay studies in septic patients have attempted to correlate the plasma levels of TNF and related cytokines with a clinical course and prognosis [1–7]. Although these studies reveal a general trend to an association, the results are somewhat patchy and may be assay-dependent. In a few cases of acute illness, a clear clinical picture emerges. In less acute cases, it may be that plasma TNF is either below assay-detection limits or absent, is associated with cells, soluble receptors, or nonspecific serum binding proteins, or is produced and/or sequestered into other body microenvironments. Direct administration of low doses of endotoxin to human volunteers elicits a measurable rise in plasma TNF [8]. This is an important observation, as endotoxin is thought to be a primary pathogenic driver in gram-negative septic shock.

A septic challenge causes a very wide variety of effects that can broadly be put into three categories. First, there are physiological changes associated mainly with deterioration in function in most vital organ systems. Second, there are cellular effects, characterized by changes in adhesion and activation states of endothelial and circulating inflammatory cells. Third, a large number of host mediators are elaborated which are likely to be orchestrating the physiological and cellular effects. The use of Mabs as specific TNF antagonists in rodent and nonhuman primate models of septic shock demonstrates a role for TNF in this complex cascade of events.

TNF administered to experimental animals and humans elicits many of the physiological and tissue disturbances characteristic of clinical septic shock [9]. The earliest demonstration that *in vivo* endotoxic shock may be treated by TNF blockade was in a mouse model of endotoxaemia using a rabbit polyclonal antiserum to murine TNF [10]. The antiserum was shown to cause an increase in the median lethal dose of endotoxin in Balb/c mice. A similar protective effect was later shown using the TNF receptor immunoadhesin [11].

More clinically relevant than endotoxaemia is the use of bacterial challenges. Studies in rodents [12, 13], pigs [14, 15] and non-human primates [16–19] have shown protection by neutralizing antibodies against the lethal effects of shock induced by sepsis. In a further attempt to mirror the genesis of clinical septic shock, Opal and colleagues developed a rat model in which the gut is colonized with

Pseudomonas aeruginosa after which the animals are rendered neutropenic with cyclophosphamide [13]. The neutropenia allows the bacteria to infect systemically, leading to death of all animals within 7 to 10 days. Treatment with an anti-murine TNF Mab (TN3 19.12), which cross-neutralizes rat TNF, resulted in a 50% survival rate. When combined with ciprofloxacin, all animals survived. In this system, the protective effects of anti-TNF were seen not only with prophylactic administration, but also when treatment was initiated after the appearance of clinical signs of disease [20].

A seminal study of anti-TNF in lethal bacteraemia in the baboon was reported by Tracey et al. [16]. Animals were passively immunized against TNF with a Fab'$_2$ fragment of a cross-reactive Mab to human TNF. This protected against an LD$_{100}$ dose of intravascular (i.v.) *Escherichia coli* given 2 h later. A later study by Hinshaw et al. [18], using a full immunoglobulin in a model similar to that of Tracey et al. [16], showed protective effects against the LD$_{100}$ challenge when antibody was given 30 min after the start of the bacterial infusion. Although this does not define a treatment window that would be of use clinically, it does remove the apparent need for preconditioning of the animals for the protective effect to be shown. These animal systems use very large bolus challenges, which do not accurately mirror the clinical course of the development of sepsis and shock, apart, perhaps, from the neutropenic rat. They do, however, define a clear pathophysiological role for TNF in gram-negative shock and endotoxaemia.

The role of TNF in morbidity and mortality associated with gram-positive infection and shock is more controversial. These infections are very important, accounting for 40 to 50% of clinical septic shock. Like endotoxin, cell wall components of gram-positive organisms are capable of inducing TNF secretion both *in vitro* and *in vivo* [21, 22]. In a baboon model of lethal *Staphylococcus aureus*-induced shock, anti-TNF Mab was highly effective in preventing multiple organ failure and preventing death [18]. These results were achieved in the presence of the antibiotic ceftriaxone, which on its own had only modest effects at prolonging survival. Results in a study of murine gram-positive challenge were less clear-cut [23].

A second area of controversy surrounding the clinical potential of TNF blockade in infectious disease concerns differences in response to anti-TNF Mab depending on the site of infection. All models in which beneficial effects have been seen primarily involve systemic challenge. Bagby and workers [24] compared the efficacy of anti-TNF Mab treatment on survival following i.v. or intraperitoneal (i.p.) instillation of *Escherichia coli*. The i.v. challenge gave high TNF plasma levels in controls and responded to anti-TNF Mab treatment. The treatment was not effective in the i.p. animals. Echtenacher et al. [25] induced peritonitis by caecal ligation and puncture (CLP). In this case anti-TNF Mab treatment was deleterious, suggesting a requirement for endogenous TNF for recovery from CLP. Indeed, Alexander and co-workers [26] have demonstrated a protective effect of exogenous recombinant TNF in a similar model of CLP in rats.

These studies raise an important issue relating to the balance between the protective and destructive effects of the endogenous TNF response to infection. TNF probably plays a role in primary host defense against infection. In circumstances where the aetiology of disease is related to infection *per se*, removing TNF may be deleterious in the absence of covering antimicrobial therapy. The responses in the intravascular models may be dominated by a systemic inflammatory response, which results, at least in part, from an inappropriate overproduction of TNF. The possibility of compromising host defense by TNF blockade has been a key safety issue in clinical studies, although to date there have been no reports of adverse events relating to infection from studies involving nearly 2000 patients treated with anti-TNF antibodies.

All of the above results provided a reasonable basis for postulating that a TNF antagonist may be of therapeutic benefit in clinical septic shock [27]. Three small clinical studies have been reported in which a murine anti-TNF Mab has been administered to septic patients [28–30]. These gave some limited support for the concept. Large clinical studies have been performed. Early studies compared a single administration of a murine Mab to TNF with placebo in patients with septic shock, with a prospective stratification of patients into either a shock or no shock stratum [31, 32]. In patients not in shock at study entry, no effect of the antibody was seen, either positive or negative. In patients in shock at entry, there was a clear trend to a reduction in mortality at both dose levels (7.5 and 15 mg/kg) used in the study. The antibody was well tolerated with very few adverse events that were judged antibody-related.

The fact that efficacy was limited to the 50% of patients with frank shock (haemodynamic instability) at entry compares well with the results of the animal model studies. These are patients, selected from a very heterogeneous group, who are in an acute phase of inflammatory degeneration during which the processes involving mediators such as TNF are likely to be active. This is the cohort likely to be most similar to the experimental animal experiencing an acute septic/inflammatory challenge. Indeed, in the most comprehensive study to date, again using a mouse anti-TNFα Mab, patients with septic shock only were included. Patients received either a single i.v. infusion of 7.5 mg/kg anti-TNF Mab (n = 949) or placebo (n = 930) and outcome was assessed as the rate of all cause mortality at 28 days [33]. There was no association between therapy with anti-TNFα Mab and reversal of initial septic shock or prevention of subsequent shock. Debates continue as to the most appropriate design of sepsis trials for this type of reagent in which there is a rationale so clearly supported by *in vitro* and *in vivo* studies.

Asthma/allergy

Allergy and asthma are characterised by an inflammatory response associated with cells of the immune system such as T cells, B cells, mast cells and eosinophils, the production of various cytokines and the predominance of a TH2-type immune response [34, 35]. Interleukin-4 (IL-4) has been shown to be the central driver in the differentiation of naive T lymphocytes into TH2 lymphocytes [36]. These TH2 lymphocytes, in turn, produce other cytokines such as IL-3 and IL-5, which have direct effects on mast cells, eosinophils and B cells. In addition to its effects on T lymphocytes, IL-4 induces B-cell proliferation and the production of antigen-specific IgE that is required to trigger an allergic response. Antigen-specific IgE induces mast cells and eosinophils to release inflammatory mediators such as histamine, prostaglandins and interleukins (IL-3, 4, 5 and 6) which further regulate the differentiation and proliferation of a variety of lymphoid and myeloid cells. Of these cytokines, IL-4 and IL-5 have been shown to be particularly important, IL-4 because of its central role in the development and potentiation of the immune response [36] and IL-5 because of its paramount importance in the control of eosinophil differentiation, maturation and function [37].

Based on this, a number of strategies have been devised to block the function of these two cytokines. These include antibodies directed against either the cytokine itself or its receptor, or by the use of soluble receptor antagonists. Neutralising antibodies against IL-4 have been shown to inhibit both primary *in vitro* responses of B cells and *in vivo* polyclonal IgE responses in mice infected with *Nippostrongylus brasiliensis* or injected with anti-IgD antibodies [38, 39]. Although these studies indicated an essential role for IL-4 in the generation of IgE antibody responses, much of the therapeutic effort in this area has been directed towards targeting the production of IL-4 using recombinant soluble receptor antagonists (sIL-4R) [40–43]. Sato and co-workers [41] have shown that the sIL-4R inhibited IgE production by up to 85% *in vivo* using an anti-IgD model in mice. Anti IL-4 Mab administration in this model resulted in comparable levels of inhibition. Further studies have shown that administration of recombinant sIL-4R to mice resulted in prolonged survival of heterotopic cardiac allografts, decreased popliteal lymph node enlargement in response to allogeneic cells and inhibition of IgE production in response to anti-IgD treatment [40, 42]. Neutralising antibodies against IL-4 or against the IL-4R were also effective inhibitors of the allogeneic response when given at 100- to 1000-fold higher concentrations than sIL-4R [40]. Interestingly, co-administration of IL-4/sIL-4R or IL-4/anti-IL-4 Mab complexes caused a superinduction of IgE production, possibly by altering the biodistribution of the cytokine [41, 42]. Additional studies have shown that sIL-4R could inhibit both IL-4 induced IgE synthesis and modulate expression of CD23 and release of soluble CD23 by human peripheral blood mononuclear cells [44]. The strategy of using sIL-4R to target diseases such as asthma and allergy in humans is being pursued by Immunex Corporation, who is carrying out world-wide Phase II trials on a genetically engineered sIL-4R (Nuvance) administered as a nebulized liquid. This receptor antagonist may also be of use in the treatment of organ transplant rejection, graft-versus-host disease and infectious disease.

Interleukin-5 (IL-5) has been shown to be crucial for eosinophil recruitment into the lungs in animal models of respiratory inflammation and to play a prominent role in the pulmonary eosinophilic inflammation associated with the pathology of human asthma and other allergic diseases [35, 45]. IL-5 therefore provides an interesting target for therapeutic intervention, particularly in diseases involving eosinophilic inflammation such as bronchial asthma. A number of strategies have been devised to modulate IL-5 function, which include inhibiting signal transduction of IL-5, blocking the IL-5 receptor (IL-5R), preventing synthesis of IL-5 or IL-5R, and neutralising IL-5 with monoclonal antibodies or soluble receptors [46, 47]. The therapeutic approach using Mabs against IL-5 has been used successfully in animal models of eosinophilia and airway hyper-responsiveness [48–52]. Experiments utilising a rat anti-murine IL-5 (TRFK-5) antibody have shown significant suppression of allergen-induced eosinophilia in the bronchoalveolar lavage of actively sensitised mice when administered up to 5 days after challenge [50]. In this study, TRFK-5 also reduced ovalbumin-induced eosinophil infiltration into the peribronchial and alveolar regions while reducing the depletion of bone marrow eosinophils occurring after antigen challenge. In addition, TRFK-5 at low doses of 0.3 mg/kg, administered 1 h before challenge, inhibited Ascaris-induced eosinophilia by about 77% 24 h after challenge which persisted for 3 months [53]. However, in a study of actively sensitised guinea-pigs, the suppression of allergen-induced pulmonary eosinophilia by TRFK-5 was unaccompanied by reduced airway hyperresponsiveness to the constrictor peptide, Substance P [51]. In contrast, TRFK-5 abolished the Ascaris-induced increase in airway responsiveness to histamine in monkeys [53]. Following on from the data in the animal models, Schering-Plough and Celltech have developed a humanised antibody to human IL-5 – Sch 55700 [47]. Sch 55700 has been shown to be well tolerated with a plasma half-life of 25 days at 0.3 mg/kg in man. It inhibited IL-5-mediated proliferation of human erythroleukemic TF-1 cells and *in vivo* inhibited the influx of eosinophils into the lungs of allergic mice. In guinea pigs, Sch 55700 inhibited pulmonary eosinophilia, acute bronchoconstriction and, less effectively, airway hyperresponsiveness in allergic guinea pigs. In allergic rabbits, the influx of eosinophils into the skin and in allergic cynomolgus monkeys, the pulmonary eosinophilia following antigen challenge was inhibited by Sch 55700. Further development of this antibody for use in human asthma is being pursued as a potential prophylactic therapeutic.

Rheumatoid arthritis

Rheumatoid arthritis (RA) is a disease characterised by inflammation and cellular proliferation of the synovial lining (pannus formation), resulting in the invasion of the adjacent cartilage matrix and its subsequent erosion. Cells of both the immune and inflammatory systems infiltrate the pannus and cartilage tissue with predominantly activated T cells infiltrating the subintimal layer of the synovium, and PMNs, macrophages, fibroblasts, and plasma cells infiltrating the pannus/cartilage junction. In parallel to the cellular infiltrate, many soluble mediators, such as IL-1, TNFα, TNFβ, IL-6, can be detected within the rheumatoid joint, all of which may contribute to the disease process [54, 55]. Because of the presence of cytokines within the synovial joint, anti-cytokine treatment as a therapy for RA has warranted much interest, in particular the neutralisation of TNF.

Leukocytes, postcapillary endothelial cells, chondrocytes, osteoclasts, and fibroblasts, all of which are found in the rheumatoid joint, produce TNF. In addition, TNF can be detected both in synovial fluid and synovial tissue from patients [54] and in cells within the synovial lining and at the cartilage pannus junction, especially around the area of cartilage erosion [55]. Apart from its effects on the inflammatory process, TNF has been shown to have many other effects that may affect the events occurring within the joint. In particular, it has been shown to increase synoviocyte proliferation, which is prevented by anti-TNF treatment [56], to stimulate chondrocytes and osteoclasts to break down cartilage and bone [57] and *in vitro* to enhance matrix breakdown and inhibit proteoglycan synthesis in cultured cartilage [58]. However, TNF does play an important role in the normal autocrine control of cartilage homeostasis [55, 57] leading to the suggestion that the balance of cytokine regulation of cartilage and bone turnover may be disturbed in RA [59].

A pivotal role for TNF in RA has been extensively supported by studies in animals. An infusion of TNF into rats with collagen II-induced polyarthritis results in the development of a more severe disease than in noninfused arthritic rats [60]. Furthermore, a transgenic mouse, which expresses a 3'-modified human TNF transgene, develops chronic inflammatory polyarthritis associated with hyperplasia of the synovial membrane, polymorphonuclear and lymphocytic infiltration into the synovium, pannus formation, articular cartilage destruction and fibrous tissue deposition [61]. Further evidence for the role of TNF in arthritis has been demonstrated with the use of antagonists of TNF. Treatment of mice with either anti-TNF Mabs or a recombinant soluble TNF receptor prevents the development of polyarthritis in a model of collagen II arthritis in mice [62], seen as both a reduction in footpad swelling and in synovial inflammation. Since no effect of anti-TNF Mab treatment was seen when given at 2 months after disease onset, the authors concluded that TNF was important in the early inflammatory phase of the disease but not during the erosion phase. Similarly, a short course of anti-TNF Mab treatment has been reported to reduce the extent of proteoglycan loss from articular cartilage by about 50% in a model of antigen-induced arthritis in rabbits and to delay onset of disease and inflammatory damage in rat collagen II arthritis [63]. Further evidence for a beneficial effect of neutralizing TNF has been shown when treatment is delayed until disease onset in mice treated with collagen II [64], where paw swelling and number and severity of joints involved was arrested. Furthermore, if the human

TNF transgenic mouse which develops polyarthritis is treated with antihuman TNF Mab, then all symptoms are ameliorated [61, 65].

These animal studies have supported the development of therapeutic agents that neutralise TNF as an approach to reduce the disease process in RA. Several of these products are now in clinical development and include antibodies against TNF and soluble TNFR fusion proteins [66–69]. Three antibodies against TNF are currently in clinical trials for use in RA patients: a mouse/human chimeric Mab (cA$_2$/inflix-imab/Remicade), an engineered human Mab (CDP571) and a human recombinant antibody (D2E7) being developed by Centocor, Celltech plc and BASF Bioresearch/Cambridge Antibody Technology Group, respectively. In 20 patients given 20 mg/kg infliximab/cA$_2$ in an open phase study, significant improvements were seen in the Ritchie score, the swollen joint count, and in serum levels of C-reactive protein [66]. In a further double-blind study improvements in Paulus index were noted using this antibody at 1 and 10 mg/kg [70]. Similar results have been observed using the engineered antibody, CDP571, in a double-blind study using doses of 0.1, 1 and 10 mg/kg. Dose-related improvements in pain, numbers of swollen joints, disease activity scores and serum levels of C-reactive protein were observed [71, 67]. D2E7 has completed formal preclinical investigation and is currently in clinical development for treatment of RA. Further strategies for neutralising TNF make use of receptor fusion proteins of which the most advanced is Enbrel (Etanercept), a recombinant human TNFR:Fc consisting of soluble p80 TNF receptor dimer fused to Fc portion of IgG1 being developed by Immunex. Two clinical studies showed that 59% of patients treated with Enbrel experienced a significant reduction in symptoms after 6 months of treatment, with a clinical response observed as early as 1 to 2 weeks from start of therapy [72, 69]. Another receptor fusion protein being developed by Amgen Inc. is a high-affinity soluble TNF receptor Type 1 (sTNFR1). sTNFR1 is PEGylated and may be designed for less frequent administration with a subcutaneous route for chronic inflammatory diseases. It is currently being tested in a multicenter, double-blind, placebo-controlled clinical trial in RA.

TNF appears to play a key role in the inflammatory disease process associated with RA and other diseases such as Crohn's disease and inflammatory bowel disease. The early clinical trials with anti-TNF therapies have shown a rapid, effective and prolonged effect and suggest that drugs capable of blocking this cytokine may have an important impact on the progression of the disease.

Transplantation

Many of the antibody therapies used in solid organ transplantation and bone marrow transplantation have concentrated on preventing rejection or graft-versus-host disease (GvHD) by targeting the cells involved in the immune process, such as the Mab OKT3. These strategies have utilised both polyclonal and monoclonal anti-lymphocyte antibodies; however, their use has been limited due to the lack of specificity, in the case of OKT3 Mab targeting all T cells, and the antibodies' immunogenicity. The development of specific Mabs with reduced foreign epitopes through chimerisation or humanisation offers further approaches for the prevention of rejection or GvHD and a reduction in side-effects. However it remains a goal to target the cells, particularly T cells, which are directly activated during the immune response. Recently two Mabs have been developed which target the alpha chain of the human interleukin-2 receptor (IL-2R or Tac) [73, 74]. This receptor is upregulated on activated T cells, and expressed on monocytes and natural killer during an ongoing immune response and, along with its ligand, IL-2, plays a pivotal role in the activation of the immune system [75]. In addition, Mabs that block the high affinity IL-2R expressed on alloantigen-reactive T lymphocytes may cause selective immunosuppression by blocking the proliferation/activation of T cells reactive to donor antigens. The first of the two anti-IL-2R Mabs, daclizumab (Zenapax), is being developed by PDL and Hoffman-La Roche. This is a genetically engineered humanised antibody and has been shown, when added with other rejection drugs, to reduce the incidence of kidney rejection episodes without increasing serious side-effects [73]. Daclizumab, when given at 1 mg/kg intravenously once a week for 5 weeks in renal transplant patients, reduced the incidence of rejection episodes from 35% to 22% [76]. Patients receiving daclizumab were more likely to have a functioning kidney six months after surgery than patients on standard therapy alone. In a further study, kidney rejection episodes were reduced from 49% to 28%, with a significant improvement in patient survival in antibody-treated patients compared to standard therapy alone [77]. This antibody is currently being evaluated for other indications such as kidney-pancreas, liver, lung, heart, paediatric kidney and intestinal-liver transplants, uveitis, leukaemia and GvHD. The second anti-IL-2R Mab, basiliximab (Simulect), is a selective high-affinity chimeric antibody being developed by Novartis AG. Novartis claims that basiliximab has a 10-fold greater activity than daclizumab and a simpler and more convenient dosing regimen. Results from clinical trials have shown that a two-dose regimen of basiliximab (20 mg per dose on days 0 and 4) reduced the incidence of biopsy-proven acute rejection to 31%, compared to placebo 45%, during the first six months post-transplant in renal transplant patients on standard immunosuppressive regimens [74, 78–80]. A follow-up of these patients showed that this effect was maintained for at least twelve months. Both of the anti-IL-2R antibodies are effective at preventing rejection of renal allografts, neither causes first-dose reactions and neither are associated with increased incidence of infectious complications, including cytomegalovirus. There is a low incidence of anti-antibody responses that appears not to alter efficacy or serum levels. The development of these antibodies has allowed the achievement of specific, safe, effective and increased efficacy in renal transplants with minimal toxicity and will undoubtedly play an important role in the clinical management of rejection, both in solid organ and bone marrow transplantation.

Elevated serum or plasma levels of TNF have been shown to be associated with allografts of kidney [81], liver [82], heart [83], pancreas [84], and bone marrow [85]. In each case, the appearance of TNF correlated with the onset of rejection or, in the case of bone marrow transplantation, GvHD. Certainly, in mouse models of GvHD, anti-TNF Mab has been shown to reduce both mortality and tissue damage caused by the reaction [86, 87]. However, a small clinical study using a murine Mab against TNF in 19 patients with severe acute GvHD refractory to conventional (steroid) therapy, demonstrated only partial responses in 14 patients during the period of treatment [88]. In all cases GvHD recurred after discontinuation, suggesting that long-term efficacy cannot be achieved, at least in this particular subset of patients. In the field of solid organ transplantation, some prolongation of graft survival has been seen in a rat cardiac transplant model using anti-TNF antibodies [89]. In addition, there are at least two reports of positive interactions between anti-TNF Mab treatment and cyclosporine [90, 91]. Stevens et al. [92] observed a modest prolongation of graft survival when an anti-TNF antibody was combined with an antibody to interferon-γ (IFNγ) in a skin graft model in rhesus monkey. In summary, it is not clear to what extent TNF blockade will be of use in transplantation. It appears to be important in the effector arm of the rejection episode; however, it does not prevent the immunological process occurring and therefore would appear not to be a pivotal cytokine in transplantation.

Inflammatory bowel disease

The inflammatory bowel diseases (IBD), which include Crohn's disease (CD) and ulcerative colitis (UC), are debilitating diseases characterized by chronic inflammation of intestinal mucosa and *lamina propria*. Although the events that trigger their onset (such as dietary, bacterial, viral, or environmental) remain unknown, they are associated with a marked leukocyte infiltrate and local production of soluble mediators. CD is a condition of chronic cell-mediated immunopathology with an infiltration of activated (CD25+) CD4+ T cells into the intestinal *lamina propria* [93]. The antigen against which these T cells are directed is obscure but is likely to be the intestinal epithelial cell [94] which stains strongly positive for HLA-DR in patients with active but not passive disease. However, although lymphocyte-mediated cytotoxicity against epithelial cells has been demonstrated *in vitro* [95], it remains unclear whether such a mechanism operates *in vivo* [94]. Like CD, there is also an influx of immune cells into the submucosa of UC patients, although there is a lower T cell (CD25+, CD3+, CD4+) and higher macrophage (CD25+, CD3−, CD4+, HLA-DR+) presence in the latter [93]. Indeed, the pathology of UC may also be autoimmune in origin, as it is often associated with the production of anti-colon antibodies by cells within the colonic mucosa [96]. Despite these differences in T cell infiltrate and in the distribution and localization of the inflammation, both are characterized by the presence of inflammatory cells such as neutrophils and macrophages [97]. As such, the potential array of soluble mediators produced by immune and inflammatory cells is large and may vary depending on the type and severity of the disease [98]. It is not surprising, therefore, that increased TNF immunoreactivity has been demonstrated in the *lamina propria* in both CD and UC patients [99]. The cells are most likely of macrophage origin and specific TNF immunoreactivity, mRNA levels (assessed by PCR [98]) or *in situ* hybridization [100] and TNF as well as IFNγ, IL-13 and IL-6 secretion by cultured *lamina propria* MNC [101] are all increased in active disease compared to normal levels. Hence, the effects of such high local concentrations of TNF are likely to be adverse.

Further indirect evidence for the involvement of TNF in the pathology of IBD has been obtained by measuring TNF activity in the stools from diseased individuals. In the stools of children with active CD or UC, TNF concentrations were 994 pg/g (CD) or 276-5982 pg/g (UC) compared to around 50 pg/g in normal children, those with nonspecific diarrhoea, or those with inactive disease [102]. Indeed, it has been suggested that routine assessment of stool TNF levels may represent a useful marker of disease activity in these patients.

TNF therefore may be a key mediator in the pathogenesis of IBD, either by a direct cytotoxic action or as an orchestrator of the inflammatory cascade. Blood-borne TNF itself can reduce intestinal blood flow and limit fuel availability to the gut mucosa [103], but it is more likely that its effect due to local production is the most important. The release of other inflammatory cytokines thought to be important in IBD, such as IL-1β [100, 104, 105], IL-6 [102, 106] and IL-8 [107] can be stimulated by TNF. The recruitment of inflammatory cells by these soluble mediators and the expression of the appropriate adhesion molecules that have been reported to be associated with IBD, such as E-selectin, VCAM-l [108], and ICAM-1 [109] can all be influenced by local TNF production. TNF- mediated increases in vascular thrombosis and inducible NOS expression [110] in IBD may represent other effector mechanisms of this cytokine.

Hence, TNF may represent a credible therapeutic target for intervention for the treatment of CD and UC. There are a number of animal models in which the hypothesis has been tested. In an acetic acid/immune complex-mediated model of colonic inflammation in rats, the maximum increase in faecal water content (a measure of disease severity) coincided with an influx of inflammatory cells into the colonic mucosa [111]. Continuous treatment with a specific anti-rodent TNF Mab (chimeric mouse/hamster TN3 19.12) prevented the diarrhoeic response and abrogated the inflammatory cell infiltrate.

The cotton-top tamarin (CTT) provides a very good model of human UC. The disease occurs spontaneously in a proportion of these animals and is characterized by an acute colonic inflammation with infiltrating neutrophils seen in the *lamina propria* and the development of eccentric crypt abscesses. The animals develop diarrhoea and an acute wasting disease, resulting in the animals' demise if not treated.

Treatment of colitic animals with a short course of humanized anti-human TNF Mab (CDP57l) [112] reverses the weight loss and diarrhoea with a rapid (< 6 days) onset of action. Moreover, there is a marked reduction in the disease severity as assessed histologically on colonic biopsies, with improvements persisting in some animals 12 months after the last treatment. In addition, some CTT become resistant to treatment with both olsalazine and prednisone. Treatment of these animals with CDP571 also has been shown to induce remission in some animals. Hence, data from animal models would seem to support the rationale for a therapeutic study of human IBD aimed at reducing the effect of TNF.

The clinical use of anti-TNF Mabs has been studied both in Europe and the USA. Two anti-TNF antibodies are currently under evaluation for CD: Infliximab (formerly called cA2, Centocor, Malvern, PA, USA), and CDP571 (Celltech, Slough, UK). Both of these Mabs strongly bind free and membrane-bound TNF. Infliximab is a mouse/human chimeric Mab with a human $\gamma 1$ heavy chain. This Mab is efficient at killing TNF-bearing cells through complement activation or antibody-dependent cytotoxicity. In contrast, CDP571 is a CDR-grafted Mab of the $\gamma 4$ isotype and can bind TNF (either soluble or cell surface) without invoking target cell lysis.

To date four published randomized, controlled trials of anti-TNF antibody in adult CD have been performed, one with CDP571, and three using infliximab. The first published trial with CDP571 studied the effect of a single 5 mg/kg i.v. dose [113]. Patients were randomised to CDP571 (n = 21) or placebo (n = 10). The median CDAI fell significantly in the treated group from 263 to 167 at 2 weeks (primary endpoint). Six of the 21 treated patients achieved remission (CDAI of less than 150), and 3 others had CDAI counts of 156 or lower. However, effects were short-lived with no significant differences between the groups seen at 8 weeks. Two larger trials of this drug have recently been completed and are due to report soon.

The earliest study with infliximab in CD was a small open-label study but the data indicated a marked therapeutic potential [114]. There followed a double-blind, placebo-controlled, dose-ranging study in patients with moderately severe CD who were randomised to receive placebo, 5 mg/kg, 10 mg/kg or 20 mg/kg as a single infusion [115]. In the infliximab-treated patients, 65% had a clinical response (reduction in CDAI of > 70 points) compared to 17% in the placebo group (p < 0.001) at 4 weeks. A third of the infliximab-treated patients were in remission at week 4, significantly more than the 4% on placebo (p = 0.005). The reduction in clinical response persisted to week 12 (41% versus 12% placebo, p = 0.008), although there was no significant difference in the percentage of patients in remission (24% infliximab, 8% placebo, p = 0.31). Patients who did not respond to the initial infusion were given a second infusion of 10 mg/kg. Four weeks later, 34% had responded and 17 % were in remission (p = 0.14 and 0.05 compared with initial placebo response), confirming a subgroup of relatively resistant patients.

This reagent is the first in its class to receive a license for clinical use in chronic indications such as CD.

Summary

Recent years have seen an upsurge in the use of Mabs in the validation of potential therapeutic targets. The approach to their use as therapeutic agents has progressed from administration of murine antibodies as single-dose therapies, such as OKT3 in transplantation, to their current use as repeat-dose therapies in chronic indications like CD or rheumatoid arthritis. This has gone in tandem with the advances in technology allowing the immunogenic potential of these reagents to be reduced through "humanisation" procedures like CDR grafting. The next years should see the development and marketing of more Mabs for use in acute and chronic disease.

References

1   Waage A, Halstensen A, Espevik T (1987) Association between tumor necrosis factor in serum and fatal outcome in patients with meningococcal disease. *Lancet* 1: 355–357

2   Waage A, Brandtzaeg P, Halstensen A, Kierulf P, Espevik T (1989) The complex pattern of cytokines in serum from patients with meningococcal septic shock. *J Exp Med* 169: 333–338

3   Girardin E, Grau GE, Dayer JM, Roux-Lombard P, The J5 Study Group, Lambert P-H (1988) TNF and IL-1 in the serum of children with severe infectious purpura. *N Engl J Med* 319: 397–400

4   Debets JMH, Kampmeijer R, van der Linden M P, Buurman WA, Van der Linden CJ (1989) Plasma TNF and mortality in critically ill sepsis patients. *Crit Care Med* 17: 489–494

5   Girardin EP, Berner ME, Grau GE, Suter S, Lacourt G, Pauvier L (1990) Serum TNF in newborns at risk for infections. *Eur J Pediatr* 149: 645–647

6   Cannon JG, Tompkins RG, Gelfand JA, Michie HR, Stanford GG, van der Meer JW, Endres S, Lonnemann G, Corsetti J, Chernow B et al (1990) Circulating IL-1 and TNF in septic shock and experimental endotoxin fever. *J Infect Dis* 161: 79–84

7   Calandra T, Baumgartner J-D, Grau GE, Wu MM, Lambert PH, Schellekens J, Verhoef J, Glauser MP (1990) Prognostic values of TNF/cachectin, IL-1, interferon-alpha and Interferon-γ in the serum of patients with septic shock. Swiss-Dutch J5 Immunoglobulin Study Group. *J Infect Dis* 161: 982–987

8   Michie HR, Manogue KR, Spriggs DR, Revhaug A, O'Dwyer S, Dinarello CA, Cerami A, Wolff SM, Wilmore DW (1988) Detection of circulating TNF after endotoxin administration. *N Engl J Med* 318: 1481–1486

9   Tracey KJ, Beutler B, Lowry SF, Merryweather J, Wolpe S, Milsark IW, Hariri RJ, Fahey TJ, Zentella A, Albert JD et al (1986) Shock and tissue injury induced by recombinant human cachectin. *Science* 234: 470–474

10 Beutler B, Milsark IW, Cerami AC (1985) Passive immunisation against cachectin/tumor necrosis factor protects mice from lethal effect of endotoxin. *Science* 229: 869–871

11   Ashkenazi A, Marsters SA, Capon DJ, Chamow SM, Figari IS, Pennica D, Goeddel DV, Palladino MA, Smith DH (1991) Protection against endotoxic shock by a tumor necrosis factor receptor immunoadhesin. *Proc Natl Acad Sci USA* 88: 10535–10539

12   Silva AT, Bayston KF, Cohen J (1990) Prophylactic and therapeutic effects of a monoclonal antibody to TNF in experimental gram-negative shock. *J Infect Dis* 162: 421–427

13   Opal SM, Cross AS, Kelly NM, Sadoff JC, Bodmer MW, Palardy JE, Victor GH (1990) Efficacy of a monoclonal antibody directed against tumor necrosis factor in protecting neutropenic rats from lethal infection with Pseudomonas aeruginosa. *J Infect Dis* 161: 1148–1152

14   Walsh CJ, Sugerman HJ, Mullen PG, Carey PD, Leeper-Woodford SK, Jesmok GJ, Ellis EF, Fowler AA (1992) Monoclonal antibody to tumor necrosis factor alpha attenuates cardiopulmonary dysfunction in porcine gram-negative sepsis. *Arch Surg* 127: 138–444; discussion 144–145

15   Windsor AC, Mullen PG, Walsh CJ, Fisher BJ, Blocher CR, Jesmok G, Fowler AA, Sugarman HJ (1994) Delayed tumor necrosis factor alpha blockade attenuates pulmonary dysfunction and metabolic acidosis associated with experimental gram-negative sepsis. *Arch Surg* 129: 80–89

16   Tracey KJ, Fong Y, Hesse DG, Manogue KR, Lee AT, Kuo GC, Lowry SF, Cerami A (1987) Anti-cachectin/TNF monoclonal antibodies prevent septic shock during lethal bacteraemia. *Nature* 330: 662–664

17   Hinshaw LB, Tekamp-Olson P, Chang ACK, Lee PA, Taylor FB, Murray CK, Peer GT, Emerson TE, Passey RB, Kuo GC (1990) Survival of primates in LD_{100} septic shock following therapy with antibody to tumor necrosis factor (TNFα). *Circ Shock* 30: 279–292

18   Hinshaw LB, Emerson TE, Taylor FB, Chang ACK, Duerr M, Peer GT, Flournoy DJ, White GL, Kosanke SD, Murray CK et al (1992) Lethal *Staphylococcus aureus*-induced shock in primates: prevention of death with anti-TNF antibody. *J Trauma-Injury Infect Crit Care* 33: 568–573

19   Fiedler VB, Loot I, Sander E, Voehringer V, Galanos C, Fournel MA (1992) Monoclonal antibody to tumor necrosis factor-alpha prevents lethal endotoxin sepsis in adult rhesus monkeys. *J Lab Clin Med* 120: 574–88

20   Cross AS, Opal SM, Palardy JE, Bodmer MW, Sadoff JC (1993) The efficacy of combination immunotherapy in experimental *Pseudomonas sepsis*. *J Infect Dis* 167: 112–118

21   Wayte J, Silva AT, Krausz T, Cohen J (1993) Observations on the role of tumour necrosis factor-alpha in a murine model of shock due to *Streptococcus* pyogenes. *Crit Care Med* 21: 1207–1212

22   Zanetti G, Heumann D, Gerain J, Kohler J, Abbet P, Barras C, Lucas R, Glauser MP, Baumgartner JD (1992) Cytokine production after intravenous or peritoneal gram-negative bacterial challenge in mice. *J Immunol* 148: 1890–1897

23   Martin RA, Silva AT, Cohen J (1993) Effect of anti-TNF treatment in an antibiotic treated murine model of shock due to *Streptococcus* pyogenes. *FEMS Microbial Lett* 110: 175–178

24   Bagby GJ, Plessala KJ, Wilson LA, Thompson JJ, Nelson S (1991) Divergent efficacy of

antibody to tumor necrosis factor-α in intravascular and peritonitis models of sepsis. *J Infect Dis* 163: 83–88

25   Echtenacher B, Falk W, Mannel DN, Krammer PH (1990) Requirement of endogenous tumor necrosis factor/cachectin for recovery from experimental peritonitis. *J Immunol* 145: 3762–3766

26   Alexander HR, Sheppard BC, Jensen JC, Langstein HN, Buresh CM, Venzon D, Walker EC, Fraker DL, Stovroff MG, Norton JA (1991) Treatment with recombinant human tumor necrosis factor-alpha protects rats against the lethality, hypotension and hypothermia of gram-negative sepsis. *J Clin Invest* 88: 34–39

27   Glauser MP, Zanetti G, Baumgartner JD, Cohen J (1991) Septic shock: pathogenesis. *Lancet* 338: 732–736

28   Exley AR, Cohen J, Buurman W, Owen R, Hanson G, Lumley J, Aulakh JM, Bodmer M, Riddell A, Stephens S et al (1990) Monoclonal antibody to TNF in severe septic shock. *Lancet* 335: 1275–1277

29   Vincent JL, Bakker J, Marecaux G, Schandene L, Kahn RJ, Dupont E (1992) Administration of anti-TNF antibody improves left ventricular function in septic shock patients. Results of a pilot study. *Chest* 101: 810–815

30   Fisher CJ, Opal SM, Dhainaut JF, Stephens S, Zimmerman JL, Nightingale P, Harris SJ, Schein RM, Panacek EA, Vincent JL et al (1993) Influence of anti-TNF monoclonal antibody on cytokine levels in patients with sepsis. *Crit Care Med* 21: 318–327

31   Abraham E, Wunderink, R, Silverman H, Perl TM, Nasraway S, Levy H, Bone R, Wenzel RP, Balk R, Alfred R et al (1995) Efficacy and safety of monoclonal antibody to human tumor necrosis factor alpha in patients with sepsis syndrome. A randomized, controlled, double-blind, multicenter clinical trial. TNF-alpha Mab Sepsis Study Group. *JAMA* 273: 934–941

32   Cohen J, Carlet T (1996) INTERSEPT: an international, multicenter placebo-controlled trial of monoclonal antibody to human tumor necrosis factor-α in patients with sepsis. Crit Care Med 24: 1431–1440

33   Abraham E, Anzueto A, Gutierrez G, Tessler S, San Pedro G, Wunderink R, Dal Nogare A, Nasraway S, Berman S, Cooney R et al (1998) Double blind randomised controlled trial of monoclonal antibody to human tumour necrosis factor in treatment of septic shock. NORASEPT II Study Group. *Lancet* 351: 929–933

34   Robinson DS, Hamid Q, Ting S, Tsicopoulos A, Barkans J, Bentley AM, Corrigan C, Durham SR, Kay AB (1992) Predominate TH2-like bronchoalveolar T-lymphocyte population in atopic asthma. *N Engl J Med* 326: 298–304

35   Kay AB Asthma and Inflammation (1991) *J Allergy Clin Immunol* 87: 893–910

36   Abbas AK, Murphy KM, Sher A (1996) Functional diversity of helper T lymphocytes. *Nature* 383: 787–793

37   Gleich GJ, Kita H, Adolphson CR (1995) Eosinophils. In: MN Frank, KF Austen, HN Cloman, ER Unanue (eds): *Samters immunologic diseases*. 5th ed. Little Brown Co, Boston, 205–245

38   Finkelman FD, Katona IM, Urban JF, Snapper CM, Ohara J, Paul WE (1986) Suppres-

sion of *in vivo* polyclonal IgE responses by monoclonal antibody to the lymphokine B-cell stimulatory factor 1. *Proc Nat Acad Sci USA* 83: 9675–9678

39   Finkelman FD, Katona IM, Urban JF, Holmes J, Ohara J, Tung AS, Sample JG, Paul WE (1988) IL-4 is required to generate and sustain *in vivo* IgE responses. *J Immunol* 141: 2335–2341

40   Fanslow WC, Clifford KN, Park LS, Rubin AS, Voice RF, Beckmann MP, Widmer MB (1991) Regulation of alloreactivity *in vivo* by IL-4 and the soluble IL-4 receptor. *J Immunol* 147: 535–540

41   Sato TA, Widmer MB, Finkelman FD, Madani H, Jacobs CA, Grabstein KH, Maliszewski CR (1993) Recombinant soluble murine IL-4 receptor can inhibit or enhance IgE responses *in vivo. J Immunol* 150: 2717–2723

42   Maliszewski CR, Sato TA, Davidson B, Jacobs CA, Finkelman FD, Fanslow WC (1994) *In vivo* biological effects of recombinant soluble IL-4 receptor. *Proc Soc Exp Biol Med* 206: 233–237

43   Renz H, Enssle K, Lauffer L, Kurrle R, Gelfand EW (1995) Inhibition of allergen-induced IgE and IgG1 production by soluble IL-4 receptor. *Int Arch Allergy Immunol* 106: 46–54

44   Konig B, Fischer A, Konig W (1995) Modulation of cell-bound and soluble CD23, spontaneous and ongoing IgE synthesis of human peripheral blood mononuclear cells by soluble IL-4 receptor and the partial antagonistic IL-4 mutant protein IL-4 (Y124D). *Immunology* 85: 604–610

45   Egan RW, Umland SP, Cuss FM, Chapman RW (1996) Biology of IL-5 and its relevance to allergic disease. *Allergy* 51: 71–81

46   Egan RW, Cuss FM, Umland SP, Chapman RW (1996) Anti-interleukin-5 antibodies as therapeutic agents in asthma and other eosinophilic diseases. In: B Henderson, M Bodmer (eds): *Therapeutic modulation of cytokines*. CRC Press, Boca Raton, Florida. 237–264

47   Egan RW, Athwal D, Bodmer MW, Carter JM, Chapman RW, Chou C-C, Cox MA, Emtage JS, Fernandez X, Foran S et al (2000) Sch 55700, a humanised monoclonal antibody to human interleukin-5, blocks eosinophilic responses and hyperreactivity. *J Exp Med; in press*

48   Gulbenkian AR, Egan RW, Fernandez X, Jones H, Kreutner W, Kung TT, Payvandi F, Sullivan L, Zurcher JA, Watnick AS (1992) Interleukin-5 modulates eosinophil accumulation in allergic guinea pig lung. *Am Rev Respir Dis* 146: 262–265

49   Kung TT, Stelts D, Zurcher JA, Watnick AS, Jones H, Mauser PJ, Fernandez X, Umland S, Kreutner W, Chapman RW, Egan RW (1994) Mechanism of allergic pulmonary eosinophilia in the mouse. *J Allergy Clin Immunol* 94: 1217–1224

50   Kung TT, Stelts DM, Zurcher JA, Adams III GK, Egan RW, Kreutner W, Watnick AS, Jones H, Chapman RW (1995) Involvement of IL-5 in a murine model of allergic pulmonary inflammation: prophylactic and therapeutic effect of anti-IL-5 antibody. *Am J Respir Cell Mol Biol* 13: 360–365

51   Mauser PJ, Pitman A, Witt A, Fernandez X, Zurcher J, Kung T, Jones H, Watnick AS,

Egan RW, Kreutner W, Adams III GK (1993) Inhibitory effect of the TRFK-5 anti-IL5 antibody in a guinea pig model of asthma. *Am Rev Respir Dis* 148: 1623–1627

52   Van Oosterout AJM, Ladenious ARC, Savelkoul HFJ, Van Ark I, Delsman KC, Nijkamp FP (1993) Effect of anti-IL-5 and IL-5 on airway hyperreactivity and eosinophils in guinea pigs. *Am Rev Respir Dis* 147: 548–552

53   Mauser PJ, Pitman A, Fernandez X, Foran SK, Adams III GK, Egan RW, Chapman RW (1995) Effect of an antibody to IL-5 in a monkey model of asthma. *Am J Respir Crit Care Med* 152: 467–472

54   Di Giovine FS, Nuki G, Duff G (1988) Tumour necrosis factor in synovial exudates. *Ann Rheum Dis* 47: 768–772

55   Chu CQ, Field M, Allard S, Ahney E, Feldmann M, Maini RN (1992) Detection of cytokines at the cartilage/pannus junction in patients with rheumatoid arthritis: implications for the role of cytokines in cartilage destruction and repair. *Br J Rheumatol* 31: 653–661

56   Alvaro-Gracia JM, Zvaifler NJ and Firestein G (1990) Cytokines in chronic inflammatory arthritis. V. Mutual antagonism between interferon-gamma and tumour necrosis factor-alpha on HLA-DR expression, proliferation, collagenase production and granulocyte macrophage colony-stimulating factor production by rheumatoid arthritis synoviocytes. *J Clin Invest* 86: 1790–1798

57   MacDonald BR, Gowen M (1992) Cytokines and bone. *Br J Rheumatol* 31: 149–155

58   Saklatvala J (1986) Tumour necrosis factor α stimulates resorption and inhibits synthesis of proteoglycan in cartilage. *Nature* 322: 547–549

59   Brennan FM, Maini RN and Feldmann M (1992) TNFα – A pivotal role in rheumatoid arthritis? *Br J Rheumatol* 31: 293–298

60   Brahn E, Peacock DJ, Banquerigo MN and Lui DY (1992) Effects of TNFα on collagen arthritis. *Lymphokine Cytokine Res* 11: 253–256

61   Keffer J, Probert L, Cazlaris H, Georgopoulos S, Kaslaris E, Kioussis D, Kollias G (1991) Transgenic mice expressing human tumour necrosis factor: a predictive genetic model of arthritis. *EMBO J* 10: 4025–4031

62   Piguet PF, Grau GE, Vesin C, Loetscher H, Gentz R, Lesslauer W (1992) Evolution of collagen arthritis in mice is arrested by treatment with anti-TNF antibody or a recombinant soluble TNF receptor. *Immunology* 77: 510–514

63   Henderson B, Blake S, Lewthwaite J, Staines N, Harper N, Andrew D, Bodmer M, Higgs G (1991) The effects of monoclonal antibodies to tumour necrosis factor in animal models of chronic arthritis. *Br J Pharmacol* 104: 442P

64   Williams RO, Feldmann M, Maini RN (1992) Anti-tumor necrosis factor ameliorates joint disease in murine collagen-induced arthritis. *Proc Natl Acad Sci USA* 89: 9784–9788

65   Kollias G, Douni E, Kassiotis G, Kontoyiannis D (1999) On the role of tumor necrosis factor and receptors in models of multiorgan failure, rheumatoid arthritis, multiple sclerosis and inflammatory bowel disease. *Immunol Rev* 169: 175–194

66   Elliott MJ, Maini RN, Feldmann M, Long-Fox A, Charles P (1993) Treatment of

rheumatoid arthritis with chimeric monoclonal antibodies to tumor necrosis factor α. *Arthritis Rheum* 36 (12): 1681–1690

67 Heath PK (1997) Use of anti-TNFα therapies as potential treatments for rheumatoid arthritis. *Rheum Arthritis* 1: 235–240

68 O'Dell JR (1999). Anticytokine therapy – a new era in the treatment of rheumatoid arthritis? *N Engl J Med* 340: 310–312

69 Weinblatt ME, Kremer JM, Bankhurst AD, Bulpitt KJ, Fleischmann RM, Fox RI, Jackson CG, Lange M, Burge DJ (1999) A trial of etanercept, a recombinant tumor necrosis factor receptor:Fc fusion protein, in patients with rheumatoid arthritis receiving methotrexate. *N Engl J Med* 340: 253–259

70 Elliott MJ, Maini RN, Feldmann M, Kalden JR, Antoni C, Smolen JS, Leeb B, Breedveld FC, Macfarlane JD, Bijl H et al (1994) Randomised double-blind comparison of chimeric monoclonal antibody to tumour necrosis factor α (cA2) versus placebo in rheumatoid arthritis. *Lancet* 344: 1105–10

71 Rankin EC, Choy EH, Kassimos D, Kingsley GH, Sopwith AM, Isenberg DA, Panayi GS (1995) The therapeutic effects of an engineered human anti-tumour necrosis factor alpha antibody (CDP571) in rheumatoid arthritis. *Br J Rheumatol* 34: 334–342

72 Moreland LW, Baumgartner SW, Schiff MH, Tindall EA, Fleischmann RM, Weaver AL, Ettlinger RE, Cohen S, Koopman WJ, Mohler K et al (1997) Treatment of rheumatoid arthritis with a recombinant human tumor necrosis factor receptor (p75)-Fc fusion protein. *N Engl J Med* 337: 141–147

73 Queen C, Schneider WP, Selick HE, Payne PW, Landolfi NF, Duncan JF, Avdalovic AN, Levitt M, Junghans RP, Waldmann TA (1989) A humanised antibody that binds to the interleukin 2 receptor. *Proc Natl Acad Sci USA* 86: 10029–10033

74 Amlot PL, Rawlings E, Fernando ON, Griffin PJ, Heinrich G, Schreier MH, Castaigne JP, Moore R and Sweny P (1995) Prolonged action of a chimeric interleukin-2 receptor (CD25) monoclonal antibody used in cadaveric renal transplantation. *Transplantation* 60: 748–756

75 Theze J, Alzari PM, Bertoglio J (1996) Interleukin 2 and its receptors: recent advances and new immunological functions. *Immunology Today* 17: 481–486

76 Vincenti F, Kirkman R, Light S, Bumgardner G, Pescovitz M, Halloran P, Neylan J, Wilkinson A, Ekberg H, Gaston R et al (1998) Interleukin-2-receptor blockade with daclizumab to prevent acute rejection in renal transplantation. Daclizumab Triple Therapy Study Group. *N Engl J Med* 338: 161–165

77 Nashan B, Light S, Harde IR, Lin A, Johnson JR (1999) Reduction of acute renal allograft rejection by daclizumab. Daclizumab Double Therapy Study Group. *Transplantation* 67: 110–115

78 Kovarik JM, Rawlings E, Sweny P, Fernando O, Moore R, Griffin PJ, Fauchald P, Albrechtsen D, Sodal G, Nordal K et al (1996) Prolonged immunosuppressive effect and minimal immunogenicity from chimeric (CD25) monoclonal antibody SDZ CHI 621 in renal transplantation. *Transpl Proc* 28: 913–914

79 Kovarik J, Wolf P, Cisterne JM, Mourad G, Lebranchu Y, Lang P, Bourbigot B, Can-

tarovich D, Girault D, Gerbeau C et al (1997) Disposition of basiliximab, an inter-leukin-2 receptor monoclonal antibody, in recipients of mismatched cadaver renal allo-grafts. *Transplantation* 64: 1701–1705

80   Kahan BD, Rajagopalan PR, Hall M (1999) Reduction of the occurrence of acute cellu-lar rejection among renal allograft recipients treated with basiliximab, a chimeric anti-interleukin-2-receptor monoclonal antibody. United States Simulect Renal Study Group. *Transplantation* 67: 276–284

81   Maury CPJ, Teppo AM (1987) Raised serum levels of cachectin/tumor necrosis factor alpha in renal allograft rejection. *J Exp Med* 166: 1132–1137

82   Tilg H, Vogel W, Aulitzky WE, Herold M, Konigsrainer A, Margreiter R, Huber C (1990) Evaluation of cytokines and cytokine-induced secondary messages in sera of patients after liver transplantation. *Transplantation* 49: 1074–1080

83   Chollet-Martin S, Depoix JP, Uvass U, Pansard Y, Vizzuzaine C, Gougerot-Pocidalo MA (1990) Raised plasma levels of tumor necrosis factor in heart allograft rejection. *Trans-plant Proc* 22: 283–286

84   Grewal, HP, Kotb M, Salem A, el Din AB, Novak K, Martin J, Gaber LW, Gaber AO (1993) Elevated tumor necrosis factor levels are predictive for pancreas allograft trans-plant rejection. *Transplant Proc* 25: 132–135

85   Holler E, Kolb HJ, Moller A, Kempeni J, Liesenfeld S, Pechumer H, Lehmacher W, Ruckdeschel G, Gleixner B, Riedner C et al (1990) Increased serum levels of tumor necrosis factor α precede major complications of bone marrow transplantation. *Blood* 75: 1011–1016

86   Piguet PF, Grau GE, Allet B, Vassalli P (1987) Tumor necrosis factor/cachectin is an effector of skin and gut lesions of the acute phase of graft-vs.-host disease. *J Exp Med* 166: 1280– 1289

87   Shalaby MR, Fendly B, Sheehan KC, Schreiber RD, Ammann A J (1989) Prevention of the graft-versus-host reaction in newborn mice by antibodies to tumor necrosis factor-alpha. *Transplantation* 47: 1057–1061

88   Herve P, Flesch M, Tiberghien P, Wijdenes J, Racadot E, Bordigoni P, Plouvier E, Stephan JL, Bourdeau H, Holler E et al (1992) Phase I-II trial of a monoclonal anti-tumor necrosis factor α antibody for the treatment of refractory severe acute graft-ver-sus-host disease. *Blood* 79: 3362–3368

89   Imagawa DK, Millis JM, Seu P, Olthoff, KM, Hart J, Wasef E, Dempsey RA, Stephens S, Busuttil RW (1991) The role of tumor necrosis factor in allograft rejection. *Trans-plantation* 51: 57–62

90   Seu P, Imagawa DK, Wasef E, Olthoff KM, Hart J, Stephens S, Dempsey RA, Busuttil RW (1991) Monoclonal anti-tumor necrosis factor-α antibody treatment of rat cardiac allografts: synergism with low-dose cyclosporine and immunohistological studies. *J Surg Res* 50: 520–528

91   Bolling SF, Kunkel SL, Lin H (1992) Prolongation of cardiac allograft survival in rats by anti-TNF and cyclosporin combination therapy. *Transplantation* 53: 283–286

92   Stevens HPJD, Van der Kwast TH, Van der Meide PH, Vuzevski VD, Buurman WA,

Jonker M (1990) Synergistic immunosuppressive effects of monoclonal antibodies specific for interferon-gamma and tumor necrosis factor alpha. A skin transplantation study in the rhesus monkey. *Transplantation* 50: 856–861

93　Choy MY, Walker-Smith JA, Williams CB, MacDonald TT (1990) Differential expression of CD25 (interleukin-2 receptor) on *lamina propria* T cells and macrophages in the intestinal lesions in Crohn's disease and ulcerative colitis. *Gut* 31: 1365–1370

94　Targan SR, Deem RL, Shanahan F (1990) Immune-mediated cytotoxicity in inflammatory bowel disease. *Trends Inflam Bowel Dis Ther Falk Symp* 29–40

95　Shorter RG, Cardoza MR, ReMilne SG, Spencer RJ, Huizinga KA (1970) Modification of *in vitro* cytotoxicity of lymphocytes from patients with chronic ulcerative colitis or chronic granulomatous colitis for allogenic colon epithelial cells. *Gastroenterology* 58: 692–698

96　Hibi T, Ohara M, Toda K, Ogata H, Iwao Y, Watanabe N, Watanabe M, Hamada Y, Kobayashi K et al (1990) *In vitro* anticolon antibody production by mucosal or peripheral blood lymphocytes from patients with ulcerative colitis. *Gut* 31: 1371–1376

97　Glebmann CM, Barrett KE (1993) Role of inflammatory cell types. In: J Scholmerich, W Kruis, H Goebell, W Hohenberger, V Gross (eds): *Inflammatory bowel diseases. Pathophysiology as basis of treatment.* Kluwer Academic, Dordrecht, Netherlands, 62–79

98　Plevy SE, Targan SR, Andus T, Toyoda H (1994) Differentiation of Crohn's disease from ulcerative colitis mucosal inflammation by levels of TNF-α mRNA. *Gastroenterology* 104: part 2, A764

99　Murch SH, Braegger CP, Walker-Smith JA, MacDonald TT (1993) Location of tumour necrosis factor α by immunohistochemistry in chronic inflammatory bowel disease. *Gut* 34: 1705–1709

100　Cappello M, Keshav S, Prince C, Jewell DP, Gordon S (1992) Detection of mRNAs for macrophage products in inflammatory bowel disease by *in situ* hybridisation. *Gut* 33: 1214–1219

101　Steffen M, Reinecker HC, Witthoft T, Voss A, Schreiber S, Raedler A (1993) Enhanced secretion of TNF-α, IFN-γ, IL-1β and IL-6 by *lamina propria* mononuclear cells from patients with inflammatory bowel disease In: J Scholmerich, W Kruis, H Goebell, W Hohenberger, V Gross (eds): *Inflammatory bowel diseases. Pathophysiology as basis of treatment.* Kluwer Academic, Dordrecht, Netherlands, 490

102　Nicholls S, Stephens S, Braegger CP, Walker-Smith J-A, MacDonald TT (1993) Cytokines in stools of children with inflammatory bowel disease or infective diarrhoea. *J Clin Pathol* 46: 757–760

103　van Lanschot JJB, Mealy K, Wilmore DW (1990) The effects of tumour necrosis factor on intestinal structure and metabolism. *Ann Surg* 212: 663–670

104　Ligumsky M, Simon PL, Karmeli F, Rachmilewitz D (1990) Role of interleukin-1 in inflammatory bowel disease – enhanced production during active disease. *Gut* 31: 686–689

105 Pullman WE, Elsbury S, Kobayashi M, Hapel AJ, Doe WF (1992) Enhanced mucosal cytokine production in inflammatory bowel disease. *Gastroenterology* 102: 529–537

106 Stevens C, Walz G, Singaram C, Lipman, ML, Zanker B, Muggia A, Antonioli D, Peppercorn MA, Strom TB (1992) Tumor necrosis factor-alpha, interleukin-1 beta and interleukin-6 expression in inflammatory bowel disease. *Dig Dis Sci* 37: 818–826

107 Yahida YR, Ceska M, Effenberger F, Kurlak L, Lindley I, Hawkey CJ (1992) Enhanced synthesis of neutrophil-activating peptide-1/interleukin-8 in active ulcerative colitis. *Clin Sci* 82: 273–275

108 Koizumi M, King N, Lobb R, Benjamin C, Podolsky DK (1992) Expression of vascular adhesion molecules in inflammatory bowel disease. *Gastroenterology* 103: 840–847

109 Kvale D, Krajci P, Brandzaeg P (1992) Expression and regulation of adhesion molecules ICAM-1 (CD45) and LFA-3 (CD58) in human intestinal epithelial cell lines. *Scand J Immunol* 35: 669–676

110 Middleton SJ, Shorthouse M, Hunter J0 (1993) Increased nitric oxide synthesis in ulcerative colitis. *Lancet* 341: 465–466

111 Ward PS, Woodger SR, Bodmer M, Foulkes R (1993) Anti-tumour necrosis factor-α monoclonal antibodies (anti-TNF Mab) are therapeutically effective in a model of colonic inflammation. *Br J Pharmacol* 110: 77

112 Watkins PE, Warren BF, Stephens S, Ward P, Foulkes R (1997) Treatment of ulcerative colitis in the cottontop tamarin using antibody to tumour necrosis factor alpha. *Gut* 40: 628–633

113 Stack WA, Mann SD, Roy AJ, Heath P, Sopwith M, Freeman J, Holmes G, Long R, Forbes A, Kamm MA (1997) Randomised controlled trial of CDP571 antibody to tumour necrosis factor-alpha in Crohn's disease [see comments]. *Lancet* 349: 521–524

114 van Dullemen HM, van Deventer SJ, Hommes DW, Bijl HA, Jansen J, Tytgat GN, Woody J (1995) Treatment of Crohn's disease with anti-tumor necrosis factor chimeric monoclonal antibody (cA2). *Gastroenterology* 109: 129–135

115 Targan SR, Hanauer SB, van Deventer SJ, Mayer L, Present DH, Braakman T, DeWoody KL, Schaible TF, Rutgeerts PJ (1997) A short-term study of chimeric monoclonal antibody cA2 to tumor necrosis factor alpha for Crohn's disease. Crohn's Disease cA2 Study Group. *N Engl J Med* 337: 1029–1035

The debut of anti-TNF therapy of rheumatoid arthritis in the clinic

Ravinder N. Maini

The Kennedy Institute of Rheumatology, 1 Aspenlea Road, London W6 8LH, UK

Introduction: The concept and preclinical evaluation

Tumour necrosis factor α (TNFα) was cloned and expressed as a recombinant protein in the middle 1980s and shown to be identical to "cachectin" [1, 2]. Shortly afterwards nucleic acid and monoclonal antibody probes became available that permitted analysis of the expression of TNF and its receptors in human tissue from patients and controls. The demonstration in *in vitro* systems that both TNF and IL-1, individually and in combination, stimulated cartilage degradation [3], bone loss [4] and production of prostaglandin E2 and collagenase [5], promoted interest in the role that these cytokines might play in rheumatoid arthritis (RA). In 1988 and 1989 experiments by Buchan et al. [6] and Brennan et al. [7] demonstrated the expression of TNFα at messenger RNA and production of TNFα protein by disaggregated cells obtained from rheumatoid synovial membranes in tissue culture. Specific antibodies were then applied to microscopic sections of synovial membrane from rheumatoid joints and showed that TNFα was present in cells in the lining layer as well as in the deeper parts of the synovium, predominantly in CD68+ cells (macrophage phenotype) but also in a minority of CD3+ T cells in perivascular areas [8]. More interestingly, it was possible to demonstrate the presence of TNFα in cells at the cartilage/pannus junction, the site of the destructive and invasive tissue that leads to the loss of cartilage and bone in rheumatoid joints [8, 9].

Rheumatoid synovitis is also characterised by enhanced expression of TNF receptors [10] and monoclonal antibody reagents directed against the two TNF receptors when applied to rheumatoid tissue showed that these co-localised in the areas in which TNF was present [11]. TNF receptors were not only increased in synovial cells but could be detected in their soluble form in significant amounts in the synovial fluid, thus providing evidence of their dual function as both signalling molecules of TNFα when cell-associated, and as an inhibitor when shed in soluble form into the joint tissues [12]. Further investigations suggested that the concentration of shed soluble TNF receptors was insufficient to neutralise the biological activity of TNF and hence it seemed likely that pro-inflammatory activity predominated. Of immense importance to our understanding of the role of TNFα in RA, it was demonstrated in an *in vitro* culture system that the addition of anti-TNF antibodies abolished the production of a number of other key pro-inflammatory cytokines over-produced by inflamed synovium such as IL-1, GM-CSF and IL-8 [7, 13, 14]. The hypothesis that arose from this work was that TNF had the capacity to drive the production of other pro-inflammatory cytokines of pathophysiological importance, and that its neutralisation could have far-reaching effects in modulating inflammatory mediators in joints [15].

In experimental models which were investigated next, it became apparent that inhibitors of TNF – either specific monoclonal anti-TNFα antibodies or soluble TNF receptor fusion proteins – could ameliorate rheumatoid-like features of arthritis in collagen-induced disease in DBA1 mice [16]. The arthritrogenic potential of TNF was further shown in a transgenic model in which the development of arthritis could be suppressed by the administration of monoclonal antibodies from birth [17]. Thus by 1992 there was a considerable body of evidence for a critical role of TNF in rheumatoid disease.

Anti-TNF agents in clinical trials

A number of anti-TNF biological agents have been developed by the biotechnology and the pharmaceutical industry and many of these have been in clinical trials in the treatment of RA (see Tab. 1). The first of these that entered clinical trials in rheumatoid arthritis was infliximab (cA2, Remicade™). In an open-label trial conducted in 1992–93 20 patients with RA who had long-standing active disease, which had not responded to conventional therapies available at that time, received intravenous injections of infliximab at a total dose of 20 mg/kg body weight, in 2 or 4 divided doses over a 2-week period. This led to a rapid and impressive improvement in individual disease features and a reduction of erythrocyte sedimentation rate (ESR) and acute phase proteins (CRP) and serum interleukin-6 concentrations [18]. The striking clinical result was confirmed in a randomised double-blind placebo-controlled trial undertaken in 1993–1994 on 73 patients from 4 European centres (Fig. 1). In this second study the primary endpoint was a response as defined by the Paulus criteria [19], which required an improvement in 4 out of 6 clinical and laboratory characteristics. Two doses of infliximab were compared with placebo (0.1% human serum albumin) and showed that 79% of patients responded to the higher dose (10 mg/kg) and 44% to the lower dose (1 mg/kg), whereas 20% responded to placebo [20]. That the response was substantial was evident from secondary assessments, such as the magnitude of the response of individual clinical features such as swollen and tender joint counts, duration of morning stiffness, pain assessment, grip strengths, patient and physician's assessment of disease activity and measurements

Table 1 - Anti-TNF agents in clinical trials in RA

Name	Manufacturer
Monoclonal antibodies	
cA2 (Infliximab, Remicade™)	Centocor
CDP571	Celltech
D2E7	CAT and BASF (Knoll)
TNFR:Fc fusion proteins	
p75TNFR:Fc (Etanercept, Enbrel™)	Immunex/AHP
p55TNFR:Fc (Ro-45-2081, Lenercept™)	Roche

of ESR and CRP. On an average there was a reduction of 60-70% from the value at baseline in these parameters.

When these trials began in 1992, monoclonal antibody therapy had not been used extensively in the treatment of RA and one cause for concern was the possibility that high concentrations of the antibody might prove to be toxic. However, this was not the case and patients tolerated infusions without any significant immediate side-effects. The safety profile in the short term at least seemed to be unremarkable.

In an extension to the above study, all patients were followed up until disease relapse. Disease activity gradually returned and a clear dose response relationship was seen (see Tab. 2). From the single infusion study it could be concluded that anti-TNF therapy had a salutary effect on disease activity and symptoms.

The next critical question that needed to be addressed was whether repeated administration of infliximab, a chimeric (mouse x human) antibody, would continue to show efficacy upon repeated administration. Preliminary experience had shown that at least some patients developed anti-chimeric antibodies which might be associated with a loss of efficacy over time [21]. Encouraged by animal model studies in which the co-administration of a T-cell depleting anti-CD4 monoclonal antibody with anti-TNF antibody had reduced the formation of antibodies to xenogeneic immunoglobulins [22], we reasoned that methotrexate, which is extensively used in treating RA, and has both anti-inflammatory and immunosuppressive effects [23], might act similarly in preventing an antiglobulin response to infliximab in RA patients. This concept was incorporated in a multicentre European trial on 101 patients, and the efficacy of infliximab alone, or infliximab in combination with a low weekly dose of methotrexate (7.5 mg/week), given repeatedly in 5 cycles over 14 weeks was evaluated against placebo treatment. The result of the trial demonstrated that infliximab given at a dose of 1 mg/kg without methotrexate was asso-

Placebo
1 mg/kg
10 mg/kg
8%
44%
79%
p = 0.0083
p < 0.0001
non-responders
responders
A
Tender Joint Count (0-60)
Swollen Joint Count (0-58)
Pain Score (0-10cm)
Patients' Assessment of disease severity (1–5)
CRP (mg/L)
ESR /mm/hour)
Normal Range
Normal Range
Weeks
Weeks
B

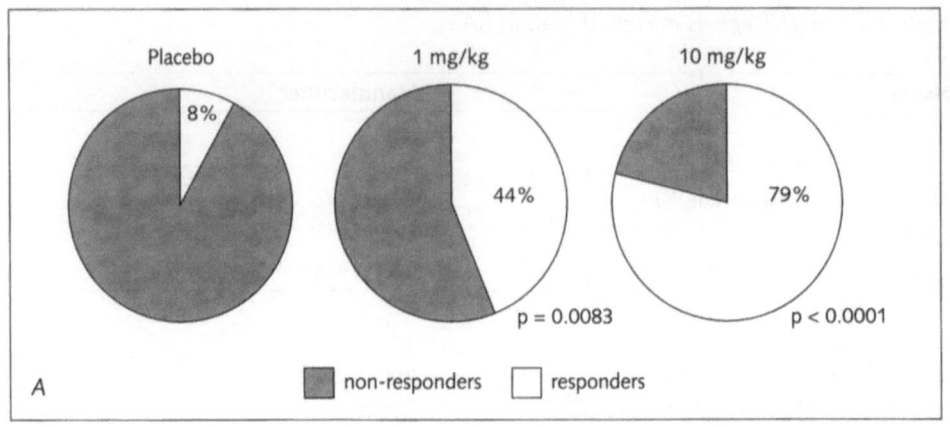

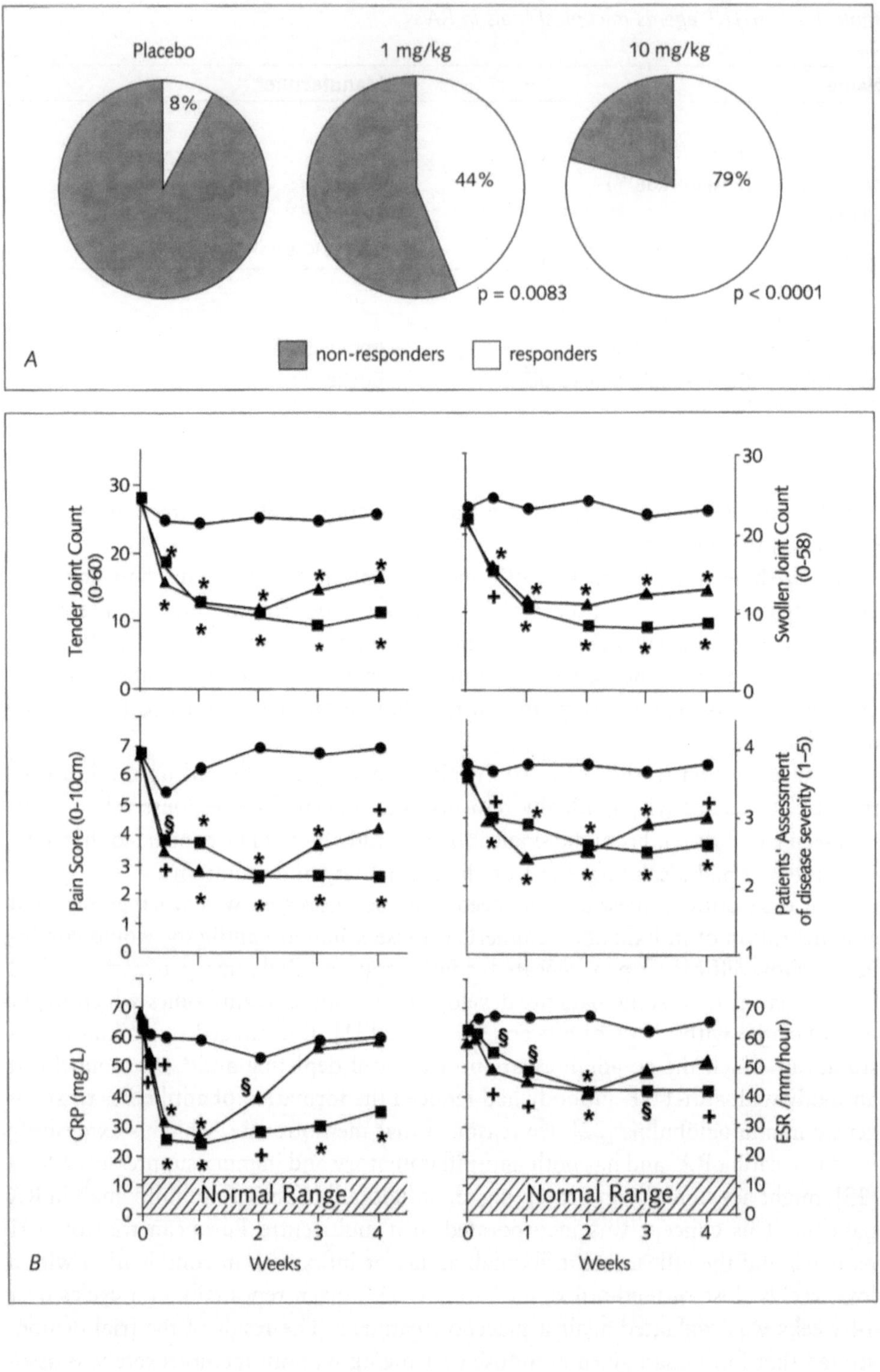

Table 2 - Duration of response to single infusion of anti-TNFα monoclonal antibody (Inflix-imab)

Dose	Duration of Paulus 20% response (median value)
1 mg/kg	3 weeks
3 mg/kg	6 weeks
10 mg/kg	8 weeks

Source: Maini et al. (1995) Arthritis Rheum 38 (Suppl 9): S186

ciated with a rapid loss of efficacy beyond the second infusion. However, in patients who continued with methotrexate there was an expectedly significant enhancement of the effect of infliximab, which was associated with a reduction in the incidence of antiglobulin production [24]. At the 2 higher doses of infliximab (3 mg and 10 mg/kg), methotrexate exerted less additional benefit, although there was some enhancement with methotrexate, both in terms of the proportion of responding patients and the duration of response [24]. At the higher doses, response rates on infliximab exceeded 60% compared with a maximum of 20% responders on place-bo (Fig. 2).

Infliximab is currently being evaluated when co-administered with methotrexate, in an ongoing trial of 2 years duration on a cohort of 428 RA patients. In this trial the patient population being treated had failed to respond to multiple disease-mod-ifying agents and had evidence of disease activity despite continuing treatment on methotrexate at a median dose of 15 mg/week. Aside from seeking to establish effi-cacy over a prolonged treatment period, this study will ascertain whether the fre-quency and dose of infliximab can be reduced without compromising efficacy. At the 6-month point it has been observed that 3 mg/kg every 8 weeks is as efficacious as its administration every 4 weeks with a response rate of 50–60% compared with 20% on placebo treatment [25]. In this trial, the placebo (control) group consists of

Figure 1
Results of a placebo-controlled randomized trial of infliximab in rheumatoid arthritis.
(A) The percent of patients responding according to Paulus criteria at the end of 4 weeks fol-lowing a single intravenous injection of infliximab 10 mg/kg or 1 mg/kg versus placebo;
(B) Changes in individual clinical and laboratory measurements.
Values are means of 24 patients at each point (25 for 1 mg/kg group).
*● = placebo, ▲ = 1 mg/kg infliximab, and ■ = 10 mg/kg infliximab. Significance versus placebo: § p < 0.05 ✚ p < 0.01, * p < 0.001 (Reproduced with permission from Elliott et al (1994) The Lancet 344: 1105–1110, © The Lancet.)*

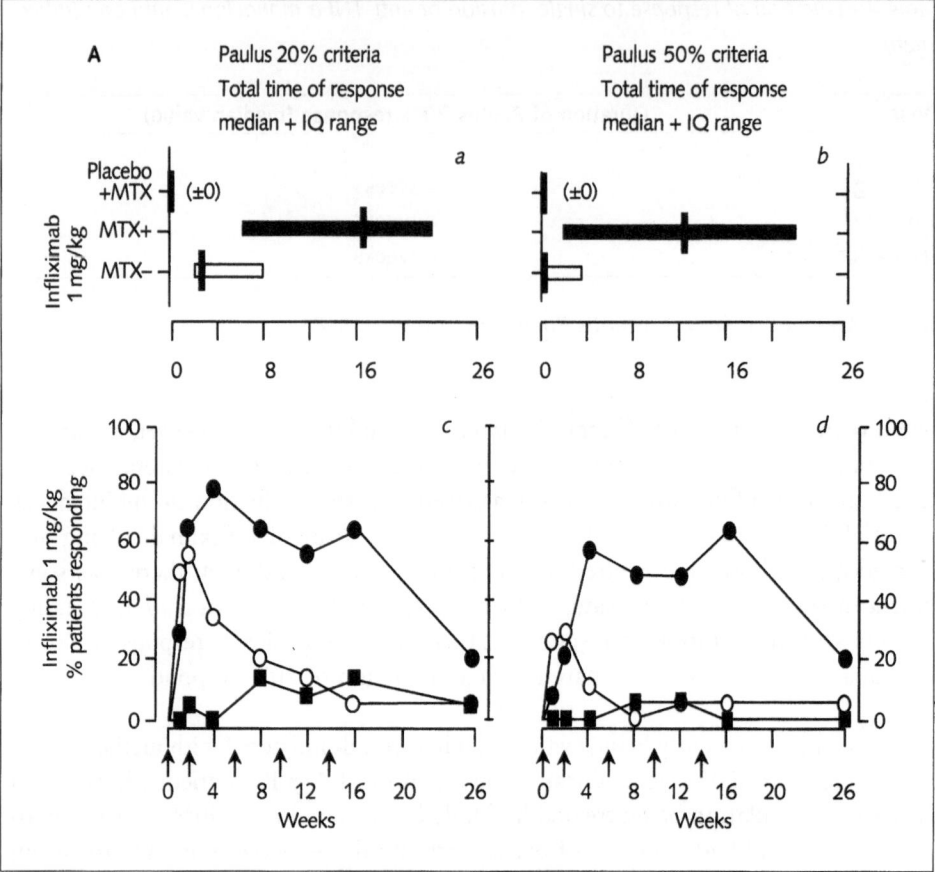

Figure 2
Efficacy of infliximab therapy based on Paulus 20% and 50% criteria.
(A) Results obtained with 1 mg/kg of infliximab with (+) and without (–) methotrexate (MTX) and placebo plus MTX.
(B) Results obtained with 3 mg/kg and 10 mg/kg of infliximab with and without MTX and placebo plus MTX. Results shown are the total time of response, as the median and interquartile (IQ) range, for each group (a, b, e, and f) and the proportion of (%) patients responding at weeks 1, 2, 4, 8, 16 and 26 (c, d, g, h, i, and j), all patients being included at each time point.
● = infliximab with MTX; ○ = infliximab without MTX; ■ = placebo plus MTX. Arrows indicate the times of infusions. For statistical analysis, see text and Table 2. The Paulus response is achieved by improvement in 4 of 6 of the following: 20% or 50% improvement in tender joint scores, swollen joint scores, duration of morning stiffness, or erythrocyte sedimentation rate, or a 2-grade improvement in the patient's and observer's assessment of disease severity. (Reproduced with permission from Lippincott Williams and Wilkins from Maini et al. (1998) Arthritis Rheum 41:1552–1563, © American College of Rheumatology).

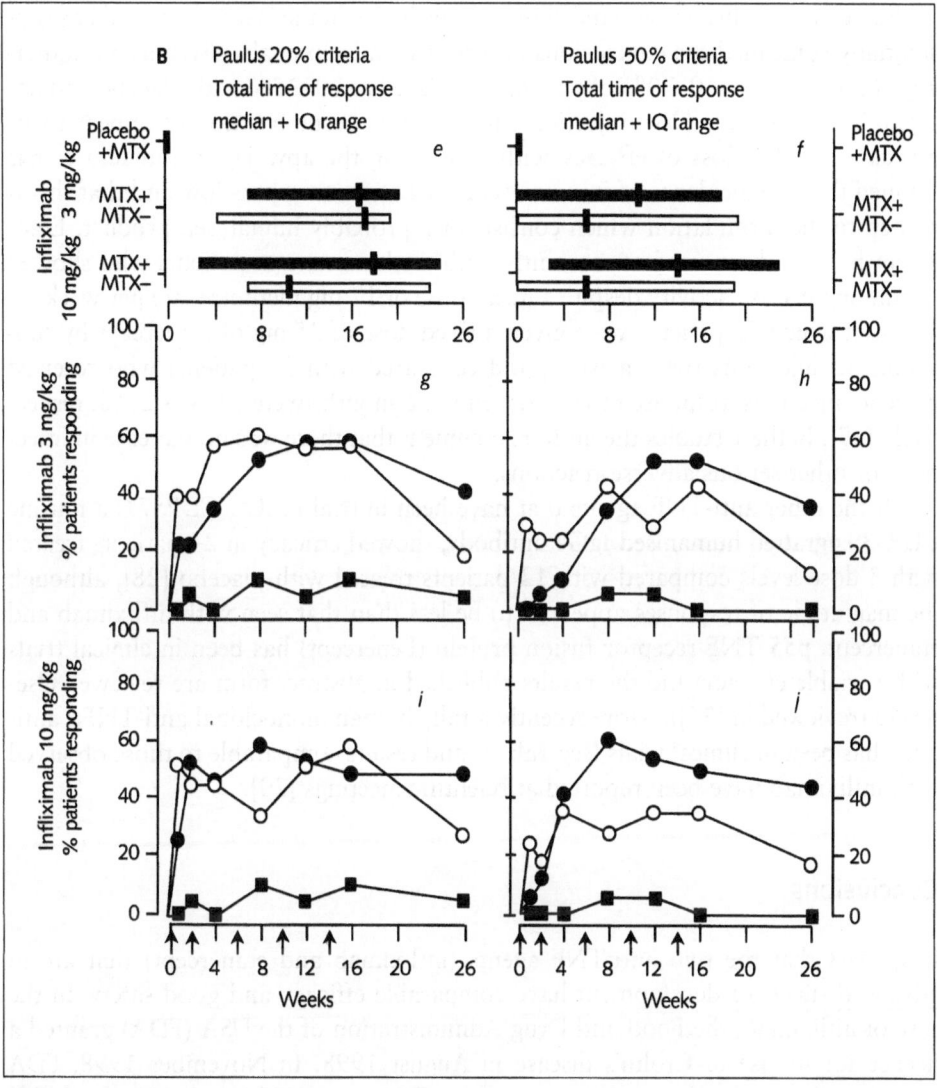

patients who have continued to receive methotrexate and therefore represent a significantly large cohort of patients that can provide comparative information on the relative safety of infliximab in this clinical setting. The preliminary 6-month analysis shows that infliximab is well tolerated, especially at the 3 mg/kg dose every 8 weeks, thus providing a regimen with a good safety and efficacy profile for continuation with longer-term studies and translation into a clinical practice. The possibility that TNF blockade might impair host defence against infection remains a concern, but observations to date have not revealed an increased incidence of serious infections compared with methotrexate treatment.

Etanercept (Enbrel™) entered clinical trials in RA in the last 4 years and has substantially confirmed the conclusions observed with infliximab. Given as monotherapy, it is effective in 60–70% of patients compared to 20–25% with placebo and has a good safety profile [26]. Injection site reactions occur but do not appear to be associated with a loss of efficacy with continuing therapy. The manufacturer has claimed that the incidence of immune reactions to etanercept is low and that this is related to the formulation which consists of a probably humanised product. Etanercept has also been used concurrently with methotrexate in patients who showed continuing disease activity despite a mean dose of 17 mg methotrexate per week. In a trial in which 59 patients were given a fixed dose of 25 mg of etanercept by subcutaneous injections twice a week, and compared with 25 patients who received placebo injections, response rates at the end of 6 months were 71% vs 27%, respectively [27]. In these studies the authors comment that there was no increase in infections or other serious adverse reactions.

Of the other anti-TNF agents that have been in trial in RA, CDP571, a murine CDR-3 engrafted humanised IgG4 antibody, showed efficacy in 24 patients treated with 3-dose levels compared with 12 patients treated with placebo [28], although the magnitude of responses appeared to be less than that seen with infliximab and etanercept. p55 TNF receptor fusion protein (Lenercept) has been in clinical trials with variable efficacy, and the results published in abstract form are reviewed elsewhere (reviewed in [29]). More recently a fully human monoclonal anti-TNFα antibody has been in clinical trials (see Tab. 1) and results comparable to those observed with infliximab have been reported at scientific meetings [30].

Conclusions

It appears that the two anti-TNF agents (infliximab and etanercept) that are in advanced stages of development have comparable efficacy and good safety. In the case of infliximab, the Food and Drug Administration of the USA (FDA) granted a licence for its use in Crohn's disease in August 1998. In November 1998, FDA granted a licence for the use of etanercept in RA patients whose disease continued to show disease activity despite current therapy. The efficacy of these two agents appears to be similar, although it should be observed that infliximab inhibits TNFα exclusively whereas etanercept, a soluble TNF receptor fusion protein, will inhibit TNFα and lymphotoxin. Whether the different biological properties will eventually result in the emergence of differences in efficacy or safety in the long term, will require considerably greater experience and a direct comparison. The duration of studies so far with any anti-TNF agent have not been conducted for a long enough period to provide evidence of protection of joints from structural damage. Similarly, whether anti-TNF therapy introduced at earlier stages of disease will be associated with more marked and longer-lasting effects, or even remission of disease, is not yet known but deserves investigation. Ultimately, the pharmacoeconomic issues will require analysis of the longer-term benefits of these agents. It has been argued that, provided anti-TNF therapy is associated with preservation of locomotor function and a high quality of physical and psycho-social health, the reduction in cumulative costs of healthcare may outweigh the higher cost of the drugs.

Addendum

Since submission of the manuscript, the results of a placebo-controlled trial of infliximab at 3 and 10 mg/kg every 4 or 8 weeks in patients with continuing disease activity, despite methotrexate treatment, has been published [25]. Methotrexate was continued at a median dose of 15 mg/week in all groups. At 30 weeks, the American College of Rheumatology (ACR) response criteria, representing a 20% improvement from baseline, were achieved in 53, 50, 58, and 52% of patients receiving 3 mg/kg every 4 or 8 weeks or 10 mg/kg every 4 or 8 weeks, respectively, compared with 20% of patients receiving placebo plus methotrexate ($p < 0.001$ for each of the four regimens vs placebo). Individual clinical response parameters improved by 50–70% from baseline. The one year analysis is now available and shows that in the infliximab treated groups there was arrest of progression of erosions and joint space narrowing assessed radiographically compared with the placebo group.

In a placebo controlled trial with etanercept, durable responses versus placebo have been reported at 6 months (Moreland et al (1999) *Ann Int Med* 130: 478–486). In a study presented at the ACR in November 1999 the results of one year of etanercept treatment showed a significant retardation of structural damage assessed by radiographs.

Acknowledgement

I wish to thank my colleagues Marc Feldmann, Fionula Brennan, Richard Williams, Michael Elliott and others for the co-operation with research carried out at The Kennedy Institute of Rheumatology, London. The Institute is supported by the Arthritis Research Campaign of Great Britain.

References

1   Pennica D, Nedwin GE, Hayflick JS, Seeburg PH, Derynck R, Palladino MA, Kohr WJ, Aggarwal BB, Goeddel DV (1984) Human tumour necrosis factor: precursor structure, expression and homology to lymphotoxin. *Nature* 312: 742–729

2   Beutler B, Greenwald D, Hulmes JD, Chang M, Pan YC, Mathison J, Ulevitch R, Cerami A (1985) Identity of tumour necrosis factor and the macrophage-secreted factor cachectin. *Nature* 316: 552–554

3   Saklatavala J, Sarsfield SJ, Townsend Y (1985) Purification of two immunologically different leucocyte proteins that cause cartilage resorption lymphocyte activation and fever. *J Exp Med* 162: 1208–1215

4   Thomas BM, Mundy GR, Chambers TJ (1987) Tumour necrosis factor α and β induce osteoblastic cells to stimulate osteoclast bone resorption. *J Immunol* 138: 775–780

5   Dayer J-M, Beutler B, Cerami, A (1985) Cachectin/tumour necrosis factor stimulates collagenase and prostaglandin E2 production by human synovial cellls and dermal fibroblasts. *J Exp Med* 162: 2163–2168

6   Buchan G, Barrett K, Turner M, Chantry D, Maini RN, Feldmann M (1988) Interleukin-1 and tumour necrosis factor mRNA expression in rheumatoid arthritis: prolonged production of IL1α. *Clin Exp Immunol* 73: 449–455

7   Brennan FM, Chantry D, Jackson A, Maini R, Feldmann M (1989) Inhibitory effect of TNFα antibodies on synovial cell interleukin-1 production in rheumatoid arthritis. *Lancet* 2:244–247

8   Chu CQ, Field M, Feldmann M, Maini RN (1991) Localization of tumor necrosis factor alpha in synovial tissues and at the cartilage-pannus junction in patients with rheumatoid arthritis. *Arthritis Rheum* 34:1125–1132

9   Chu CQ, Field M, Allard S, Abney E, Feldmann M, Maini RN (1992) Detection of cytokines at the cartilage/pannus junction in patients with rheumatoid arthritis: implications for the role of cytokines in cartilage destruction and repair. *Br J Rheumatol* 31: 653–661

10   Brennan FM, Gibbons DL, Mitchell T, Cope AP, Maini RN, Feldmann M (1992) Enhanced expression of TNF receptor mRNA and protein in mononuclear cells isolated from rheumatoid arthritis synovial joints. *Eur J Immunol* 22: 1907–1912

11   Deleuran BW, Chu CQ, Field M, Brennan FM, Mitchell T, Feldmann M, Maini RN (1992) Localization of Tumour Necrosis Factor receptors in the synovial tissue and cartilage/pannus junction in rheumatoid arthritis: Implication for local actions of TNFα. *Arthritis Rheum* 35: 1170–1178

12   Cope A, Aderka D, Doherty M, Engelmann H, Gibbons D, Jones AC, Brennan FM, Maini RN, Wallach D, Feldmann M (1992) Increased levels of soluble tumour necrosis factor (TNF) receptors in the sera and synovial fluids of patients with rheumatic diseases. *Arthritis Rheum* 35: 1160–1169

13   Haworth C, Brennan FM, Chantry D, Turner M, Maini RN, Feldmann M (1991) Expression of granulocyte-macrophage colony-stimulating factor in rheumatoid arthritis: regulation by tumor necrosis factor-α. *Eur J Immunol* 21: 2575–2579

14   Butler DM, Maini RN, Feldmann M, Brennan FM (1995) Modulation of proinflammatory cytokine release in rheumatoid synovial membrane cell cultures. Comparison of monoclonal anti-TNFα antibody with the IL-1 receptor antagonist. *Eur Cytokine Ntwk* 6: 225–230

15   Brennan FM, Maini RN, Feldmann M (1992) TNFα – a pivotal role in rheumatoid arthritis? *Br J Rheumatol* 31: 293–298

16   Williams RO, Feldmann M, Maini RN (1992) Anti-tumor necrosis factor ameliorates

joint disease in murine collagen-induced arthritis. *Proc Nat Acad Sci USA* 89: 9784–9788

17   Keffer J, Probert L, Cazlaris H, Georgopoulos S, Kaslaris E, Kioussis D, Kollias G (1991) Transgenic mice expressing human tumour necrosis factor: a predictive genetic model of arthritis. *EMBO J* 10: 4025–4031

18   Elliott MJ, Maini RN, Feldmann M, Long-Fox A, Charles P, Katsikis P, Brennan FM, Walker J, Bijl H, Ghrayeb J et al (1993) Treatment of rheumatoid arthritis with chimeric monoclonal antibodies to TNFα. *Arthritis Rheum* 36: 1681–1690

19   Paulus HE, Egger MJ, Ward JR, Willams HJ, Cooperative Systemic Studies of Rheumatic Disease Group (1990) Analysis of improvement in individual rheumatoid arthritis patients treated with disease-modifying anti-rheumatic drugs based on the findings in patients treated with placebo. *Arthritis Rheum* 33: 477–484

20   Elliott MJ, Maini RN, Feldmann M, Kalden JR, Antoni C, Smolen JS, Leeb B, Breedveld FC, Macfarlane JD, Bijl H, et al (1994) Randomised double-blind comparison of chimeric monoclonal antibody to tumour necrosis factor α (cA2) versus placebo in rheumatoid arthritis. *Lancet* 344: 1105–1110

21   Elliott MJ, Maini RN, Feldmann M, Long-Fox A, Charles P, Bijl H, Woody JN (1994) Repeated therapy with a monoclonal antibody to tumour necrosis factor α in patients with rheumatoid arthritis. *Lancet* 344: 1125–1127

22   Williams RO, Mason LJ, Feldmann M, Maini RN (1994) Synergy between anti-CD4 and anti-tumor necrosis factor in the amelioration of established collagen-induced arthritis. *Proc Natl Acad Sci USA* 91: 2762–2766

23   Kremer JM (1994) The mechanism of action of methotrexate in rheumatoid arthritis: The search continues. *J Rheumatol* 21: 1–5

24   Maini RN, Breedveld FC, Kalden JR, Smolen JS, Davis D, Macfarlane JD, Antoni C, Leeb B, Elliott MJ, Woody JN et al (1998) Therapeutic efficacy of multiple intravenous infusions of anti-tumor necrosis factor α monoclonal antibody combined with low-dose weekly methotrexate in rheumatoid arthritis. *Arthritis Rheum* 41: 1552–1563

25   Maini RN, St Clair EW, Breedveld FC, Furst D, Kalden JR, Weisman M, Smolen JS, Emery P, Harriman G, Feldmann M, Lipsky P and the ATTRACT study group (1999) Randomised phase III trial of infliximab (chimeric anti-TNFα monoclonal antibody) versus placebo in rheumatoid arthritis patients receiving concomitant methotrexate. *Lancet* 354: 1932–1939

26   Moreland LW, Baumgartner SW, Schiff MH, Tindall EA, Fleischmann RM, Weaver AL, Ettlinger RE, Cohen S, Koopman WJ, Mohler K et al (1997) Treatment of rheumatoid arthritis with a recombinant human tumor necrosis factor receptor (p75)-Fc fusion protein. *N Eng J Med* 337: 141–147

27   Weinblatt ME, Kremer JM, Bankhurst AD, Bulpitt KJ, Fleischmann RM, Fox RI, Jackson CG, Lange M, Burge DJ (1999) A trial of etanercept, a recombinant tumor necrosis factor receptor:Fc fusion protein, in patients with rheumatoid arthritis receiving methotrexate. *N Eng J Med* 340: 253–259

28   Rankin ECC, Choy ESH, Kassimos D, Kingsley GH, Sopwith AM, Isenberg DA, Panayi

GS (1995) The therapeutic effects of an engineered human anti-tumour necrosis factor alpha antibody (CD571) in rheumatoid arthritis. *Br J Rheumatol* 34:334–342

29   Feldmann M, Elliott MJ, Woody JN, Maini RN (1997) Anti-tumor necrosis factor-α therapy of rheumatoid arthritis. *Adv Immunol* 64: 283–350

30   Rau R, Sander O, den Broeder A, van Riel PLCM, van der Putte L, Kruger K, Schattenkirchner M, Fenner H, Lassmann A, Kupper H et al (1998) Long-term efficacy and tolerability of multiple i.v. doses of the fully human anti-TNF-antibody D2E7 in patients with rheumatoid arthritis. *Arthritis Rheum* 41 (Suppl 9): S55

Blockade of cytokine activity by soluble cytokine receptors

Anthony Meager

Division of Immunobiology, National Institute for Biological Standards and Control, Blanche Lane, South Mimms, Potters Bar, Hertfordshire EN6 3QG, UK

Introduction

A great number and variety of specific cytokine receptors have been characterised. One feature they have in common is that their extracellular domains (ectodomains) are universally required for cytokine binding. While in general receptors, following cytokine binding, are internalised, many shed their ectodomains, largely in their entirety, from the cell membrane due to enzymatic cleavage (Tab. 1). These intact ectodomains are known as soluble cytokine receptors (sCRs) [1]. More rarely, sCRs are generated by alternative splicing of mRNAs (Tab. 1), where such modification leads to receptors lacking transmembrane- and cytoplasmic domains. In addition, certain viruses, especially those of the poxvirus family, contain genes encoding cytokine-binding proteins which are structurally similar in most cases to the sCR for particular cytokines. Such "viral sCRs" enhance the virulence of the virus by competing for binding of the cognate cytokine with its specific cell-membrane receptor, thus reducing any antiviral effects of the cytokine. The role of naturally produced, non-viral, sCRs seems less obvious, but since they retain their ability to fully bind their respective cognate cytokine, they too could effectively neutralise cytokine action by preventing the normal binding to cell surface receptors. Such sCR activity appears to be important during acute and chronic inflammatory conditions, where raised levels of certain sCRs have been reported [2, 3].

That sCRs can potentially act as cytokine antagonists has led to research studies to provide evidence of their efficacy in disease models. In turn these have led to the development of recombinant sCRs as potentially therapeutic agents for the treatment and management of human disease. In particular, sCRs are indicated in inflammatory disorders, where excessive amounts of proinflammatory cytokines are generated by activated leukocytes. These cytokines, which include IL-1β, IL-6, IL-8, and TNFα, have often been correlated with symptoms and pathology [126]. Therefore, injection of sCRs which block the actions of these cytokines may alleviate or

Novel Cytokine Inhibitors, edited by Gerry A. Higgs and Brian Henderson
© 2000 Birkhäuser Verlag Basel/Switzerland

Table 1 - Soluble cytokine receptors

Molecule	Mechanism of generation	References
IL-1RII (CDw121b)	Enzymatic cleavage (ec)	[3, 66]
IL-2Rα (CD25)	ec	[71, 76]
IL-4Rα (CDw124)	Alternative splicing of mRNA (as)	[67]
IL-5Rα	as	[105]
IL-6Rα (CD126)	ec	[86]
IL-6Rβ (gp130) (CDw130)	ec	[89, 91]
IL-7Rα (CD127)	as	[106]
TNF-R1 (CD120a)	ec	[2, 15]
TNF-R2 (CD120b)	ec	[2]
LIF-Rα	as	[107]
G-CSF-Rα	as	[108]
GM-CSF-Rα	as	[109]
Type I IFN-R (CD118)	not known	[110]
Type II IFN-R (CD119)	not known	[110]
CNTF-Rα	ec	[111]
SCF-R (c-kit)	ec	[112, 113]

prevent symptoms and pathology. Much of the drive for the development of recombinant sCRs has thus come from studies of the inhibitory actions of sCRs that block proinflammatory cytokine activities, especially those of TNFα and IL-1β.

Soluble cytokine receptors

Soluble TNF receptors

Cytokines of the TNF superfamily are mostly mediators of host defence and immune regulation. Members of this superfamily (Tab. 2) act either locally through direct cell-to-cell contact or as soluble mediators capable of diffusing to more distant targets. Many of these mediators are synthesized as type-II transmembrane proteins with the C-terminal ectodomains mediating binding to the receptors of the TNF receptor (TNF-R) superfamily (Tab. 2). Some of the type-II transmembrane proteins are cleaved *in situ* by proteinases to release active soluble cytokines. TNFα is one of four such cytokines (the others are LTα or TNFβ, Fas ligand, and TWEAK) [4, 5]. TNFα acts primarily as an inflammatory cytokine involved in the coordination of host defence mechanisms against invading pathogens, especially bacteria, by

Table 2 - TNF ligand and receptor superfamily members

Ligands	Receptors
TNFα	TNF-R1, TNF-R2, (T2, Va53/SalF19R, M-T2, soluble homologues of TNF-R specified by Shope fibroma, vaccinia and myxoma viruses, respectively)
LTα (TNFβ)	TNF-R1 (TNF-R2 and viral soluble homologues, as above) and herpesvirus entry mediator (HVEM)
LTβ (found in LTα 2b complex)	TNF-receptor related protein (TNF-Rrp or LTβR)
LIGHT	TNF-Rrp/LTβR and HVEM
TWEAK (LTγ)	TWEAK-R (?)
Fas.L or APO-1L	Fas antigen or APO-1 (CD95)
CD27.L (CD70)	CD27
CD30.L (CD153)	CD30
CD40.L (gp39;CD154)	CD40
4-1BB.L	4-1BB (CD137)
OX40.L	OX40 (CD134)
DR3.L (?)	DR3 (also known as APO-3/LARD/TRAMP/WSL-1)
TRAIL (APO-2.L)	DR4, DR5 (functional), TRID, TRUNDD (non-functional)
RANK.L (TRANCE/OPGL)	RANK and osteoprotegerin (OPG)
(NGF, not a TNF ligand)	LNGF-R (low affinity p75 NGF-receptor)
VEGI (vascular endo-thelial growth factor)	VEGI.R (?)

References: [4, 5, 8–10, 114–122].

activating a wide range of immunological and non-immunological cellular responses [6]. LTα appears to have a similar range of activities to TNFα, but may have more involvement in the development of peripheral lymphoid organs and splenic architecture [7].

The activities of TNFα are mediated through two related TNF-Rs [8–10]. These two TNF-R types are either known as TNF-R1 and TNF-R2, or p55 and p75, respectively, according to their molecular masses. The amino acid sequences of TNF-R1 and TNF-R2 are 28% homologous. They each contain four characteristic cysteine-rich repeat units of approximately 40 amino acids in their ectodomains. TNF-R1 and TNF-R2 ectodomains show structural similarity to those of other members of the TNF-R superfamily (Tab. 2), and can be shown to contain topologically distinct types of modules [11].

The majority of TNFα actions are mediated by the smaller of the two TNF-Rs, TNF-R1. This is also the major receptor for LTα [12]. The role of TNF-R2 is less well established, although it has been reported to be strongly stimulated by membrane-bound TNFα [13]. However, the smaller TNF-R1 has recently been shown to have a significantly higher affinity for soluble TNFα at 37 °C than TNF-R2, thus providing a rationale for the predominant role of TNF-R1 in cellular responses stimulated by soluble TNFα [12].

The ectodomains of both TNF-R1 and TNF-R2 can be shed, following limited proteolytic cleavage [14], from the cell membrane and were originally isolated as TNF-binding proteins (TNF-BP) from human urine [15–17] and sera from cancer patients [2, 18]. TNF-BP were found to have an approximate molecular weight of 27–35 kDa and, except for the variable loss of several N-terminal amino acids due to proteolytic degradation, to contain the four cysteine-rich repeat domains of membrane-bound receptors [2]. They bound TNF with an affinity (kd) of approximately 5 nM, i.e., about 100-fold and 10-fold lower, respectively, than membrane-bound TNF-R1 and TNF-R2. Subsequently, it has been shown that recombinant soluble (rs) human TNF-R1 containing all 236 amino acids of the ectodomain, produced as a secreted protein from baculovirus-infected insect cells or Chinese hamster ovary cells [19], bound TNF with a similar affinity to TNF-BP. The relatively low affinities of TNF-BP and rsTNF-R1 compared to the membrane-bound TNF-R1 and TNF-R2 counterparts are probably largely accounted for by the fact the TNF-BP and rsTNF-R1 are monomeric binding proteins, whereas TNF forms a stable complex with membrane-bound TNF-R by a trimeric interaction [20]. To gain increased affinity, rsTNF-R1 has been fused to the human IgG Fc to form a recombinant chimeric homodimer [21–23]. Similarly, a rsTNF-R2.IgG Fc chimeric homodimer has been shown to bind TNF with a much higher affinity than the rsTNF-R2 monomer [24].

In normal healthy persons, circulating serum levels of TNF-BP/sTNF-R are 1–2 ng/ml [25, 26]. Similar levels have also been found in baboons [27]. These levels are significantly increased (3–5-fold) in individuals treated with lipopolysaccharide (LPS), or in response to natural bacterial and HIV-1 infections [28–30]. They can also be raised in certain types of cancer, arthritis and renal failure [31, 32]. The loss of cell-membrane TNF-R by increased shedding may lead to a transient decreased responsiveness of cells to TNF. Complexes of TNF-TNF-BP may serve as a reservoir that liberates small amounts of TNF over prolonged periods. On the other hand, it has been speculated that TNF-BP/sTNF-R could play a role in the paracrine and autocrine regulatory system by potentially neutralising the toxic and other stimulatory effects of TNF. However, the TNF "neutralising capacity " of TNF-BP/sTNF-R is relatively low and a 30-fold and 300-fold molar excess of TNF-BP55/sTNF-R1 and TNF-BP75/sTNF-R2, respectively, are required to inhibit the cytotoxic action of TNF [33]. Release of sTNF-R2 from malignant cells can, however, protect against TNF-mediated cytotoxicity [34]. Furthermore, rsTNF-R.IgG Fc chimeric homodimers, which have increased affinity for TNF over the monomeric soluble receptors, have been shown to be much more effective in inhibiting TNF-mediated cytotoxicity *in vitro*. For example, a rsTNF-R1.IgG fusion protein (p55-sf2), containing a CH1 domain to be more "IgG-like", neutralised the cytotoxicity of TNF in a human rhabdomyosarcoma cell line (KYM-1D4) on an equimolecular basis with TNF (based on monomeric mass of TNF) and was over 3,000-fold more effective than rsTNF-R1 [35]. Interestingly, rsTNF-R1.IgG/p55-sf2 was approximately 9-fold more effective than rsTNF-R2.IgG/p75-sf2 [35]. Such chimeric fusion proteins are as, or more, effective in neutralising TNF-mediated cytotoxicity as strong neutralising anti-TNF monoclonal antibodies. A potential advantage of rsTNF-R.IgG over monoclonal antibodies is that they also neutralise the activity of LTα (TNFβ), which is also bound by them.

Although strong neutralisation of TNF has been demonstrated *in vitro*, the neutralising capacity of rsTNF-R and rsTNF-R.IgG fusion proteins needs to be assessed *in vivo* in relevant inflammation and disease models. In a murine model of gram-negative bacterial sepsis (endotoxic shock), pre-treatment of animals with 200 μg of a rsTNF-R1.IgG was found to give good protection against death with complete neutralisation of circulating TNF [36]. A rsTNF-R2.IgG was less effective in this model; low levels of bioactive TNF persisted and death was delayed, but not prevented. The lower efficacy of rsTNF-R2.IgG appears thus to mirror its poorer effectiveness *in vitro* compared with rsTNF-R1.IgG constructs. Similar results for rsTNF-R1.IgG have been generated in a rat endotoxic shock model, but even at doses as high as 5.0 mg/kg, not all animals ultimately survived [37]. It has also recently been shown that baboons can be protected against lethal *E. coli* bacteraemia by pre-treatment with rsTNF-R1.IgG and that free plasma TNFα was neutralised even though inactive TNFα/rsTNF-R1.IgG complexes remained in the circulation for long periods [51]. While in septic shock models administration of rsTNF-R.IgG within the first three hours of a lethal endotoxic or gram-negative bacteraemic insult could protect animals, such pro-active strategies are not possible in human severe sepsis or septic shock (SS). Patients admitted to hospital with SS are already many hours into overwhelming bacterial infection, caused by a diversity of pathogenic agents, where the continuous stimulation of leukocytes has already generated dangerously excessive levels of proinflammatory mediators; these cause or contribute to hypotensive shock, organ failure, and death [38]. TNF has been strongly implicated as a contributory cause of SS or endotoxic shock, leading to mortality [39–41]. However, the course of SS is complex and variable and it has not been possible to pinpoint when TNF levels peak [60]. Nevertheless, this uncertainty has not stopped clinical trials of TNF inhibitors going ahead. These were first carried out in SS patients using an anti-TNF monoclonal antibody (mAb). A large randomised, controlled, double-blind, multicentre trial, in which 971 SS patients were given 7.5 mg/kg, 15.0 mg/kg or placebo as a single infusion, indicated that, overall, anti-TNF did not cause a decrease in mortality at 28 days between placebo and anti-

TNF in all infused patients [42]. However, in a subset of SS patients infused with anti-TNF mAb, a significant reduction in mortality was found at 3 days after infusion. Although a trend towards reduced mortality continued at 28 days following treatment with anti-TNF mAb, the difference in mortality among SS patients treated with placebo or anti-TNF was not significant. Other trials involving the use of anti-TNF mAbs have similarly been unsuccessful [55–57]. These trials have been followed by others in which rsTNF-R.IgG fusion proteins have been given to SS patients in order to reduce mortality. The first of these trials, conducted by the "Soluble TNF receptor sepsis group", was a randomised, double-blind, placebo-controlled, multicentre trial in which a total of 141 SS patients were randomly assigned to receive either placebo or a single intravenous infusion of one of three doses of rsTNF-R2.FcIgG (0.15, 0.45, or 1.5 mg/kg of body weight) [43]. The primary endpoint was mortality from all causes at 28 days. The results indicated that treatment with rsTNF-R2.FcIgG did not reduce mortality compared with placebo, and, worryingly, at the highest dose appeared to be associated with increased mortality. A volunteer study, in which a similar rsTNF-R2.FcIgG construct was used to counter inflammatory responses following intravenous (i.v.) endotoxin administration, has indicated inflammatory responses can be maintained independently of high circulating levels of TNF-blocking activity, thus questioning the rationale for using inhibitors of TNF in SS [44]. A further clinical trial where a rsTNF-R1.IgG fusion protein was used for the treatment of sepsis has, however, produced a more hopeful outcome [45]. While there was no decrease in mortality between placebo and rsTNF-R1.IgG in all infused patients, a trend towards reduced mortality was observed at day 28 in a prospectively defined population with severe sepsis. This trend became significant when predicted mortality and plasma IL-6 levels were included in a logistic regression analysis.

The reason for the apparent better efficacy of rsTNF-R1.IgG compared to rsTNF-R2.IgG may stem from a difference in the exchange rates of TNF with these two constructs. While they both bind TNF with high affinity, there is a large difference in the binding kinetics, the exchange rate of TNF bound to rsTNF-R2.IgG being 65-fold faster than that of the rsTNF-R1.IgG construct [36]. In mice treated with rsTNF-R2.IgG, circulating bioactive TNF persisted, whereas rsTNF-R1.IgG neutralised all bioactive TNF completely [36]. However, a similar persistence of bioactive serum TNF was not seen in endotoxin-challenged volunteers treated with rsTNF-R2.IgG, but (enhanced) inflammatory responses still developed and were maintained irrespective of the presence of rsTNF-R2.IgG [44]. Alternatively, the apparent better efficacy of rsTNF-R1.IgG may have something to do with the dose of this protein. In comparison with rsTNF-R2.IgG, which was used at 1.5 mg/kg (highest dose), much lower amounts of rsTNF-R1.IgG, 0.083 mg/kg (i.e., about 20-fold lower), were used for treatment of sepsis [43, 45]. There is evidence from the INTERSEPT trial [42] that increasing doses of anti-TNF mAb, with resultant more extensive neutralisation of bioactive TNF, may be accompanied by increased mortality in septic patients. By analogy, complete removal of TNF activity by high-dose rsTNF-R2.IgG may also be detrimental by removing one of a network of bactericidal mediators [43]. Another variable could be that the patients in these "anti-TNF" trials were not equivalent, and for some unknown reason septic patients were more responsive to treatment in the rsTNF-R1.IgG trial [45].

Whatever the reason for the reduced mortality in severe sepsis patients in the rsTNF-R1.IgG trial, further more extensive trials are required to confirm the beneficial effect of rsTNF-R1.IgG in sepsis. Possibly, the inflammatory responses during severe sepsis and shock are too overwhelming for one agent alone to be effective, and perhaps combinations of anti-cytokine molecules could produce more beneficial effects. In future studies it will, however, be important to find ways that do not compromise host defense mechanisms [52, 53].

Although rsTNF-R.IgG treatments of septic patients has proved rather uncertain, perhaps such therapies will prove more efficacious in certain chronic disorders, such as rheumatoid arthritis (RA) and Crohn's disease, an inflammatory condition of the bowel, in which TNF has also been implicated in pathology. Anti-TNF mAb treatment of RA has beneficial effects, including reduction of joint pain and increased mobility [46]. A double-blind, placebo-controlled trial of a chimeric murine anti-TNF mAb/ human IgG fusion protein, cA2, in patients with severe RA has provided evidence that TNF blockade was achieved and correlated with impressive reductions in disease activity [47, 48]. This same fusion mAb (cA2) is also proving of some benefit in Crohn's disease [49, 50]. Trials involving rsTNF-R.IgG are already underway in RA and Crohn's disease patients [54].The potential beneficial effect of rsTNF-R.IgG in Crohn's disease may possibly be linked to inhibition of TNF-induced matrix metalloproteinase production by gut mucosal mesenchymal cells [58].There is also growing evidence for inflammatory mediators in heart failure, and thus a rationale for the use of rsTNF-R.IgG has become apparent and Phase I clinical trials are being done [59]. Possibly rsTNF-R.IgG will prove more effective than anti-TNF mAb because it has the capacity to neutralise both TNF and LTα.

The emerging field of gene therapy is opening further avenues for targetted treatment of RA. A recent study of the direct adenovirus-mediated gene transfer of sTNF-R1.IgG and sIL-1RI.IgG to rabbit knees with experimental arthritis has shown that such treatment induces both local and distal anti-arthritic effects [125].

The potential of the soluble counterparts of other members of the TNF-R superfamily (Tab. 2) to act as inhibitors of their interactions with cognate TNF-ligands has yet to be studied. There may well be several opportunities to use sCR in a clinical setting, but further basic research to establish the biology of these interactions is first required to provide rationales for the treatment of particular diseases. An area of growing interest is that of T-cell activation; several members of the TNF- and TNF-R-superfamilies are involved as co-stimulators. For example, the interaction of

CD40 and its T-cell-based ligand CD40.L appears to play a key role in activation of T killer cells [61, 62] and many other immune responses [95]. Inhibition of this interaction by application of anti-CD40.L, in combination with CTLA4.Ig fusion protein, has been shown to prolong allograft survival in primates [63, 69, 70]. Anti-CD40L has also been recently shown to reduce atherosclerosis in hyperlipidaemic mice by inhibiting inflammatory processes triggered by CD40 signalling [127]. Soluble CD40 may therefore offer potential as a "blocker" of allograft rejection and atherogenesis. This field has yet to be fully developed and exploited.

Soluble interleukin receptors

Many soluble interleukin receptors have been characterised (Tab. 1).These all have some potential for treating various diseases, but as yet little clinical work has been carried out. For the purposes of this review discussion will be limited to soluble interleukin-1 receptor (sIL-1R), sIL-2R and sIL-6R.

In the case of IL-1, two types of cell surface receptors, designated IL-1RI and IL-1RII, have been characterised [64]. The larger of these two receptors, the 80 kDa IL-1RI, acts as a functional receptor for IL-1α and IL-1β, but the smaller 60 kDa IL-1RII probably acts as a "non-functional", decoy receptor, which serves as a negative regulator of IL-1 action [65, 66]. Both receptors have ectodomains containing three Ig-like domains, the amino acid sequences of which are 28% related. However, only the ectodomain of IL-1RII is shed from the cell surface and behaves as a soluble receptor. A variety of agents and mediators, including IL-4, TNF and glucocorticoids, appear to stimulate shedding of sIL-1RII [64, 66, 67]. Normally, circulating levels of sIL-1RII are low, but they can be dramatically increased in acute illness, particularly in severe sepsis [67]. Soluble IL-1RII has a higher binding affinity for IL-1β than for IL-1α or IL-1ra and, since circulating IL-1 is mostly IL-1β, sIL-1RII could therefore, in principle, act as an antagonist of IL-1β-mediated activity. However, only a recombinant soluble form of IL-1RI, which has equally high affinity for IL-1α, IL-1β and IL-1ra, has been developed as a potential therapeutic agent for the use in, for example, cutaneous allergy [68] and HIV-infected individuals [128]. IL-1 up-regulates HIV expression *in vitro* and thus inhibition of IL-1 activity may be expected to suppress HIV activation. A phase I/II trial has been completed [128] the results of which have suggested possible beneficial effects of sIL-1RI.

In the case of IL-2, high affinity receptors are comprised of three distinct sub-units, a 55 kDa α-chain (also known as the Tac antigen, CD25), a 75 kDa β-chain (CD122), and a 65 kDa γ-chain [71, 72]. The IL-2Rα chain is inducible and expressed on T lymphocytes following activation, whereas IL-2Rβ and IL-2Rγ are normally constitutively expressed. The latter are members of the cytokine (or haematopoietin) receptor superfamily [73] and have large intracellular domains, whereas IL-2Rα is related to complement control protein and has only a short cytoplasmic tail [71]. The IL-2Rα binds IL-2 with low affinity (kd = 10 nm) and is incapable of signalling, except when associated with the β- and γ-subunits [72]. It appears to be the most readily shed of the three chains, giving rise to soluble (s) IL-2Rα of molecular weight 35–40 kDa which also binds IL-2 with low affinity (kd = 10 nm) [74, 75]. Circulating levels of sIL-2Rα have been found to be raised in a wide variety of diseases states, including those of infectious or neoplastic nature, e.g., AIDS, hairy cell leukaemia, and those conditions involving T cells, e.g., systemic lupus erythematosus, allograft transplantation [76–80]. When sIL-2Rα levels increase in the microenvironment around cells, the concentration of sIL-2Rα may be sufficient to effectively compete with (high-affinity) cell surface IL-2 receptors for binding of IL-2, and thus render the immune system unresponsive [75]. There are several reports of the inhibitory action of sIL-2Rα on IL-2-induced activations of peripheral mononuclear cells and cytotoxic T-lymphocyte line (CTLL) *in vitro* [81–85]. However, inhibition of IL-2-driven CTLL proliferation can be partial [81] and requires very high concentrations of sIL-2Rα.

In the case of the high-affinity receptor for IL-6, an IL-6-binding subunit (80 kDa α-chain) associates with a larger affinity converter and signalling subunit (130 kDa β-chain) known as gp130 [86, 87]. The ectodomain of IL-6Rα contains an outer N-terminal Ig-like domain, that is a signature of the Ig superfamily, and two inner domains that are related to those found in the cytokine (or haematopoietin) receptor superfamily [73]. The Ig-like domain is not involved in IL-6 binding. IL-6Rβ or gp130 is also a member of the cytokine receptor superfamily [73]. This gp130 is shared by several other cytokine receptors, including leukaemia inhibitory factor (LIF), oncostatin M (OSM), ciliary neurotrophic factor (CNTF) and interleukin-11 (IL-11), for which it serves as a converter/signal transducer [87]. The occupied high-affinity IL-6R appears to be a hexameric complex consisting of two molecules of IL-6, two of IL-6Rα and two of IL-6Rβ (gp130) [88].

A soluble form of IL-6Rα (50–60 kDa) has been found in the serum and urine of healthy volunteers [89]. However, unlike the potential antagonistic capability of other sCRs, sIL-6Rα appears to enhance cellular responses to IL-6. This is because the complex of IL-6/sIL-6Rα can still associate with IL-6Rβ (gp130) and thus induce biological activities [87, 90]. Therefore, sIL-6Rα could potentially act as a carrier for IL-6 or enhance the biological effects of locally released IL-6. Nevertheless, the regulation of IL-6 actions may be even more complicated since a soluble form of IL-6Rβ/gp130 (90–110 kDa) can be generated and this, by binding to circulating IL-6/sIL-6Rα complexes, has the potential to inhibit binding and therefore signalling through cell-bound IL-6Rβ/gp130 [91]. Circulating levels of sIL-6Rβ/sgp130 in the serum of healthy individuals has been reported to be much higher [91] than circulating levels of IL-6Rα [89]. Raised levels of sIL-Rα have been found in the serum of blood donors seropositive for HIV [92], and increased levels of both sIL-6Rα and -β have been reported in patients with multiple myelomatosis

[93, 94]. Recombinant forms of sIL-6Rα or -β have not been tested *in vivo* to see whether they can have inhibitory action against IL-6 and could be potentially used as therapeutic agents in diseases where IL-6 has adverse effects, e.g., multiple myeloma.

While the biological roles of other soluble interleukin receptors remain to be fully established, some may have potentially useful clinical applications. For example, sIL-4Rα was shown to delay cardiac allograft rejection in mice suggesting that IL-4 is critical for the generation of alloreactivity [96]. Therefore, sIL-4Rα may be helpful for allotransplantation of organs in man.

Soluble viral cytokine receptors

It has emerged that several of the larger viruses, e.g., herpesviruses and poxviruses, encode proteins with similar structures to sCR and which appear to mimic the action of specific sCRs. In particular, poxviruses may express several soluble viral cytokine receptors (svCR). For example, *vaccinia* virus has among its ca. 150 proteins svCR-specific for TNF, IL-1, type I IFN and type II IFN [97–102], all of which could potentially act as cytokine-response modifiers. With the exception of the svCR for type I IFN, these svCRs are structurally related to known CRs, i.e., TNF-R2, IL-1RII, IFNGR1. The *vaccinia* soluble type-I IFN receptor is unrelated to either IFNAR1 or IFNAR2, but is instead a member of the immunoglobulin superfamily [100, 101]. Other poxviruses, e.g., cowpoxvirus, myxoma and shope fibroma, also encode similar svCRs (Tab. 3). In contrast, herpesviruses are svCR-specific for certain chemokines (Tab. 3) and viral counterparts of IL-6, IL-10 and IL-17. However, it has recently been shown that poxviruses too express a secreted soluble chemokine-binding protein which is specific for CC, but not CXC or C, chemokines [103]. This protein has no sequence homology with known chemokine receptors.

All of the svCRs and binding proteins bind cytokines with high affinity and block their activity by preventing interaction with cellular receptors. Removal of the viral genes encoding these cytokine inhibitors or their inactivation has profound effects on viral pathogenesis; their virulence is significantly reduced. For example, *myxoma* virus usually causes a fatal disease in rabbits, but if rabbits are infected with a mutant virus lacking the T2 gene, which encodes a soluble TNF-R2-like protein, they are able to shake off the infection [104]. This finding indicates that the TNF-R2-like protein probably acts as an inhibitor of TNF and is therefore critical for the virus's usual virulence.

Although normally associated with viral virulence, svCRs and virus-specified soluble cytokine-binding proteins could be developed as therapeutic agents. For example, the *vaccinia*-specified chemokine-binding protein was shown to block eotaxin-induced eosinophil infiltration, a feature of allergic inflammatory reactions [103].

Table 3 - Soluble viral cytokine receptors and binding proteins

Virus family	Virus	Viral protein	Cytokine receptor
Poxviruses	Shope fibroma	T2	TNF-R2
	Myxoma	M-T2	TNF-R2
		T7	Type II IFN-R
	Vaccinia (some strains) (some strains)	CrmB, CrmC, CrmD	TNF-R2
	Vaccinia/Cowpox	B15R	IL-1IIR
		B18R	Type I IFN-R
		B9R	Type II IFN-R
		vCKBP	CCR
Herpesviruses	Herpesvirus-6	U12	Chemokine receptor?
	Herpesvirus-7	U12	Chemokine receptor?
	Herpesvirus-8	ORF74	CXCR1-like receptor
	Cytomegalovirus	US28	CCR1 (MIP-1aR)
		US27	Chemokine receptor?
		UL33	Chemokine receptor?
	Saimiri (monkey)	ECRF3	CXCR1 (IL-8R)

References: [96–104, 123, 124]

Thus, it should be possible to design cytokine/chemokine antagonists on the basis of such viral proteins. They could potentially also be used to neutralise pathogens that have cytokine or chemokine receptors as docking sites on cells, e.g., HIV or the malarial parasite.

References

1   Rose-John S, Heinrich PC (1994) Soluble receptors for cytokines and growth factors: generation and biological function. *Biochem J* 300: 281–290

2   Olsson I, Gatanaga T, Gulberg U, Lantz M, Granger GA (1993) Tumour necrosis factor (TNF) binding proteins (soluble TNF receptor forms) with possible roles in inflammation and malignancy. *Eur Cytokine Netw* 4: 169–180

3   Giri JG, Wells J, Dower SK, McCall CE, Guzman RN, Slack J, Bird TE, Shaneback K, Grabstein KH et al (1994) Elevated levels of shed type II IL-1 receptor in sepsis: potential role for the type II receptor in regulation of IL-1 responses. *J Immunol* 153: 5802–5809

4   Gruss H-J, Dower S (1995) Tumour necrosis factor ligand superfamily: involvement in the pathology of malignant lymphomas. *Blood* 85: 3378–3404

5   Chicheportiche Y, Bourdon PR, Xu H, Hsu Y-M, Scott H, Hession C, Garcia I, Browning JL (1997) TWEAK, a new secreted ligand in the tumour necrosis factor family that weakly induces apoptosis. *J Biol Chem* 272: 32401–32410

6   Jaatela M (1991) Biologic activities and mechanisms of action of tumour necrosis factor-α/cachectin. *Lab Invest* 64: 724–742

7   Liu Y-J, Banchereau J (1996) Mutant mice without B lymphocyte follicles. *J Exp Med* 184: 1207–1211

8   Loetscher H, Pan YC-E, Lahm H-W, Gentz R, Brockhaus M, Tabichi H, Lesslauer W (1990) Molecular cloning and expression of the human 55kd tumour necrosis factor receptor. *Cell* 61: 351–359

9   Schall TJ, Lewis M, Koller KJ, Lee A, Rice GC, Wong GHW, Gatanaga T, Granger GA, Lentz R et al (1990) Molecular cloning and expression of a receptor for tumour necrosis factor. *Cell* 61:361–370

10   Smith CA, Davis T, Anderson D, Solam L, Beckmann MP, Jerzy R, Dower SK, Cosman D, Goodwin RG (1990) A receptor for tumour necrosis factor defines an unusual family of cellular and viral proteins. *Science* 248: 1019–1023

11   Naismith JH, Sprang SR (1998) Modularity in the TNF-receptor family. *Trends Biochem Sci* 23: 74–79

12   Grell M, Wajant H, Zimmermann G, Scheurich P (1998) The type I receptor (CD120a) is the high-affinity receptor for soluble tumour necrosis factor. *Proc Natl Acad Sci USA* 95: 570–575

13   Grell M, Douni E, Wajant H, Lohden M, Clauss M, Maxeiner B, Georgopoulos S, Lesslauer W, Kollias G et al (1995) The transmembrane form of tumour necrosis factor is the prime activating ligand of the 80 kDa tumour necrosis factor receptor. *Cell* 83: 793–802

14   Brakebusch C, Varfolomeev EE, Batkin M, Wallach D (1994) Structural requirements for the inducible shedding of the p55 tumour necrosis factor receptor. *J Biol Chem* 269: 32488–32496

15   Engelmann H, Aderka D, Rubinstein M, Rotman D, Wallach D (1989) A tumour necrosis factor-binding protein purified to homogeneity protects cells from tumour necrosis factor toxicity. *J Biol Chem* 264: 11974–11980

16   Olsson I, Lantz M, Nilsson E, Peetre C, Thysell H, Grubb A, Adolf G (1989) Isolation and characterisation of a tumour necrosis factor binding protein from urine. *Eur J Haematol* 42: 270–275

17   Lantz M, Gulberg U, Nilsson E, Olsson I (1990) Characterisation *in vitro* of a human tumour necrosis factor-binding protein. A soluble form of tumour necrosis factor receptor. *J Clin Invest* 86: 1396–1402

18   Gatanaga T, Hwang C, Kohr W, Cappucini F, Lucci JA, Jeffes EWB, Lentz R, Tomich J, Yamamoto RS et al (1990) Purification and characterisation of an inhibitor (soluble

tumour necrosis factor receptor) for tumour necrosis factor and lymphotoxin from the serum ultrafiltrate of human cancer patients. *Proc Natl Acad Sci USA* 87: 8781–8784

19 Loetscher H, Gentz R, Zulauf M, Lustig A, Tabichi H, Schlaeger EJ, Brockhaus M, Gallati H, Manneberg M et al (1991) Recombinant 55-kDa tumour necrosis factor (TNF) receptor. *J Biol Chem* 266: 18324–18329

20 Banner DW, D'Arcy A, Janes W Gentz R, Schoenfeld H-J, Broger C, Loetscher H, Lesslauer W (1993) Crystal structure of the soluble 55 kd TNF receptor-human TNFβ complex: implications for TNF receptor activation. *Cell* 73: 431–445

21 Ashkenazi A, Marsters SA, Capon DJ, Chamow SM, Figari IS, Pennica D, Goeddel DV, Palladino MA, Smith DH (1991) Protection against endotoxic shock by a tumour necrosis factor receptor immunoadhesin. *Proc Natl Acad Sci USA* 88: 10535–10539

22 Peppel K, Crawford D, Beutler B (1991) A tumour necrosis factor (TNF) receptor-IgG heavy chain chimeric protein as a bivalent antagonist of TNF activity. *J Exp Med* 174: 1483–1489

23 Haak-Frendscho M, Marsters SA, Mordenti J, Brady S, Gillett NA, Chen SA, Ashkenazi A (1994) Inhibition of TNF by a TNF immunoadhesin. *J Immunol* 152:1347–1353

24 Mohler KM, Torrance DS, Smith CA, Goodwin RG, Stremler KE, Fung VP, Madini H, Widmer MB (1993) Soluble tumour necrosis factor (TNF) receptors are effective therapeutic agents in lethal endotoxemia and function simultaneously as both TNF carriers and TNF antagonists. *J Immunol* 151:1548–15561

25 Aderka D, Engelman H, Hornik V, Skornick Y, Levo Y, Wallach D, Kushtai G (1991) Increased levels of soluble receptors for tumour necrosis factor in cancer patients. *Cancer Res* 51: 5602–5607

26 Aderka D, Engelman H, Shermer-Avni Y, Hornik Y, Galil A, Sarov B, Wallach D (1992) Variation in serum levels of the soluble TNF receptors among healthy individuals. *Lymphokine Cytokine Res* 11: 157–16–59

27 Redl H, Schlag G, Adolf G, Natmessnig B, Davies J (1995) Tumour necrosis factor (TNF)-dependent shedding of the p55 TNF receptor in a baboon model of bacteremia. *Infect Immun* 63: 297–300

28 Girardin E, Roux-Lombard P, Grau GE, Suter P, Gallati H and the J5 Study Group (1992) Imbalance between tumour necrosis factor-alpha and soluble TNF receptor concentrations in severe meningococcaemia. *Immunology* 76: 20–23

29 Spinas GA, Keller U, Brockhaus M (1992) Release of soluble receptors for tumour necrosis factor (TNF) in relation to circulating TNF during experimental endotoxaemia. *J Clin Invest* 90: 533–536

30 Zangerle R, Gallati H, Sarcletti M, Wachter H, Fuchs D (1995) Tumour necrosis factor alpha and soluble tumour necrosis factor receptors in individuals with immunodeficiency virus infection. *Immunol Lett* 41: 229–234

31 Cope AP, Aderka D, Doherty M, Engelmann H, Gibbons D, Jones AC, Brennan FM, Maini RN, Wallach D, Feldman M (1992) Soluble tumour necrosis factor (TNF) receptors are increased in the serum and synovial fluids of patients with rheumatic diseases. *Arthritis Rheum* 35: 1160–1169

32   Diez-Ruiz A, Tilz GP, Zangerle R, Baier-Bitterlich G Wachter H, Fuchs D (1995) Soluble receptors for tumour necrosis factor in clinical laboratory diagnosis. *Eur J Haematol* 54: 1–8

33   Van Zee KJ, Kohno T, Fischer E, Rock CS, Moldawer LL, Lowry SF (1992) Tumour necrosis factor soluble receptors circulate during experimental and clinical inflammation and can protect against excessive tumour necrosis factor-α *in vitro* and *in vivo*. *Proc Natl Acad Sci USA* 89: 4845–4849

34   Neuner P, Klosner G, Pourmojib M, Trautinger F, Knobler R (1994) Selective release of tumour necrosis factor binding protein II by malignant human epidermal cells reveals protection from tumour necrosis factor-α cytotoxicity. *Cancer Res* 54: 6001–6005

35   Butler DM, Scallon B, Meager A, Kissonerghis M, Corcoran A, Chernajovsky Y, Feldmann M, Ghrayeb J, Brennan FM (1994) TNF receptor fusion proteins are effective inhibitors of TNF-mediated cytotoxicity on human KYM-1D4 rhabdomyosarcoma cells. *Cytokine* 6: 616–623

36   Evans TJ, Moyes D, Carpenter A, Martin R, Loetscher H, Lesslauer W, Cohen J (1994) Protective effect of 55- but not 75-kD soluble tumour necrosis factor receptor-immunoglobulin G fusion proteins in an animal model of gram-negative sepsis. *J Exp Med* 180: 2173–2179

37   Jin H, Yang R, Marsters SA, Bunting SA, Wurm FM, Chamow SM, Ashkenazi A (1994) Protection against rat endotoxic shock by p55 tumour necrosis factor (TNF) receptor immunoadhesin. Comparison with anti-TNF monoclonal antibody. *J Infect Dis* 170: 1323–1326

38   Glauser MP, Zanetti G, Baumgartner J-D, Cohen J (1991) Septic shock: pathogenesis. *Lancet* 338: 32–736

39   Tracey KJ, Beutler B, Lowry SF, Merryweather J, Wolpe S, Milsark I W, Harri RJ Fahey TJ, Zentella A (1986) Shock and tissue injury induced by recombinant human cachectin. *Science* 234: 470–474

40   Silva AT, Bayston KF, Cohen J (1990) Prophylactic and therapeutic effects of a monoclonal antibody to TNF in experimental Gram-negative shock. *J Infect Dis* 162: 421–427

41   Bodmer MW, Foulkes R (1996) TNFα neutralisation by biological antagonists. In: B Henderson, MW Bodmer (eds): *Therapeutic modulation of cytokines*. CRC Press, Boca Raton, FL, 221–236

42   Abraham E, Wunderink R, Silverman H, Perl TM, Nasraway S, Levy H, Bone R, Wenzel RP, Balk R, Allred R et al (1995) Efficacy and safety of monoclonal antibody to human tumour necrosis factor-α in patients with sepsis syndrome. *JAMA* 273: 934–941

43   Fisher CJ, Agosti JM, Opal SM, Lowry SF, Balk RA, Sadoff JC, Abraham E, Schein RMH, Benjamin E for the Soluble TNF Receptor Sepsis Study Group (1996) Treatment of septic shock with the tumour necrosis factor: Fc fusion protein. *N Engl J Med* 334: 1697–1702

44   Suffredini AF, Reda D, Bank SM, Tropea M, Agosti JM, Miller R (1995) Effects of

recombinant dimeric TNF receptor on human inflammatory responses following intravenous endotoxin administration. *J Immunol* 155: 5038–5045

45 Abraham E, Glauser MP, Butler T, Garbino J, Gelmont D, Laterre PF, Kudsk K, Bruining HA, Otto C, Tobin E et al (1997) p55 tumour necrosis factor receptor fusion protein in the treatment of patients with severe sepsis and septic shock. *JAMA* 277: 1531–1538

46 Feldmann M, Elliot MJ, Woody JN, Maini RN (1997) Anti-tumour necrosis factor-alpha therapy of rheumatoid arthritis. *Adv Immunol* 64: 238–308

47 Maini RN, Elliot MJ, Brennan FM, Feldman M (1995) Beneficial effects of tumour necrosis factor-alpha (TNFα) blockade in rheumatoid arthritis. *Clin Exp Immunol* 101: 207– 212

48 Elliott MJ, Maini RN, Feldmann M, Long-Fox A, Charles P, Bijl H, Woody JN (1994) Repeated therapy with monoclonal antibody to tumour necrosis factor a (cA2) in patients with rheumatoid arthritis. *Lancet* 344: 1125–1127

49 Derkx B, Taminiau J, Radema S, Stronkhorst A, Wortel C, Tytgat G, van Deventer S (1993) Tumour necrosis factor antibody treatment in Crohn's disease. *Lancet* 342: 173–174

50 Van Dullemen HM, van Deventer SJH, Hommes DW, Bijl H, Tytgat GN, Woody J (1995) Treatment of Crohn's disease with anti-tumour necrosis factor chimeric monoclonal antibody (cA2). *Gastroenterol* 109: 129–135

51 Van Zee K, Moldawer L, Oldenberg H, Thompson W, Stackpole A, Montegut W, Rogy M, Meschter C, Ashkenazi A, Chamow SM et al (1996) Protection against lethal *Escherichia coli* bacteremia in baboons (*Papio anubis*) by treatment with a 55 kDa TNF receptor (CD 120a)-IgG fusion protein, Ro 45-2081. *J Immunol* 156: 2221–2230

52 Lynn W, Cohen J (1995) Adjunctive therapy for septic shock: a review of experimental approaches. *Clin Infect Dis* 20: 143–158

53 Zanetti G, Glauser M (1997) Prevention and treatment of sepsis and septic shock. Curr Opin Infect Dis 10: 139–143

54 Moreland K, Baumgartner S, Schiff M, Tindall E, Fleishmann R, Weaver A, Ettlinger R, Cohen S, Koopman W et al (1997) Treatment of rheumatoid arthritis with a recombinant human tumour necrosis factor receptor (p75)-Fc fusion protein. *N Engl J Med* 337: 141–147

55 Cohen J, Carlet J, The INTERSEPT Study Group (1996) INTERSEPT: an international, multicenter, placebo-controlled trial of monoclonal antibody to human tumour necrosis factor-a in patients with sepsis. *Crit Care Med* 24: 1431–1440

56 Reinhart K, Weigand-Lohnert C, Grimminger F, Kaul M, Withington S, Treacher D, Eckhart J, Wilats S, Bouza C, Krausch D et al (1996) Assessment of the safety and efficacy of the monoclonal anti-tumour necrosis factor antibody fragment, MAK 195F, in patients with sepsis and septic shock: a multicenter, randomised, placebo-controlled, dose-ranging study. *Crit Care Med* 24: 733–742

57 Dhainaut JF, Vincent JL, Richard C, Lejeune P, Martin C, Fierobe L, Stephens S, Ney UM (1995) CDP571, a humanised antibody to human tumour necrosis factor-a: safety,

pharmacokinetics, immune response, and influence of the antibody on cytokine concentrations in patients with septic shock. *Crit Care Med* 23: 1461–1469

58 Pender SLF, Fell JME, Chamow SM, Ashkenazi A, MacDonald TT (1998) A p55 TNF receptor immunoadhesin prevents T cell-mediated intestinal injury by inhibiting matrix metalloproteinase production. *J Immunol* 160: 4098–4103

59 Mann DL (1998) The emerging role of tumour necrosis factor-α in the pathogenesis of heart failure: Mechanisms and current therapies (Abstract). *J Interferon Cytokine Res* 18: A–115

60 Grau G, Maennel D (1997) TNF inhibition and sepsis – sounding a cautionary note. Conflicting results from anti-TNF trials in sepsis patients suggest that timely administration of TNF inhibitor drugs is crucial. *Nature Med* 3: 1193–1195

61 Bennett SRM, Carbone FR, Karamalis F, Flavell RA, Miller JFAP, Heath WR (1998) Help for cytotoxic-T-cell responses is mediated by CD40 signalling. *Nature* 393: 478–480

62 Schoenberger SP, Toes REM, van der Voort EIH, Offringa R, Melief CJM (1998) T-cell help for cytotoxic T lymphocytes is mediated by CD40-CD40L interactions. *Nature* 393: 480–483

63 Kirk AD, Harlan DM, Armstrong NN, Davis TA, Dong Y, Gray GS, Hong X, Thomas D, Fechner JH Jnr, Knechtle SJ (1997) CTLA4-Ig and anti-CD40 ligand prevent renal allograft rejection in primates. *Proc Natl Acad Sci USA* 94: 8789–8794

64 Dinarello C (1996) Biologic basis for interleukin-1 in disease. Blood 87: 2095–2147

65 Collotta F, Re F, Muzio M, Bertini R, Polentarutti N, Sironi M, Giri JG, Dower SK, Sims JE, Mantovani A (1993) Interleukin-1 type II receptor: a decoy target for IL-1 that is regulated by IL-4. *Science* 261: 472–475

66 Sims JE, Giri JG, Dower SK (1994) The two interleukin-1 receptors play different roles in IL-1 actions. *Clin Immunol Immunopathol* 72: 9–14

67 Mosley B, Beckmann MP, March CJ, Idzerda RL, Gimpel SD, Vanden Bos T, Frizand D, Alpert A, Anderson D, Jackson J et al (1989) The murine interleukin-4 receptor: Molecular cloning and characterization of secreted and membrane bound isoforms. *Cell* 59: 335–344

68 Mullarkey MF, Leferman KM, Peters MS, Caro I, Roux ER, Hanna RK, Rubin AS, Jacobs CA (1994) Human cutaneous allergic late-phase response is inhibited by soluble IL-1 receptor. *J Immunol* 152: 2033–2041

69 Larsen CP, Elwood ET, Alexander DZ, Ritchie SC, Hendrix R , Tucker-Burden C, Cho HR, Aruffo D, Hollenbaugh D, Linsley PS et al (1996) Long-term acceptance of skin and cardiac allografts after blocking CD40 and CD28 pathways. *Nature* 381: 434–436

70 Saito K, Sakurai J, Ohata J, Kohsaka T, Hashimoto H, Okumura K, Abe R, Azuma M (1998) Involvement of CD40 ligand-CD40 and CTLA4-B7 pathways in murine acute graft-versus-host disease induced by allogeneic T cells lacking CD28. *J Immunol* 160: 4225–4231

71 Minami Y, Kono T, Yamada K, Taniguchi T (1992) The interleukin-2 receptors: insights into a complex signalling mechanism. *Biochim Biophys Acta* 1114: 163–177

72   Taniguchi T, Minami Y (1993) The IL-2/IL-2 receptor system: A current overview. *Cell* 73: 5–8

73   Cosman D (1993) The haematopoietin receptor superfamily. *Cytokine* 5: 95–106

74   Rubin LA, Kurman CC, Fritz ME, Biddison WE, Boutin B, Yarchoan R, Nelson DL (1985) Soluble interleukin-2 receptors are released from activated human lymphoid cells *in vitro*. *J Immunol* 135: 3172–3177

75   Rubin LA, Jay G, Nelson DL (1986) The released interleukin-2 receptor binds interleukin-2 efficiently. *J Immunol* 137: 3841–3844

76   Nelson DL (1986) Soluble interleukin-2 receptors: analysis in normal individuals and in certain disease states. *Fed Proc* 45: 377–383

77   Colvin RB, Fuller TC, MacKeen L, Kung PC, Ip SH, Cosimi AB (1987) Plasma interleukin-2 receptor levels in renal allograft recipients. *Clin Immunol Immunopathol* 43: 273–276

78   Keystone EC, Snow KM, Bombardier C, Chang C-H, Nelson DL, Rubin DL (1988) Elevated soluble interleukin-2 receptor levels in the sera and synovial fluids of patients with rheumatoid arthritis. *Arthritis Rheum* 31: 844–849

79   Steis RG, Marcon L, Clark J, Urba W, Longo DL, Nelson DL, Maluish AE (1988) Serum soluble IL-2 receptors as a tumour marker in patients with hairy cell leukaemia. *Blood* 71: 1304–1309

80   Jakobsen PH, Morris-Jones S, Theander TG, Hviid L, Hansen MB, Bendtzen K (1994) Increased plasma levels of soluble IL-2R are associated with severe *Plasmodium falciparum* malaria. *Clin Exp Immunol* 96: 98–103

81   Treiger BF, Leonard WJ, Svetlik P, Rubin LA, Nelson DE, Greene WC (1986) A secreted form of the human interleukin-2 receptor encoded by an 'anchor minus' cDNA. *J Immunol* 136: 4099–4105

82   Chilosi M, Semenzato G, Cetto A, Ambrosetti A, Fiore-Donati L, Perona G, Berton G, Lestani M, Scarpa A, Agostini L et al (1987) Soluble interleukin-2 receptors in the sera of patients with hairy cell leukaemia: relationship with the effect of recombinant a-interferon therapy on the clinical parameters and natural killer *in vitro* activity. *Blood* 70: 1530–1535

83   Symons JA, Wood NC, Di Giovine FS, Duff GW (1988) Soluble IL-2 receptor in rheumatoid arthritis. Correlation with disease activity, IL-1 and Il-2 inhibition. *J Immunol* 141: 2612–2618

84   Kondo N, Kondo S, Shimizu A, Honjo T, Hamuro J (1988) A soluble 'anchorminus' interleukin-2 receptor suppresses *in vitro* interleukin-2-mediated immune response. *Immunol Lett* 19: 299–308

85   Zorn U, Dallmann I, Grobe J, Kirchner H, Poliwoda H, Atzpodien J (1994) Soluble interleukin-2 receptors abrogate IL-2 induced activation of peripheral mononuclear cells. *Cytokine* 6: 358–364

86   Kishimoto T, Akira S, Taga T (1992) Interleukin-6 and its receptor: a paradigm for cytokines. *Science* 258: 593–597

87   Kishimoto T, Akira S, Narazaki M, Taga T (1995) Interleukin-6 family of cytokines and gp130. *Blood* 86: 1243–1254

88   Ward LD, Howlett GJ, Discolo G, Kasukawa K, Hamacher A, Moritz RL, Simpson RJ (1994) High affinity interleukin-6 receptor is a hexameric complex consisting of two molecules each of interleukin-6, interleukin-6 receptor, and gp130. *J Biol Chem* 269: 23286–23289

89   Freiling JTM, Sauerwein RW, Wijdenes J, Hendriks T, van der Linden CJ (1994) Soluble interleukin-6 receptor in biological fluids from human origin. *Cytokine* 6: 376–381

90   Mackiewicz A, Schooltink H, Heinrich PC, Rose-John S (1992) Complex of soluble human IL-6 receptor/IL-6 up-regulates expression of acute phase proteins. *J Immunol* 149: 2021–2027

91   Narazaki M, Yasukawa K, Saito T, Ohsugi Y, Fukui H, Koishihara Y, Yancopoulos GD, Taga T, Kishimoto T (1993) Soluble forms of the interleukin-6 signal-transducing receptor component gp130 in human serum possessing a potential to inhibit signals through membrane-anchored gp130. *Blood* 82: 1120–1126

92   Honda M, Yamamoto S, Cheng M, Yasukawa K, Suzuki H, Saito T, Osugi Y, Tokunaga T, Kishimoto T (1992) Human soluble IL-6 receptor: its detection and enhanced release by HIV infection. *J Immunol* 148: 2175–2180

93   Gaillard J-P, Bataille R, Brailly H, Zuber C, Yasukawa K, Attal M, Maruo N, Taga T, Kishimoto T, Klein B (1993) Increased and highly stable levels of functional soluble interleukin-6 receptor in sera of patients with monoclonal gammopaphy. *Eur J Immunol* 23: 820–824

94   Klein B, Zhang X-G, Lu Z-Y, Bataille R (1995) Interleukin-6 in human multiple myeloma. *Blood* 85: 863–872

95   Grewal IS, Flavell RA (1998) CD40 and CD154 in cell-mediated immunity. *Annu Rev Immunol* 16: 111–135

96   Maliszewski CR, Morrissey PJ, Fanslow WC, Sato TA, Willis C, Davison B (1992) Delayed allograft rejection in mice transgenic for a soluble form of the IL-4 receptor. *Cell Immunol* 143: 434–448

97   Alcami A, Smith G (1992) A soluble receptor for interleukin-1β encoded by vaccinia virus: a novel mechanism of virus modulation of the host response to infection. *Cell* 71: 153–167

98   Spriggs M, Hruby DE, Maliszewski CR, Pickup DJ, Sims JE, Buller RML, VanSlyke J (1992) Vaccinia and cowpox viruses encode a novel secreted interleukin-1 binding protein. *Cell* 71: 145–152

99   Alcami A, Smith G (1995) Cytokine receptors encoded by poxviruses: a lesson in cytokine biology. *Immunol Today* 16: 474–478

100 Symons J, Alcami A, Smith G (1995) Vaccinia virus encodes a soluble type I interferon receptor of novel structure and broad species specificity. *Cell* 81: 551–560

101 Colamonici O, Domanski P, Sweitzer SM, Larner A, Buller RML (1995) Vaccinia virus B18R gene encodes a type I interferon-binding protein that blocks interferon α transmembrane signaling. *J Biol Chem* 270: 15974–15978

102 Loparev VN, Parsons JM, Knight JC, Panus JF, Ray CA, Buller RML, Pickup DJ, Esposito JJ (1998) A third tumour necrosis factor receptor of orthopoxviruses. *Proc Natl Acad Sci USA* 95: 3786–3791

103 Alcami A, Symons JA, Collins PD, Williams TJ, Smith GL (1998) Blockade of chemokine activity by a soluble chemokine binding protein from vaccinia virus. *J Immunol* 160: 624–633

104 Barinaga M (1992) Viruses launch their own 'star wars'. *Science* 258: 1730–1731

105 Tavernier J, Tuypens T, Plaetinck G, Verhee A, Fiers W, Devos R (1992) Molecular basis of the membrane-anchored and two soluble isoforms of the human interleukin-5 receptor a subunit. *Proc Natl Acad Sci USA* 89: 7041

106 Goodwin RG, Friend D, Ziegler SF, Jerzy R, Falk BA, Gimpel S, Cosman D, Dower SK, March CJ, Namen AE et al (1990) Cloning of the human and murine interleukin-7 receptors: demonstration of a soluble form and homology to a new receptor superfamily. *Cell* 60: 941

107 Layton MJ, Cross BA, Metcalf D, Ward LD, Simpson RJ, Nicola NA (1992) A major binding protein for leukaemia inhibitory factor in normal mouse serum: identification as a soluble form of the cellular receptor. *Proc Natl Acad Sci USA* 89: 8616

108 Fukunaga R, Seto Y, Mizushima S, Nagata S (1990) Three different mRNAs encoding human granulocyte colony-stimulating factor receptor. *Proc Natl Acad Sci USA* 87: 8702

109 Raines MA, Liu L, Quan SG, Joe V, Di Persio JF, Golde DW (1991) Identification and molecular cloning of a soluble human granulocyte-macrophage colony-stimulating factor receptor. *Proc Natl Acad Sci USA* 88: 8203

110 Uze G, Lutfalla G, Mogensen KE :(1995) Alpha and beta interferons and their receptor and their friends and relations. *J Interferon Cytokine Res* 15(1): 3–26

111 Davis S, Aldrich TH, Ip NY, Stahl N, Scherer S, Farruggella T, Di Stefano PS, Curtis R, Panayotatos N, Gascan H et al (1993) Released form of CNTF receptor α component as a soluble mediator of CNTF responses. *Science* 251: 1736

112 Wypych J, Bennett LG, Schwartz MG, Clogston CL, Lu HS, Broudy VC, Bartley TD, Parker VP, Langley KE (1995) Soluble kit receptor in human serum. *Blood* 85: 66–73

113 Turner AM, Bennett LG, Lin NL, Wypych J, Bartley TD, Hunt RW, Atkins HL, Langley KE, Parker V, Martin F et al (1995) Identification and characterization of a soluble c-kit receptor produced by human hematopoietic cell lines. *Blood* 85: 2052–2058

114 Mauri D, Ebner R, Montgomery R, Kochel K, Cheung T, Yu G, Ruben S, Murphy M, Eisenberg R, Cohen G (1998) LIGHT, a new member of the TNF superfamily, and lymphotoxin alpha are ligands for herpesvirus entry mediator. *Immunity* 8: 21–30

115 Yasuda H, Shima N, Nakagawa N, Yamaguchi K, Kinosaki M, Mochizuki S-I, Tomoyasu A, Yano K, Goto M, Murakami A et al (1998) Osteoclast differentiation factor is a ligand for osteoprotegerin/osteoclastogenesis-inhibitory factor and is identical to TRANCE/ RANKL. *Proc Natl Acad Sci USA* 95: 3597–3602

116 Lacey D, Tims E, Tan H-L, Kelley M, Dunstan C, Burgess T, Elliott R, Colombero A,

Elliott G et al (1998) Osteoprotegrin ligand is a cytokine that regulates osteoclast differentiation and activation. *Cell* 93: 165–167

117 Wong B, Josien R, Lee S, Sauter B, Li H-L, Steinman R, Choi Y (1997) TRANCE (Tumour necrosis factor [TNF]-related activation-induced cytokine), a new TNF family member predominantly expressed in T cells, is a dendritic cell-specific survival factor. *J Exp Med* 186: 2075–2080

118 Pan G, Ni J, Yu G-L, Wei Y-F, Dixit V(1998) TRUNDD, a new member of the TRAIL receptor family that antagonizes TRAIL signalling. *FEBS Lett* 424: 41–45

119 Langstein J, Michel J, Fritsche J, Kreutz M, Andreesen R, Schwarz H (1998) CD137 (ILA/4-1BB), a member of the TNF receptor family, induces monocyte activation *via* bidirectional signaling. *J Immunol* 160: 2488–2494

120 Bazzoni F, Beutler B (1996) The tumour necrosis factor ligand and receptor families. *N Engl J Med* 334: 1717–1725

121 Anderson D, Maraskovsky E, Billingsley W, Dougall W, Tometsko M, Roux E, Teepe M, DuBose R, Cosman D, Galibert L (1997) A homologue of the TNF receptor and its ligand enhance T-cell growth and dendritic-cell function. *Nature* 390: 175–179

122 Tan K, Harrop J, Reddy M, Young P, Terrett J, Emery J, Moore G, Truneh A (1997) Characterization of a novel TNF-like ligand and recently described TNF ligand and TNF receptor superfamily genes and their constitutive and inducible expression in hematopoietic and non-hematopoietic cells. *Gene* 204: 35–46

123 Gooding LR (1992) Virus proteins that counteract host immune defences. *Cell* 71: 5

124 Murphy P (1993) Molecular mimicry and the generation of host defense protein diversity. *Cell* 72: 823–826

125 Ghivizzani S, Lechman E, Kang R, Tia C, Kolls J, Evans C, Robbins P (1998) Direct adenovirus-mediated gene transfer of interleukin-1 and tumour necrosis factor a soluble receptors to rabbit knees with experimental arthritis has local and distal anti-arthritic effects. *Proc Natl Acad Sci USA* 95: 4613–4618

126 Hack CE, Aarden L, Thijs L (1997) Role of cytokines in sepsis. *Adv Immunol* 66: 101–195

127 Mach F, Schonbeck U, Sukhova G, Atkinson E, Libby P (1998) Reduction of atherosclerosis in mice by inhibition of CD40 signalling. *Nature* 394: 200–203

128 Takebe N, Paredes J, Pino M, Lownsbury W, Agosti J, Krown S (1998) PhaseI/II trial of the type I solunble recombinant human interleukin-1 receptor in HIV-1-infected patients. *J Interferon Cytokine Res* 18: 321–326

Interleukin-1 receptor antagonist

Michael F. Smith Jr.

Division of Gastroenterology and Hepatology, University of Virginia School of Medicine, Charlottesville, VA 22908, USA

Introduction

Interleukin-1 (IL-1) is a proinflammatory cytokine produced by a wide variety of cell types with an equally pleiotropic range of activities including the induction of fever, hypotension, adhesion molecule expression, neutrophilia, cartilage destruction, prostanoid production, and induction of the expression of a number of other cytokines. Two forms of IL-1 (IL-1α and IL-1β) have been cloned and, although sharing only approximately 25% amino acid homology, these two molecules have essentially identical biological activities. The biological role for IL-1 in normal physiology and in the development of disease has been extensively reviewed elsewhere [1]. Support for a role of IL-1 as an important mediator of pathological events observed in a number of acute and chronic inflammatory diseases has been demonstrated in a large number of animal models. A common characteristic of many of these acute and chronic inflammatory diseases is that the production even of small amounts of IL-1 can have severe consequences in terms of tissue destruction and systemic homeostasis. The discovery of a naturally occurring IL-1 receptor antagonist (IL-1Ra) has suggested a means through which the pathological effects of IL-1 can be modulated.

Discovery of an IL-1 receptor antagonist

IL-1Ra was first identified in the mid-1980s by two independent laboratories. William Arend's laboratory described a 22-kDa IL-1 inhibitory activity in the supernatants from human peripheral blood monocytes that were cultured on a substrate of adherent IgG [2]. At about the same time, a group led by Dayer isolated a 25–35-kDa IL-1 inhibitory activity from the urine of three patients with myelomonocytic leukaemia [3]. This same group later went on to partially purify the urine-derived material and determined that it had an apparent molecular weight of 18–22 kDa, was antigenically distinct from IL-1, and could block the binding of radiolabeled

Novel Cytokine Inhibitors, edited by Gerry A. Higgs and Brian Henderson
© 2000 Birkhäuser Verlag Basel/Switzerland

IL-1 to murine EL-4-6.1 thymoma cells [4]. In 1989, Arend's group partially purified the monocyte-derived material and demonstrated that it could block IL-1β-induced collagenase production by rabbit articular chondrocytes, and PGE2 production by human foreskin fibroblasts or synovial cells. Similarly to the Dayer group, Arend et al. demonstrated that this IL-1 inhibitor could also block the binding of IL-1 to its cell surface receptor [5]. Subsequently, the IL-1 inhibitor was purified to homogeneity from monocyte supernatants and characterized as an 18-kDa protein and a 22-kDa glycoprotein [6]. The cDNA was then cloned from a human monocyte cDNA library [7]. As predicted, the recombinant protein was able to block the binding of radiolabeled IL-1α and Il-1β to EL4-6.1 cells. This protein was thus termed IL-1 receptor antagonist (IL-1Ra). By both protein and gene structures, IL-1Ra was determined to be the third member of the IL-1 gene family [8]. A second group cloned the identical protein from the human myelomonocytic cell line U937 and termed it IL-1 receptor antagonist protein (IRAP) [9].

The term IL-1Ra actually refers to two closely related proteins derived from the alternative splicing of two different first exons (outlined in Fig. 1). The originally described molecule, secretory or sIL-1Ra, encodes a protein of 177 amino acids including a 25 amino acid hydrophobic leader sequence which is subsequently cleaved resulting in a secreted 152 amino acid mature protein [7]. An alternative form of IL-1Ra, intracellular or icIL-1Ra, was cloned from an adherent monocyte cDNA library [10]. This structural variant was created when an alternative first exon was spliced into an internal acceptor site within the first exon of sIL-1Ra in the region encoding for the secretory leader sequence. Thus icIL-1Ra is identical to the mature sIL-1Ra except for seven additional amino acids at the amino terminal end and icIL-1Ra lacks the hydrophobic leader sequence required for secretion. Both forms of IL-1Ra are equally capable of inhibiting the binding IL-1 to cell surface receptors. To date, a clear understanding of a role for an intracellular IL-1 receptor antagonist in the regulation of the IL-1 system is lacking. There has been one report in the literature which indicated that expression of the icIL-1Ra molecule in ovarian epithelial cells resulted in a decrease in mRNA stability for three IL-1-inducible genes: GRO, IL-8 and A20.MAD-6 [11]. Interestingly, IL-1 signaling was not affected since the IL-1-induced transcription of these three genes was not affected.

Mechanism of action

Two distinct cell surface IL-1 receptors have been identified and cloned. All three ligands (IL-1α, IL-1β, and IL-1Ra) can bind to both types I and II IL-1 receptors, albeit with different affinities. Of the two receptors, only the type-I IL-1R is capable of transducing a signal. When bound to one of the agonist molecules, the type-I receptor forms a complex with the IL-1R accessory protein (IL-1RacP) and trans-

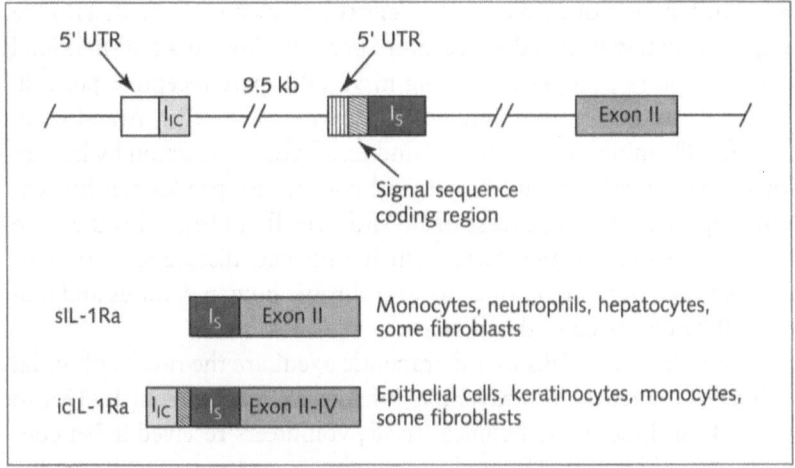

Figure 1
IL-1 receptor antagonist isoforms gene and protein structures.

mits a cytoplasmic signal [12]. IL-1Ra does not promote the recruitment of the IL-1RacP to the complex and thus a signal is not initiated. The type-II IL-1R has a very short cytoplasmic tail, does not associate with the IL-1RacP, and does not transduce a signal. This receptor thus acts as a "decoy" receptor and provides an additional means for the body to regulate the activity of the IL-1 system .

The relative binding of the three IL-1R ligands has been extensively studied. IL-1Ra binds to the type-I receptor with an affinity approximately equal to that of IL-1α and IL-1β [13] with a K_D of approximately 150 pM on EL4-6.1 cells and 213 pM on human synovial fibroblasts. IL-1Ra binds to the type-II receptor with an affinity that is approximately 100-fold lower than for IL-1R type I (15 nM on Raji cells and 8 nM on human neutrophils). These affinities are equal to those for IL-1α but approximately 15-fold lower than IL-1β. Thus IL-1Ra shows a preference for binding to the signaling receptor, which may have important implications for the biological role of IL-1Ra in regulating the activity of the IL-1 system.

One important caveat to bear in mind when assessing the potential role for IL-1Ra in regulating IL-1 activity is the exquisite sensitivity of animals and cells to the biological effects of IL-1. Both *in vitro* and *in vivo* studies have demonstrated that a 100–1000-fold molar excess of IL-1Ra over IL-1 is required for inhibition of IL-1 biological responses. This phenomenon is due primarily to two factors: the so-called "spare receptor" effect [14, 15] and the rapid clearance of IL-1Ra from the circulation [16]. Curtis et al. showed that Chinese hamster ovary (CHO) cells responded to IL-1 when only 7 receptors per cell were occupied and had a half maximal response with occupation of 22 receptors. Cells transfected with the IL-1R type I, and expressing approximately 100,000 receptors per cell, showed a half-maximal response with 500 receptors occupied, resulting in 99,000 spare receptors per cell. In some of the earliest studies on the biological properties of IL-1Ra, Arend et al. demonstrated that a 50% inhibition of the IL-1-induced PGE_2 production by human synovial cells or rabbit articular chondrocytes and collagenase production by synovial cells required up to a 100-fold excess of IL-1Ra over IL-1 [14]. Thus the ratio of IL-1 to IL-1Ra may be an important factor which influences disease severity. This concept has been explored in a variety of acute and chronic human diseases and animal models and will be discussed in detail below.

Complicating the utility of IL-1Ra as a therapeutic agent are the results of initial studies on the pharmacokinetics of intravenous infusion of recombinant IL-1Ra to healthy humans [16]. In these phase I clinical trials, volunteers received a 3-h continuous infusion of recombinant IL-1Ra at doses ranging from 1 mg/kg to 10 mg/kg. Following infusion, plasma IL-1Ra levels dropped rapidly, demonstrating an initial plasma half-life for IL-1Ra of 21 minutes with a maximum plasma concentration of 29 µg/ml in volunteers receiving the highest dose. The implications of this short *in vivo* half-life for the treatment of both acute and chronic inflammatory disorders will be discussed below.

The production of IL-1Ra and its role in the regulation of the inflammatory response have been a subject of intensive research in the ten years since its initial description. A large body of evidence has suggested that IL-1Ra may play a protective role in controlling the progression of inflammation. Indeed, in several animal models of acute and chronic inflammatory diseases, the administration of exogenous IL-1Ra has been demonstrated to decrease disease severity. Results of these animal studies have prompted a number of clinical trials to assess the therapeutic potential of IL-1Ra in the treatment of human diseases. This review will examine the data from animal models of disease for which IL-1 has been suggested to be a major contributory factor and discuss recent progress in the use of IL-1Ra as a therapeutic agent in the treatment of the corresponding human disease. Although a number of animal models have demonstrated that IL-1Ra can prove efficacious for treatment of inflammatory diseases, to date its use in the treatment of human diseases has only been reported for three diseases: sepsis, rheumatoid arthritis, and graft-versus-host disease.

Sepsis syndrome and endotoxic shock

IL-1 has been implicated to play a role in the pathogenesis of sepsis and endotoxic shock and is rapidly induced in response to injection of gram-negative bacteria or purified lipopolysaccharide (LPS). Many of the inflammatory and haemodynamic changes observed in animals or humans injected with purified gram-negative LPS (endotoxin) can be mimicked by injection of recombinant IL-1. Thus the use of IL-1Ra as a means of preventing or reversing the development of septic shock has received a large amount of attention. Indeed, as will be discussed below, the first clinical trials to be attempted with IL-1Ra were designed to examine its potential for the treatment of sepsis syndrome.

Studies in animal models

The ability of IL-1Ra to provide protection from endotoxic shock was first evaluated in a rabbit model [17]. Endotoxaemia was induced by injection of 0.5 mg/kg E. coli LPS. This dose of endotoxin killed 80% of the animals within 48 h. When IL-1Ra was injected just prior to LPS and every 2 h thereafter, up to 24 h, there was a dose-dependent decrease in mortality throughout the entire 7-day study period. Those rabbits receiving the highest dose (100 mg/kg) had a 90% survival rate after 7 days. When IL-1Ra treatment was initiated 1 or 2 h after endotoxin administration, 7 out of 8 animals survived the seven-day study period, suggesting that IL-1 toxicity can be reversed as late as 2 h following endotoxin injection. A similar study by Wakabayashi and colleagues [18] examined the haemodynamic and histological changes that occurred in rabbits following injection of live E. coli bacteria in the presence or absence of exogenously administered IL-1Ra. Rabbits received a 10 mg/kg bolus injection of IL-1Ra 15 minutes prior to E. coli infusion followed by a continuous IL-1Ra infusion for 4 h thereafter. The IL-1Ra-treated animals demonstrated a sustained mean arterial pressure and systemic vascular resistance compared to saline controls. Additionally, treated animals showed much lower levels of polymorphonuclear leukocyte (PMN) infiltration into the lungs than did controls.

Studies in IL-1Ra gene knockout and overexpressing mice provided additional confirmation of the roles of IL-1 and IL-1Ra in the response to bacterial infection [19]. Mice in which the endogenous IL-1Ra gene was disrupted displayed increased sensitivity (i.e., decreased survival) to the effects of LPS administration, while the IL-1Ra overexpressing mice were more resistant than their normal littermates to endotoxin administration. Conversely, the IL-1Ra knockout mice were less susceptible to Listeria monocytogenes infection than normal littermates. Interestingly, this effect was more pronounced in the female mice than in the males, suggesting that the IL-1 system may interact with the endocrine system. Likewise, the IL-1Ra overexpressing mice were more susceptible to Listeria infection than control mice. Two surprising findings in terms of the physiological roles of IL-1 and IL-1Ra were to come out of this study. First, IL-1Ra knockout mice had significantly lower body weights than the wild-type or hemizygous mice, suggesting that IL-1Ra may play a positive role in the regulation of weight gain, perhaps by inhibiting the IL-1-induced appetite suppression [20]. Second, in contrast to earlier studies which demonstrated that exogenous IL-1Ra administration could result in a decrease in LPS-induced IL-

1 accumulation in plasma, IL-1β levels in the IL-1Ra knockout mice were significantly lower than their normal littermates. The implication from this result is that IL-1Ra is an important participant in a regulatory feedback loop involving the IL-1 system and that levels of IL-1 in these mice may be reduced in an effort to maintain a normal homeostatic environment in the animals. Alternatively, IL-1Ra may play an important role in the development of cellular systems which are involved in the production of IL-1α and IL-1β.

As a prelude to using IL-1Ra as therapy for sepsis syndrome in humans, two preliminary studies in baboons were undertaken. The first study compared haemodynamic, metabolic, and hormonal effects of sublethal endotoxin administration to the effects of IL-1α administration [21]. This study demonstrated that many, but not all, of the effects induced by endotoxin could be mimicked by IL-1. Most notably, tumor necrosis factor α (TNFα) and IL-6 production were induced by endotoxin but not by IL-1α. Surprisingly, IL-1 production in response to endotoxin was not measured. Earlier studies had indicated that TNFα could induce both IL-1 and IL-6 production *in vivo*, thus potentially placing TNF earlier in the cytokine cascade than IL-1. These results would therefore suggest that perhaps TNF is the more critical cytokine to control in treatment of sepsis.

In the follow-up study [22], IL-1Ra was tested for its ability to control the host response to injection of either purified LPS or live *E. coli* bacteria. When animals were challenged with a sublethal dose of endotoxin (500 μg/kg), IL-1Ra did not significantly alter the haemodynamic or cytokine responses. Surprisingly, and in contrast to experimental endotoxaemia in humans [23], plasma IL-1β was not increased. However, IL-1Ra was protective when animals were treated with a lethal dose of live *E. coli*. In this model of septic shock, IL-1β was detected in the plasma of infected animals. IL-1Ra reduced IL-1β and IL-6 production and attenuated the decrease in haemodynamic parameters. 24 h following *E. coli* infection, all seven of the IL-1Ra-treated animals survived, while only 3 out of seven control animals survived. At the end of the 24 h period, in six out of the seven IL-1Ra treated animals, indices of renal, hepatic, and pulmonary function were all within normal limits, whereas only two of the control animals had normal indices. Despite some minor differences between the human and baboon endotoxaemia models, these were very encouraging results and suggested that IL-1Ra may be clinically useful in the treatment of human disease.

Studies in human disease

Phase I trials to assess the safety of IL-1Ra administration were performed on 25 healthy, male volunteers [16]. Subjects were infused with 1–10 mg/kg IL-1Ra over the course of 3 h. No adverse effects of IL-1Ra treatment were observed, indicating that administration of IL-1Ra was safe and well tolerated. Importantly, the plasma levels of IL-1Ra were observed to drop quickly at the termination of infusion with a very short plasma half-life of only 21 min. This very short plasma half-life of IL-1Ra may have profound effects on its utility as a therapeutic agent, especially in terms of treatment of chronic inflammatory diseases.

The first clinical trial of IL-1Ra was performed to assess its efficacy in the treatment of sepsis syndrome. The initial open-label, placebo-controlled, phase II trial was performed on 99 patients with sepsis syndrome or septic shock [24]. In this preliminary trial to evaluate safety and efficacy, patients received a continuous 72-h infusion of 17, 67, or 133 mg/h IL-1Ra. When evaluated 28 days following treatment initiation, there was a dose-dependent decrease in mortality in the IL-1Ra-treated subjects. Mortality rates decreased from 44% for the placebo group to 16% for the group receiving 133 mg/h. Additionally, APAHCE II scores showed a dose-dependent reduction by the end of the treatment period.

A second larger, double-blind, phase III trial was performed to further study the efficacy of IL-1Ra in the treatment of sepsis syndrome [25]. This study, involving 893 patients who received a 72-h continuous infusion of IL-1Ra at 1 or 2 mg/kg per hour, did not demonstrate a statistically significant decrease in 28-day mortality for the IL-1Ra-treated group. Secondary analysis, however, suggested that IL-1Ra could increase survival time in a subgroup of the more seriously ill patients (those with a predicated risk of mortality of ≥ 24%). A second confirmatory phase III trial of 696 patients was terminated after an interim analysis indicated that no statistically significant effect of IL-1Ra treatment would be observed [26]. The 28-day mortality rate in this study did not differ with respect to any parameters examined, including site of infection, organism, organ dysfunction, or predicted risk at the time of entry into the study.

The disparate results between the efficacy of IL-1Ra in human clinical trials and animal models of sepsis and endotoxic shock serve to underscore the complexity of the human disease and patient population. The disappointing results of the IL-1Ra trials and of trials utilizing anti-TNF monoclonal antibodies to treat sepsis demonstrate that this is really a multifactorial syndrome that is not likely to be responsive to treatment with a single "silver bullet". Perhaps future studies using combination treatments (e.g., IL-1Ra plus anti-TNF) may prove to be more effective in controlling the devastating inflammatory response observed in the septic patient.

Rheumatoid arthritis

Rheumatoid arthritis (RA) is an inflammatory disease of the joints characterized by a chronic synovitis leading to the destruction of joint tissue. The major contributory agents involved in the development of RA are thought to be cytokines released from activated macrophages and synoviocytes within the joint. Of the variety of cytokines which have been demonstrated to be present within the rheumatoid synovial fluid, a considerable amount of evidence points to IL-1 as one of the most important mediators of the pathophysiology observed in RA (reviewed in [27]). By virtue of its ability to act on a number of cell types within the joint, IL-1 can induce prostaglandin production, collagenase production by chondrocytes, bone resorption by osteoclasts, and the production of a number of proteolytic enzymes active in cartilage destruction. In fact, a recent study demonstrated that essentially all of the hallmark traits of RA could be reproduced by intra-articular expression of IL-1 following *ex vivo* transduction of synoviocytes with an IL-1-expressing retrovirus [28]. Additionally, in the collagen-induced arthritis model in mice, administration of neutralizing antibodies to IL-1 was capable of both protecting against disease development and ameliorating the progression of established arthritis [29, 30].

Studies in animal models

Several studies have demonstrated the efficacy of exogenous administration of IL-1Ra in controlling both the development and progression of disease in animal models of RA. In both LPS-induced arthritis in rabbits and LPS-induced arthritis in rats, IL-1Ra was effective in suppressing disease when given concurrently or within a few hours following disease induction [31, 32]. Likewise, neutralization of endogenous IL-1Ra by coinjection of antibodies to rabbit IL-1Ra along with LPS into rabbit knees, resulted in an exacerbation of disease activity [33]. More impressively, in collagen-induced arthritis in mice, administration of high doses of IL-1Ra by continuous infusion using osmotic pumps led to a drastic reduction in disease severity even after the establishment of full-blown arthritis [34].

Studies in human disease

In human disease, the relative ratio of IL-1Ra to IL-1 may play a critical role in the progression of joint destruction. High levels of IL-1Ra have been detected in the synovial fluid of 80% of the patients studied with active RA [35]. In Lyme arthritis, patients with high levels of IL-1Ra and low levels of IL-1β had more rapid resolution of disease than those patients with the opposite pattern of cytokine expression, suggesting that a high ratio of IL-1Ra to IL-1 is prognostic of a rapid recovery from Lyme arthritis [36]. An examination of IL-1 and IL-1Ra expression in the cells of the rheumatoid synovium suggested that the amount of IL-1Ra produced by synovial tissue cells is insufficient to overcome the amount of IL-1 produced [37]. In unfractionated fresh or cultured rheumatoid synovial tissue cells, the ratios of IL-1Ra to IL-1 ranged from 1.2 to 3.6, far below the 100–1000-fold levels suggested from earlier studies to be required for inhibition of IL-1 biological activity. These studies went on to suggest that defective production of IL-1Ra by rheumatoid synovial macrophages may be a contributing factor to the progression of the inflammatory synovitis.

Based upon the results from animal studies which demonstrated that exogenous administration of IL-1Ra can be therapeutically useful, clinical trials in humans have been initiated. In the initial clinical trial to evaluate safety and dosing regimens, recombinant human IL-1Ra was administered to 175 patients with active RA [38]. Patients received 20, 70, or 200 mg IL-1Ra, once, three times or seven times per week for an initial treatment period of three weeks. At the end of the three-week initial dosing period, treatment was continued for 4 weeks on a one dose per week maintenance phase. Administration of even the highest dose of IL-1Ra was well tolerated. Although not placebo-controlled, preliminary efficacy results suggested that daily administration of the highest dose was effective in reducing both investigator and patient assessments of disease severity. This initial study provided the impetus for the design of a more extensive clinical trial.

Results of a second, more definitive clinical study on the treatment of RA have now been reported in abstract form [39, 40]. In this study, 472 patients with active RA were enrolled in a double-blind, placebo-controlled trial in Europe. Patients were treated with subcutaneous 30, 75, or 150 mg rHuIL-1Ra daily over the course of 24 weeks. A significant clinical improvement as determined by number of swollen and tender joints, pain, and radiological assessment, was observed in patients treated with the 150 mg/day dosage. In the follow-up study, the patients who received IL-1Ra in the study above, continued to receive IL-1Ra treatment for another 24 weeks [41]. The patients maintained the improvement observed during the initial 24-week period and demonstrated that long-term treatment with IL-1Ra was well tolerated. Likewise, a smaller study demonstrated that treatment with 150 mg/day of IL-1Ra was associated with a reduction in synovial T-cell infiltration [42].

The studies described above used IL-1Ra as the sole therapeutic agent. A concept that is just beginning to be explored is the use of IL-1Ra as an adjunct in combination therapy. One study in the rat adjuvant arthritis model has indicated that combination therapy with IL-1Ra and methotrexate results in an additive effect [43]. Treatment with IL-1Ra alone resulted in a 6% decrease in paw swelling and a 53% decrease in bone resorption. Methotrexate alone resulted in 47% and 58% inhibition of paw swelling and bone resorption, respectively. However, combination therapy resulted in a dramatic 84% decrease in paw swelling and a 97% inhibition of bone resorption. Since IL-1Ra has already proven to be well tolerated by RA patients, there is little reason not to begin trials to assess its therapeutic effect in combination with other well-described arthritis treatment regimens.

One of the barriers to the effective use of IL-1Ra as a therapeutic agent is the requirement for the maintenance of sustained high levels of the protein in the circulation and tissues. The use of daily, subcutaneous injections as a means of drug delivery results in a less than ideal situation, both in terms of patient comfort and pharmacokinetics. To circumvent this limitation, researchers are taking two very different approaches. The first approach is attempting to create a slow-release formulation of IL-1Ra [44]. IL-1Ra was incorporated into a viscous solution of hyaluronic acid and injected into a rat model of collagen-induced arthritis which was unresponsive to a daily administration of 100 mg/kg IL-1Ra in aqueous formulation. In contrast, rats injected daily with 100 mg/kg hyaluronic acid-formulated IL-1Ra had a 62% inhibition of paw swelling. Every-other-day dosing resulted in a 51% inhibition and every-third-day dosing in a 27% inhibition. These results suggest that an alternative formulation of IL-1Ra in which the drug is delivered in a more controlled fashion may prove efficacious in the treatment of human disease.

One of the difficulties with systemic delivery of a protein to treat RA is that, in general, proteins are not readily diffusible into the joint. Although direct intra-articular injection of IL-1Ra has proven to be efficacious in the treatment of animal models, this method of drug delivery is not especially useful as a means of long-term treatment. The second approach to a more sustained high-level delivery of IL-1Ra is through the use of gene therapy. The general approach used in these studies is to specifically target an IL-1Ra expression vector to synoviocytes as a means of generating high local levels of IL-1Ra protein within the joint itself. This approach circumvents the problems associated with the need for repeated daily injections of extremely large amounts of recombinant protein and delivers the drug directly to the site of action.

Several groups have reported on the use of direct delivery of IL-1Ra to the synovium as a means of protection from experimentally induced arthritis in three different animal systems. In the first reported use of IL-1Ra in gene therapy by Bandara and colleagues, rabbit synoviocytes were transduced in culture with a retrovirus encoding the human IL-1Ra protein [45]. The IL-1Ra-expressing cells were then transplanted by intraarticular injection into the knee joints of rabbits. Using this method, they were able to maintain IL-1Ra expression within the joint for approximately 5 weeks. In addition, the rabbits were protected from the leukocytosis observed following direct intra-articular injection of IL-1. Makarov et al. utilized a similar approach to deliver IL-1Ra to the ankle joints of rats in a bacterial cell wall-induced arthritis model [46]. In this model, direct delivery of IL-1Ra to the synovium was approximately four orders of magnitude more effective than systemic injection in suppressing the severity of disease. In a mouse model of collagen-induced arthritis, Bakker et al. demonstrated an almost complete inhibition of disease induction in joints that were injected with IL-1Ra-transduced NIH 3T3 cells [47]. Thus the results of gene therapy experiments in animals have demonstrated that intra-articular expression of high levels of IL-1Ra can be of therapeutic value in the treatment of rheumatoid arthritis.

The encouraging results of animal studies have led to the initiation of human clinical trials to assess the safety and efficacy of transducing human synoviocytes with an IL-1Ra-expressing retrovirus. Preliminary trials are in progress on patients about to undergo total joint replacement of metacarpal joints [48]. Prior to joint replacement surgery, autologous synovial tissue was obtained, transduced in culture with an IL-1Ra-expressing retrovirus, and injected into 2 of the 4 joints to be replaced. The other two joints were injected with untransduced cells. One week later, patients underwent total joint replacement and the retrieved tissues were analyzed for IL-1Ra expression. In the first three patients treated, all joints transduced with IL-1Ra showed evidence of IL-1Ra gene expression, thus demonstrating the feasibility of this approach as a means of delivery of a protein directly to the joint tissue [49]. More expanded clinical trials are planned to determine the therapeutic value of this delivery system in the treatment of human rheumatoid arthritis.

Graft-versus-host disease

Studies in animal models

Acute graft-versus-host (GVHD) disease is a devastating inflammatory response following bone marrow transplantation that is mediated in part by activated T cells. A role for IL-1 in the pathogenesis of GVHD was demonstrated by McCarthy et al. using a murine bone marrow allograft model [50]. Administration of IL-1Ra twice daily for 10 days, beginning on the day of transplantation, resulted in a significant increase in survival when followed up to 60 days post-transplant (70% for IL-1Ra treated vs 0% for saline controls). Most importantly, there were no detrimental effects on haematopoetic reconstitution. The follow-up study by Abhyankar et al. [51] examined cytokine expression in the bone marrow allograft model. In contrast to IL-2 production in the spleen, which was evident during the first week following transplant, levels of IL-1, especially in the skin, rose more slowly and did not reach peak levels until several weeks later. This observation suggested that IL-1Ra might be beneficial if treatment was begun after disease initiation. When IL-1Ra treatment was delayed until 10 days following bone marrow engraftment, at a time when clinical GVHD was evident, a beneficial effect was still observed. A 10-day course of treatment resulted in an increase in survival from 30% in saline controls to 80% in IL-1Ra treated animals. These two studies thus provided convincing evidence that IL-1 is an effector molecule in the development of acute GVHD and that exogenous administration of IL-1Ra may prove to be efficacious in the treatment of human disease.

Studies in human disease

A phase I/II clinical trial of IL-1Ra in human bone marrow transplant patients with steroid resistant GVHD has been completed [52]. IL-1Ra was administered as a continuous infusion for 7 days at doses ranging from 400 to 3200 mg/day. Overall improvement in the severity of disease by at least one grade was observed in 10 of 16 patients (63%). The most extensive improvement was observed in the clinical grades of GVHD in skin (53% responders) and gut (82% responders). Although difficult to assess because of the advanced stage of disease when IL-1Ra therapy was initiated, no detrimental haematopoetic effects were observed. One of the most encouraging results from this study was the effect of IL-1Ra treatment on IL-1β and TNFα levels. When assessed with regard to clinical response to IL-1Ra treatment, it was observed that responders had significantly lower levels of both IL-1β and TNFα mRNAs at the end of the 7-day treatment period when compared to pre-treatment levels.

This small study indicated that IL-1Ra may be highly efficacious in the treatment of acute GVHD. Clearly, if bone marrow transplantation is to be a viable treatment for a number of leukaemic malignancies, the management of acute GVHD needs to be improved. The use of IL-1Ra either to prevent GVHD, as the first mouse model demonstrated, or to decrease the severity of disease, as in the human trial, needs to be tested further. Extrapolating from the mouse model, it is evident that earlier and more aggressive treatment, in terms of both dosage and length of treatment, with IL-1Ra needs to be explored.

Inflammatory bowel disease

Crohn's disease (CD) and ulcerative colitis (UC) represent the two major forms of chronic inflammatory bowel disease (IBD). Although clinically distinct, these two diseases share the common trait of chronic, relapsing, nonspecific, destructive inflammation of the colon. Crohn's disease commonly involves inflammation of the entire alimentary tract while UC is primarily confined to the mucosa of the rectum and left colon. The aetiological causes of UC and CD are unknown, although a variety of genetic and environmental factors have been implicated, suggesting that these syndromes are multifactorial in origin.

Studies in animal models

That IL-1 may play a role in the pathogenesis of IBD was first demonstrated in a rabbit immune complex colitis model by Cominelli and colleagues in 1990 [53]. In this model, a 0.45% formalin solution is delivered to the distal colon, followed 2 h later by an intravenous injection of immune complexes. Within 4 h following administration of immune complexes, an acute inflammation develops in the colon. Concomitant with the development of inflammation, both IL-1α and IL-1β production were demonstrated in colonic tissue. Administration of recombinant human

IL-1Ra (rhIL-1Ra) 2 h before and every 8 h following immune complex administration resulted in significant reduction in the inflammatory index, oedema, and
necrosis observed in the colon. A follow-up study assessed dosage and time of
administration of IL-1Ra in the same rabbit colitis model [54]. When administered
10 min prior to and 3, 11, 19, 27, 35, and 43 h following immune complex injection, rhIL-1Ra demonstrated a dose-dependent inhibition of essentially every index
of colonic inflammation measured. The IL-1Ra treatment was beneficial when first
administered up to 3 h after the induction of colitis; however, a delay in 12 h resulted in no significant decrease in colonic inflammation. Additionally, when rabbits
were given IL-1Ra only 10 min before and one dose 3 h after colitis induction, no
protection was observed, indicating that protection from the deleterious effects of
IL-1 was required during the later stages of the disease. The same group went on to
demonstrate that neutralization of endogenous IL-1Ra by pretreatment of the rabbits with anti-rabbit IL-1Ra antibodies resulted in an exacerbation of disease activity and increased mortality [55]. This study provided the first convincing evidence
for a role of endogenously produced IL-1Ra in controlling the inflammatory effects
of IL-1 in an animal model of disease.

Additional evidence that endogenous IL-1Ra may play a role in the development
and progression of animal models of colitis comes from a preliminary study with IL-
1Ra knockout and overexpressing mice [56]. In this model, colitis is induced by 5%
dextran suphate (DSS) in the drinking water for 7 days. Although only small numbers of mice were reported, the IL-1Ra knockout mice appeared more susceptible to
colitis with all 6 dying by day 9, whereas the majority of normal mice did not die
until day 14, with one recovering completely. Additional and more refined studies
are required to confirm the role of endogenous IL-1Ra in this disease model but the
preliminary data are encouraging.

Studies in human disease

That IL-1Ra may be an important anti-inflammatory cytokine in human IBD has
been suggested in a number of studies. Hyams and colleagues examined IL-1 and IL-
1Ra levels in colonic tissue biopsies from paediatric patients with UC or CD [57].
They observed that mucosal IL-1β levels were approximately four-fold higher in
moderately and severely inflamed specimens from CD and UC patients compared
with none or mildly inflamed tissue samples. IL-1β levels were also elevated in the
none/mildly inflamed tissue from CD and UC patients when compared to non-
inflamed control subjects. Most interestingly, and in contrast to the animal studies
described above [55], the levels of IL-1Ra were not significantly different between
the IBD groups and only modestly different between controls and severely inflamed
UC patients. This differential expression of the two cytokines resulted in a significantly decreased IL-1Ra to IL-1β ratio in all of the disease groups. Very similar

results were reported by Casini-Raggi et al. [58]. In their study of adult UC and CD patients, they also observed a decreased IL-1Ra to IL-1 ratio in freshly isolated intestinal mucosal cells that correlated with the severity of disease. Likewise, no correlation with disease severity and levels of IL-1Ra expression were observed. Taken together, these studies imply that a lack of response by the IL-1Ra gene to the inflammatory mediators responsible for the initiation and development of IBD in humans may be a critical element in disease progression. Further support for a role of IL-1Ra in the development of human disease has been suggested by an increasing number of studies indicating an association between ulcerative colitis and the carriage of polymorphic allele within the second intron of the IL-1Ra gene [59-63]. However, it must be pointed out that no clear consensus has developed as to the role of this allele in affecting the expression of IL-1Ra and it remains to be determined if this allele is simply in linkage with another gene which may be important in the development of UC.

Given the very encouraging data from both the animal and human studies described above, IBD would seem to be a good therapeutic target for IL-1Ra. However, as with other chronic inflammatory diseases, the difficulty in treating this disease with a natural product remains the inability to deliver the agent directly to the colon and the necessity of repeated, basically chronic, delivery of IL-1Ra.

Chronic and acute myelogenous leukaemia

The autonomous proliferation of malignant myeloid cells may, at least in part, be governed by endogenously produced growth factors that act as part of a paracrine loop. In acute (AML) and chronic myelogenous leukaemia (CML), IL-1 has been implicated as a spontaneously produced factor which may play a critical role in maintaining the proliferation of AML and CML cells *in vitro*. Rambaldi and colleagues demonstrated that AML cells from 20 of 23 patients examined constitutively expressed IL-1β mRNA while only 2 patients expressed IL-1Ra mRNA [64]. Inclusion of IL-1Ra in the cultures resulted in a decrease in spontaneous proliferation and reduced levels of granulocyte macrophage colony-stimulating factor (GM-CSF) and IL-1β levels in the culture media. Similar results for both AML [65] and CML [66] were reported by Estrov et al. In the CML study, it was observed that IL-1β could augment the proliferation of CML cells derived from patients with early stage, interferon α (IFNα) sensitive disease, but did not augment growth in cells from IFN-resistant disease. Strikingly, IL-1β expression was increased in cells from patients with IFN-resistant disease. Not unexpectedly, IL-1Ra was more effective in inhibiting the growth of cells derived from late-stage, IFN-resistant disease than it was on the less developed IFN-sensitive cells. Essentially the same results have been obtained in several other studies [67–69]. These results therefore suggested that cells from the IFN-resistant disease respond to endogenously produced IL-1β as a paracrine growth factor.

As with other diseases for which IL-1 has been implicated to be a critical effector molecule, it may be that the ratio of IL-1 to IL-1Ra is an important prognostic factor in the development and progression of disease. This hypothesis was in fact borne out in a study that demonstrated elevated IL-1β levels in leukocytes from patients with CML [70]. Moreover, IL-1β levels were significantly higher in patients in accelerated/blastic crisis than those in the chronic phase of disease. Levels of IL-1Ra in cells from patients in the chronic phase of disease were not significantly different than those observed in healthy volunteers. However, levels of IL-1Ra in patients in accelerated/blastic crisis were significantly lower. Thus it appears that IL-1 may act as an *in vitro* paracrine growth factor for myelogenous leukaemic cells. The effect of IL-1 may in part be through its ability to induce the expression of GM-CSF in these cultures.

Although these *in vitro* data provide hope for a very exciting new treatment regimen for AML and CML, to date there have been no reports of the use of IL-1Ra to treat clinical disease. Unfortunately, these diseases are not particularly amenable to testing in animal models and thus we may have to wait for the results of trials in humans to see if IL-1Ra therapy is efficacious in humans. Since both the sepsis and arthritis trials have shown that long-term treatment with IL-1Ra is well tolerated, there should be very few roadblocks to testing this compound in human disease.

Other diseases

In the past three years, a number of other studies have established a role for IL-1 in the pathogenesis of several other inflammatory diseases. This has raised the prospect for the use of IL-1Ra in a wide variety of clinical situations. Studies in animal models of glomerulonephritis [71–73], pancreatitis [74, 75], and corneal transplants [76] have all demonstrated a beneficial effect of exogenous administration of IL-1Ra. Whether or not exogenous administration of IL-1Ra will be effective in the treatment of the corresponding human disease remains to be determined.

A small IL-1 receptor antagonist peptide

A small 15-mer polypeptide has been identified from screening of a phage display library that specifically blocks the binding of IL-1 to the human type I IL-1R [77, 78]. This peptide (AF12198) blocked IL-1-induced IL-6 production by human fibroblasts *in vitro* and IL-1-induced IL-6 production *in vivo* in monkeys. Although an intriguing finding, this peptide acts at concentrations 1000-fold higher than IL-1Ra. Furthermore, the peptide is poorly soluble in water and is rapidly metabolized: declining from 5 nmol/ml at the end of the infusion period to approximately

Table 1 - Effective uses of IL-1Ra

Disease	Animal model	Human trials
Sepsis syndrome	+	−
Rheumatoid arthritis	+	+
Inflammatory bowel disease	+	nt
Graft-versus-host disease	+	+/−
Acute/chronic myelogenous leukaemia	nt	*
Glomerulonephritis	+	nt
Pancreatitis	+	nt
Corneal transplantation	+	nt
Premature labor	+	nt
Diabetes mellitus	+	nt

+, efficacious; +/−, potentially efficacious; −, non-effective; nt, not tested; *, effective in in vitro studies on human tissue

0.1 nmol/ml 30 min later. Thus the therapeutic value of AF12198 is very limited. However, it does demonstrate the possibility of developing a small molecule antagonist that might be clinically useful if issues of metabolism and pharmacokinetics can be addressed.

Conclusions

The studies described above have demonstrated that IL-1Ra shows some potential as a means to control the pathophysiology of a number of acute and chronic inflammatory diseases (summarized in Tab. 1). Promising results from clinical trials for RA and GVHD provide hope for a new therapeutic approach to the treatment of these diseases. Clearly, clinical studies are warranted to explore the therapeutic potential of IL-1Ra in the treatment of IBD and myelogenous leukaemia.

These studies have also served to underscore the difficulties inherent in extrapolating from animal models to human disease. In particular, the very disappointing results of the sepsis trials demonstrate that laboratory models in which a homogenous population of animals is given a very specific disease inducer do not always mimic the human situation in which the population and disease-inducing agent are both heterogeneous. The diverse clinical presentations of most diseases suggest that a variety of host factors probably play a significant role in the development, progression and, quite possibly, underlying aetiology of any given disease. Thus in future clinical trials with any biological response modifier, extra care needs to be taken in evaluating and interpreting clinical outcomes. Due to the diverse nature of human disease, such therapeutics as IL-1Ra may prove to be effective in only a sub-population of patients. Identifying these potential responders prior to administration of the drug will be a critical part of future therapies.

References

1   Dinarello CA (1996): Biologic basis for interleukin-1 in disease. *Blood* 87: 2095–2147
2   Arend WP, Joslin FG, Massoni RJ (1985) Effects of immune complexes on production by human monocytes of interleukin 1 or an interleukin 1 inhibitor. *J Immunol* 134: 3868–3875
3   Balavoine J-F, de Rochemonteix B, Williamson K, Seckinger P, Cruchaud A, Dayer J-M (1986) Prostaglandin E_2 and collagenase production by fibroblasts and synovial cells is regulated by urine-derived human interleukin 1 and inhibitors. *J Clin Invest* 78: 1120–1120
4   Seckinger P, Lowenthal JW, Williamson K, Dayer J-M, MacDonald HR (1987) A urine inhibitor of interleukin 1 activity that blocks ligand binding. *J Immunol* 139: 1546
5   Arend WP, Joslin FG, Thompson RC, Hannum CH (1989) An IL-1 inhibitor from human monocytes Production and characterization of biologic properties. *J Immunol* 143: 1851–1858
6   Hannum CH, Wilcox CJ, Arend WP, Joslin FG, Dripps DP, Heimdal PL, Armes LG, Sommer A, Eisenberg SP, Thompson RC (1990) Interleukin-1 receptor antagonist activity of a human interleukin-1 inhibitor. *Nature* 343: 336–340
7   Eisenberg SP, Evans RJ, Arend WP, Verderber E, Brewer MT, Hannum CH, Thompson RC (1990) Primary structure and functional expression from complementary DNA of a human interleukin 1 receptor antagonist. *Nature* 343: 341–346
8   Eisenberg SP, Brewer MT, Verderber E, Heimdal PL, Brandhuber BJ, Thompson RC (1991) IL-1 receptor antagonist (IL-1ra) is a member of the IL-1 gene family. Evolution of a cytokine control mechanism. *Proc Natl Acad Sci USA* 88: 5232–5236
9   Carter DB, Deibel MR Jr, Dunn CJ, Tomich C-SC, LaBorde AL, Slightom JL, Berger AE, Bienkowski MJ, Sun FF et al (1990) Purification, cloning, expression, and biological characterization of an interleukin-1 receptor antagonist protein. *Nature* 344: 633–638
10   Haskill S, Martin G, Van Le L, Morris J, Peace A, Bigler CF, Jaffe GJ, Hammerberg C, Sporn SA, Fong S, Arend WP, Ralph P (1991) cDNA cloning of an intracellular form of the human interleukin 1 receptor antagonist associated with epithelium. *Proc Natl Acad Sci USA* 88: 3681–3685
11   Watson JM, Lofquist AK, Rinehart CA, Olsen JC, Makarov SS, Kaufman DG, Haskill JS (1995) The intracellular IL-1 receptor antagonist alters IL-1-inducible gene expression without blocking exogenous signaling by IL-1β. *J Immunol* 155: 4467–4475
12   Greenfeder SA, Nunes P, Kwee L, Labow M, Chizzonite RA, Ju G (1995) Molecular

cloning and characterization of a second subunit of the interleukin 1 receptor complex. *J Biol Chem* 270: 13757–13765

13   Dripps DJ, Verderber E, Ng RK, Thompson RC, Eisenberg SP (1991) Interleukin-1 receptor antagonist binds to the type II interleukin-1 receptor on B cells and neutrophils. *J Biol Chem* 266: 20311–20315

14   Arend WP, Welgus HG, Thompson RC, Eisenberg SP (1990) Biological properties of recombinant monocyte-derived interleukin 1 receptor antagonist. *J Clin Invest* 85: 1694–1697

15   Curtis BM, Gallis B, Overell RW, McMahan CJ, DeRoos P, Ireland R, Eisenman J, Dower SK, Sims JE (1989) T-cell interleukin 1 receptor cDNA expressed in Chinese hamster ovary cells regulates functional responses to interleukin 1. *Proc Natl Acad Sci USA* 86: 3045–3049

16   Granowitz EV, Porat R, Mier JW, Pribble JP, Stiles DM, Bloedow DC, Catalano MA, Wolff SM, Dinarello CA (1992) Pharmacokinetics, safety and immunomodulatory effects of human recombinant interleukin-1 receptor antagonist in healthy humans. *Cytokine* 4: 353–360

17   Ohlsson K, Bjork P, Bergenfeldt M, Hageman R, Thompson RC (1990) Interleukin-1 receptor antagonist reduces mortality from endotoxin shock. *Nature* 348: 550–552

18   Wakabayashi G, Gelfand JA, Burke JF, Thompson RC, Dinarello CA (1991) A specific receptor antagonist for interleukin 1 prevents *Escherichia coli*-induced shock in rabbits. *FASEB J* 5: 338–343

19   Hirsch E, Irikura VM, Paul SM, Hirsh D (1996) Functions of interleukin 1 receptor antagonist in gene knockout and overproducing mice. *Proc Natl Acad Sci USA* 93: 11008–11013

20   Plata-Salaman CR, French-Mullen JM (1992) Intracerebroventricular administration of a specific IL-1 receptor antagonist blocks food and water intake suppression induced by interleukin-1 beta PhysiolBehav 51: 1277–1279

21   Fischer E, Marano MA, Barber AE, Hudson A, Lee K, Rock CS, Hawes AS, Thompson RC, Hayes TJ, Anderson TD (1991) Comparison between effects of interleukin-1 alpha administration and sublethal endotoxemia in primates. *Am J Physiol* 261: R442–R452

22   Fischer E, Marano MA, Van Zee KJ, Rock CS, Hawes AS, Thompson WA, DeForge L, Kenney JS, Remick DG, Bloedow DC (1992) Interleukin-1 receptor blockade improves survival and hemodynamic performance in *Escherichia coli* septic shock, but fails to alter host responses to sublethal endotoxemia. *J Clin Invest* 89: 1551–1557

23   Granowitz EV, Santos AA, Poutsiaka DD, Cannon JG, Wilmore DW, Wolff SM, Dinarello CA (1991) Production of interleukin-1-receptor antagonist during experimental endotoxaemia. *Lancet* 338: 1423–1424

24   Fisher CJJ, Slotman GJ, Opal SM, Pribble JP, Bone RC, Emmanuel G, Ng D, Bloedow DC, Catalano MA (1994) Initial evaluation of human recombinant interleukin-1 receptor antagonist in the treatment of sepsis syndrome: a randomized, open- label, placebo-controlled multicenter trial. The IL-1RA Sepsis Syndrome Study Group. *Crit Care Med* 22: 12–21

25 Fisher CJJ, Dhainaut JF, Opal SM, Pribble JP, Balk RA, Slotman GJ, Iberti TJ, Rackow EC, Shapiro MJ, Greenman RL (1994) Recombinant human interleukin 1 receptor antagonist in the treatment of patients with sepsis syndrome Results from a randomized, double-blind, placebo-controlled trial. Phase III rhIL-1ra. Sepsis Syndrome Study Group. *JAMA* 271: 1836–1843

26 Opal SM, Fisher CJJ, Dhainaut JF, Vincent JL, Brase R, Lowry SF, Sadoff JC, Slotman GJ, Levy H, Balk RA et al (1997) Confirmatory interleukin-1 receptor antagonist trial in severe sepsis: a phase III, randomized, double-blind, placebo-controlled, multicenter trial. The Interleukin-1 Receptor Antagonist Sepsis Investigator Group. *Crit Care Med* 25: 1115–1124

27 Arend WP, Dayer JM (1995) Inhibition of the production and effects of interleukin-1 and tumor necrosis factor alpha in rheumatoid arthritis. *Arthritis Rheum* 38: 151–160

28 Ghivizzani SC, Kang R, Georgescu HI, Lechman ER, Jaffurs D, Engle JM, Watkins SC, Tindal MH, Suchanek MK, McKenzie LR et al (1997) Constitutive intra-articular expression of human IL-1 beta following gene transfer to rabbit synovium produces all major pathologies of human rheumatoid arthritis. *J Immunol* 159: 3604–3612

29 van de Loo FA, Arntz OJ, Otterness IG, Van den Berg WB (1992) Protection against cartilage proteoglycan synthesis inhibition by antiinterleukin 1 antibodies in experimental arthritis. *J Rheumatol* 19: 348–356

30 Van den Berg WB, Joosten LA, Helsen M, van de Loo FA (1994) Amelioration of established murine collagen-induced arthritis with anti-IL-1 treatment. *Clin Exp Immunol* 95: 237–243

31 Matsukawa A, Ohkawara S, Maeda T, Takagi K, Yoshinaga M (1993) Production of IL-1 and IL-1 receptor antagonist and the pathological significance in lipopolysaccharide-induced arthritis in rabbits. *Clin Exp Immunol* 93: 206–211

32 Schwab JH, Anderle SK, Brown RR, Dalldorf FG, Thompson RC (1991) Pro- and anti-inflammatory roles of interleukin-1 in recurrence of bacterial cell wall-induced arthritis in rats. *Infect Immun* 59: 4436–4442

33 Fukumoto T, Matsukawa A, Ohkawara S, Takagi K, Yoshinaga M (1996) Administration of neutralizing antibody against rabbit IL-1 receptor antagonist exacerbates lipopolysaccharide-induced arthritis in rabbits. *Inflamm Res* 45: 479–485

34 Joosten LA, Helsen MM, van de Loo FA, Van den Berg WB (1996) Anticytokine treatment of established type II collagen-induced arthritis in DBA/1 mice. A comparative study using anti-TNF alpha, anti- IL-1 alpha/beta, and IL-1Ra. *Arthritis Rheum* 39: 797–809

35 Malyak M, Swaney RE, Arend WP (1993) Levels of synovial fluid interleukin-1 receptor antagonist in rheumatoid arthritis and other arthropathies Potential contribution from synovial fluid neutrophils. *Arth Rheum* 36: 781–789

36 Miller LC, Lynch EA, Isa S, Logan JW, Dinarello CA, Steere AC (1993) Balance of synovial fluid IL-1β and IL-1 receptor antagonist and recovery from Lyme arthritis. *Lancet* 341: 146–148

37 Firestein GS, Boyle DL, Yu C, Paine MM, Whisenand TD, Zvaifler NJ, Arend WP

(1994) Synovial interleukin-1 receptor antagonist and interleukin-1 balance in rheumatoid arthritis. *Arthritis Rheum* 37: 644–652

38 Campion GV, Lebsack ME, Lookabaugh J, Gordon G, Catalano M (1996) Dose-range and dose-frequency study of recombinant human interleukin-1 receptor antagonist in patients with rheumatoid arthritis The IL-1Ra Arthritis Study Group. *Arthritis Rheum* 39: 1092–1101

39 Bresnihan B, Lookabaugh J, Witt K, Musikic P (1996) Treatment with recombinant human interleukin-1 receptor antagonist (rhIL-1Ra) in rheumatoid arthritis (RA) Results of a randomized double-blind, placebo-controlled multicenter trial. *Arth Rheum* 39: S73 (Abstract)

40 Watt I, Cobby M (1996) Recombinant human IL-1 receptor antagonist (rhIL-1Ra) reduces the rate of joint erosion in rheumatoid arthritis (RA). *Arth Rheum* 39: S123 (Abstract)

41 Nuki G, Rozman B, Pavelka K, Emery P, Lookabaugh J, Musikic P (1997) Interleukin-1 receptor antagonist continues to demonstrate clinical improvement in rheumatoid arthritis. *Arth Rheum* 40: S224 (Abstract)

42 Cunnane G, Madigan A, Fitzgerald B, Bresnihan B (1996) Treatment with recombinant human interleukin-1 receptor antagonist (rhIL-1Ra) may reduce synovial infiltration in rheumatoid arthritis. *Arth Rheum* 34: S245 (Abstract)

43 Bendele AM, Senello G, McAbee T, Frazier J, Chlipala E (1997) Effects of interleukin-1 receptor antagonist alone and in combination with methotrexate in adjuvant arthritic rats. *Arth Rheum* 40: S181 (Abstract)

44 Collins D, McAbee T, Woodward M, Frazier J, Chipala E, Bendele AM (1997) A slow release formulation of interleukin-1 receptor antagonist (IL-1Ra) exhibits improved *in vivo* activity. *Arth Rheum* 40: S180 (Abstract)

45 Bandara G, Mueller GM, Galea-Lauri J, Tindahl MH, Georgescu HI, Suchanek MK, Hung GL, Glorioso JC, Robbins PD, Evans CH (1993) Intraarticular expression of biologically active interleukin 1 receptor antagonist protein by ex vivo gene transfer. *Proc Natl Acad Sci USA* 90: 10764–10768

46 Makarov SS, Olsen JC, Johnston WN, Anderle SK, Brown RR, Baldwin ASJ, Haskill JS, Schwab JH (1996) Suppression of experimental arthritis by gene transfer of interleukin 1 receptor antagonist cDNA. *Proc Natl Acad Sci USA* 93: 402–406

47 Bakker AC, Joosten LA, Arntz OJ, Helsen MM, Bendele AM, van de Loo FA, Van den Berg WB (1997) Prevention of murine collagen-induced arthritis in the knee and ipsilateral paw by local expression of human interleukin-1 receptor antagonist protein in the knee. *Arthritis Rheum* 40: 893–900

48 Evans CH, Robbins PD, Ghivizzani SC, Herndon JH, Kang R, Bahnson AB, Barranger JA, Elders EM, Gay S, Tomaino MM et al (1996) Clinical trial to assess the safety, feasibility, and efficacy of transferring a potentially anti-arthritic cytokine gene to human joints with rheumatoid arthritis. *Hum Gene Ther* 7: 1261–1280

49 Ghivizzani SC, Kang R, Muzzonigro T, Whalen JD, Watkins SC, Herndon JH, Robbins PD, Evans CH (1997) Gene therapy for arthritis – Treatment of the first three patients.

Arth Rheum 40: S223 (Abstract)

50 McCarthy PLJ, Abhyankar S, Neben S, Newman G, Sieff C, Thompson RC, Burakoff SJ, Ferrara JL (1991) Inhibition of interleukin-1 by an interleukin-1 receptor antagonist prevents graft-versus-host disease. *Blood* 78: 1915–1918

51 Abhyankar S, Gilliland DG, Ferrara JL (1993) Interleukin-1 is a critical effector molecule during cytokine dysregulation in graft versus host disease to minor histocompatibility antigens. *Transplantation* 56: 1518–1523

52 Antin JH, Weinstein HJ, Guinan EC, McCarthy P, Bierer BE, Gilliland DG, Parsons SK, Ballen KK, Rimm IJ, Falzarano G (1994) Recombinant human interleukin-1 receptor antagonist in the treatment of steroid-resistant graft-versus-host disease. *Blood* 84: 1342–1348

53 Cominelli F, Nast CC, Clark BD, Schindler R, Llerena R, Eysselein VE, Thompson RC, Dinarello CA (1990) Interleukin 1 (IL-1) gene expression, synthesis, and effect of specific IL-1 receptor blockade in rabbit immune complex colitis. *J Clin Invest* 86: 972–980

54 Cominelli F, Nast CC, Duchini A, Lee M (1992) Recombinant interleukin-1 receptor antagonist blocks the proinflammatory activity of endogenous interleukin-1 in rabbit immune colitis. *Gastroenterology* 103: 65–71

55 Ferretti M, Casini-Raggi V, Pizarro TT, Eisenberg SP, Nast CC, Cominelli F (1994) Neutralization of endogenous IL-1 receptor antagonist exacerbates and prolongs inflammation in rabbit immune colitis. *J Clin Invest* 94: 449–453

56 Cominelli F, Melani L, Hirsch E, Kosiewicz MM, Guanzon M, Pizarro TT, Hirsch D, Nast CC (1997) Differential susceptibility to experimental colitis in IL-1Ra mutant mice. *Cytokine* 9: 960 (Abstract)

57 Hyams JS, Fitzgerald JE, Wyzga N, Muller R, Treem WR, Justinich CJ, Kreutzer DL (1995) Relationship of interleukin-1 receptor antagonist to mucosal inflammation in inflammatory bowel disease. *J Pediatr Gastroenterol Nutr* 21: 419–425

58 Casini-Raggi V, Kam L, Chong YJT, Fiocchi C, Pizarro TT, Cominelli F (1995) Mucosal imbalance of IL-1 and IL-1 receptor antagonist in inflammatory bowel disease: A novel mechanism of chronic intestinal inflammation. *J Immunol* 154: 2434–2440

59 Heresbach D, Alizadeh M, Dabadie A, Le Berre N, Colombel JF, Yaouanq J, Bretagne JF, Semana G (1997) Significance of interleukin-1beta and interleukin-1 receptor antagonist genetic polymorphism in inflammatory bowel diseases. *Am J Gastroenterol* 92: 1164–1169

60 Mansfield JC, Holden H, Tarlow JK, di Giovine FS, McDowell TL, Wilson AG, Holdsworth CD, Duff GW (1994) Novel genetic association between ulcerative colitis and the anti- inflammatory cytokine interleukin-1 receptor antagonist. *Gastroenterology* 106: 637–642

61 Roussomoustakaki M, Satsangi J, Welsh K, Louis E, Fanning G, Targan S, Landers C, Jewell DP (1997) Genetic markers may predict disease behavior in patients with ulcerative colitis. *Gastroenterology* 112: 1845–1853

62 Bioque G, Bouma G, Crusius JB, Koutroubakis I, Kostense PJ, Meuwissen SG, Pena AS (1996) Evidence of genetic heterogeneity in IBD: 1 The interleukin-1 receptor antago-

nist in the predisposition to suffer from ulcerative colitis. *Eur J Gastroenterol Hepatol* 8: 105–110

63 Andus T, Daig R, Vogl D, Aschenbrenner E, Lock G, Hollerbach S, Kollinger M, Scholmerich J, Gross V (1997) Imbalance of the interleukin 1 system in colonic mucosa – association with intestinal inflammation and interleukin 1 receptor agonist genotype 2. *Gut* 41: 651–657

64 Rambaldi A, Torcia M, Bettoni S, Vannier E, Barbui T, Shaw AR, Dinarello CA, Cozzolino F (1991) Modulation of cell proliferation and cytokine production in acute myeloblastic leukemia by interleukin 1 recpetor antagonist and lack of its expression by leukemic cells. *Blood* 78: 3248–3253

65 Estrov Z, Kurzrock R, Estey E, Wetzler M, Ferrajoli A, Harris D, Blake M, Gutterman JU, Talpaz M (1992) Inhibition of acute myelogenous leukemia blast proliferation by interleukin-1 (IL-1) receptor antagonist and soluble IL-1 receptors. *Blood* 79: 1938–1945

66 Estrov Z, Kurzrock R, Wetzler M, Kantarjian H, Blake M, Harris D, Gutterman JU, Talpaz M (1991) Supression of chronic myelogenous leukemia colony growth by interelukin-1 (IL-1) receptor antagonist and soluble IL-1 receptors: A novel application for inhibitors of IL-1 activity. *Blood* 78: 1476–1484

67 Stosic-Grujicic S, Basara N, Milenkovic P, Dinarello CA (1995) Modulation of acute myeloblastic leukemia (AML) cell proliferation and blast colony formation by antisense oligomer for IL-1 beta converting enzyme (ICE) and IL-1 receptor antagonist (IL-1ra). *J Chemother* 7: 67–70

68 Schiro R, Longoni D, Rossi V, Maglia O, Doni A, Arsura M, Carrara G, Masera G, Vannier E, Dinarello CA (1994) Suppression of juvenile chronic myelogenous leukemia colony growth by interleukin-1 receptor antagonist [see comments]. *Blood* 83: 460–465

69 Yin M, Gopal V, Banavali S, Gartside P, Preisler H (1992) Effects of an IL-1 receptor antagonist on acute myeloid leukemia cells. *Leukemia* 6: 898–901

70 Wetzler M, Kurzrock R, Estrov Z, Kantarjian H, Gisslinger H, Underbrink MP, Talpaz M (1994) Altered levels of interleukin-1β and interleukin-1 receptor antagonist in chronic myelogenous leukemia: Clinical and prognostic correlates. *Blood* 84: 3142–3147

71 Karkar AM, Tam FW, Steinkasserer A, Kurrle R, Langner K, Scallon BJ, Meager A, Rees AJ (1995) Modulation of antibody-mediated glomerular injury *in vivo* by IL-1ra, soluble IL-1 receptor, and soluble TNF receptor. *Kidney Int* 48: 1738–1746

72 Lan HY, Nikolic-Paterson DJ, Mu W, Vannice JL, Atkins RC (1995) Interleukin-1 receptor antagonist halts the progression of established crescentic glomerulonephritis in the rat. *Kidney Int* 47: 1303–1309

73 Chen A, Sheu LF, Chou WY, Tsai SC, Chang DM, Liang SC, Lin FG, Lee WH (1997) Interleukin-1 receptor antagonist modulates the progression of a spontaneously occurring IgA nephropathy in mice. *Am J Kidney Dis* 30: 693–702

74 Norman JG, Franz MG, Fink GS, Messina J, Fabri PJ, Gower WR, Carey LC (1995)

Decreased mortality of severe acute pancreatitis after proximal cytokine blockade. *Ann Surg* 221: 625–631

75 Norman J, Franz M, Messina J, Riker A, Fabri PJ, Rosemurgy AS, Gower WRJ (1995) Interleukin-1 receptor antagonist decreases severity of experimental acute pancreatitis. *Surgery* 117: 648–655

76 Dana MR, Yamada J, Streilein JW (1997) Topical interleukin 1 receptor antagonist promotes corneal transplant survival. *Transplantation* 63: 1501–1507

77 Yanofsky SD, Baldwin DN, Butler JH, Holden FR, Jacobs JW, Balasubramanian P, Chinn JP, Cwirla SE, Peters-Bhatt E, Whitehorn EA et al (1996) High affinity type I interleukin 1 receptor antagonists discovered by screening recombinant peptide libraries. *Proc Natl Acad Sci USA* 93: 7381–7386

78 Akeson AL, Woods CW, Hsieh LC, Bohnke RA, Ackermann BL, Chan KY, Robinson JL, Yanofsky SD, Jacobs JW, Barrett RW et al (1996) AF12198, a novel low molecular weight antagonist, selectively binds the human type I interleukin (IL)-1 receptor and blocks *in vivo* responses to IL-1. *J Biol Chem* 271: 30517–30523

Therapeutic regulation of cytokine signalling by inhibitors of p38 mitogen-activated protein kinase

Raymond J. Owens and Simon Lumb

Celltech Chiroscience, 216 Bath Road, Slough, Berkshire SL1 4EN, UK

Introduction

Over the last four years, selective inhibitors of the enzyme p38 mitogen-activated protein kinase (p38 MAPK) have emerged as potential new drugs for the treatment of inflammatory diseases, such as rheumatoid arthritis. Inhibition of the kinase activity of p38 MAPK is associated with the modulation of cytokine production, in particular tumor necrosis factor (TNF) and interleukin-1 (IL-1), and also the response of cells to these pro-inflammatory agents. This profile suggests that inhibitors of p38 MAPK may be disease-modifying, hence this enzyme has become the focus of considerable activity in the pharmaceutical industry. In this chapter we will review the chemistry and biology of p38 MAPK inhibitors and discuss how blocking the activity of p38 MAPK leads to an anti-inflammatory effect.

p38 MAPK Cascade

p38 MAPK was first identified as a kinase which became tyrosine phosphorylated in mouse monocytes following treatment with lipopolysaccharide (LPS). Molecular cloning and sequencing of the p38 MAPK cDNA showed that it was related to the osmosensing gene, *HOG1*, in *Saccharomyces cerevisiae* and a member of the MAPK family of enzymes [1]. A link between p38 MAPK and the response of cells to cytokines was first established by Saklatvala and co-workers [2] who showed that IL-1 activates a protein kinase cascade that results in the phosphorylation of the small heat shock protein, Hsp27, probably by mitogen-activated protein activated protein kinase 2 (MAPKAP kinase-2). Analysis of peptide sequences derived from the purified kinase indicated that it was related to the p38 MAPK activated by LPS in mouse monocytes [1]. At the same time it was shown that p38 MAPK was itself activated by an upstream kinase in response to a variety of cellular stresses, including exposure to UV radiation and osmotic shock, and the identity of the kinase that directly phosphorylates Hsp27 was confirmed as MAPKAP kinase-2 [3]. Subsequently, workers at SmithKline Beecham showed that p38 MAPK was the molecular target of a series of pyridinylimidazole compounds that inhibited the production of TNF from LPS-challenged human monocytes [4]. This was a key discovery and has led to the development of a number of selective inhibitors of p38 MAPK and the elucidation of its role in cytokine signalling.

It is now known that multiple forms of p38 MAPK (α, β, γ, δ) each encoded by a separate gene [5–7], form part of a kinase cascade involved in the response of cells to a variety of stimuli, including osmotic stress, UV light and cytokines. A second parallel pathway leading to the activation of c-jun N-terminal kinase (JNK) is also triggered by many of the same stimuli. The activation of extracellular signal-regulated kinases 1 and 2 (ERK 1/ERK2) by growth factors forms a third and distinct kinase cascade in mammalian cells (Fig. 1). All three pathways involve two upstream levels of kinases, i.e., MAPK kinase kinase (MAPKKK) and MAPK kinase (MAPKK), both of which comprise several homologues. The MAPKK family of enzymes is unique in being dual-specificity kinases, phosphorylating MAPKs on both the threonine and tyrosine residues of a three-amino acid motif, in which the identity of the intervening residue characterises the MAPK family (for ERK Thr, *glutamate*, Tyr for p38 MAPK Thr *glycine*, Tyr for JNK Thr, *proline*, Tyr). Dual phosphorylation of p38 MAPK is required for its activation and has been demonstrated in human cell lines treated with cytokines (IL-1 and TNF on HeLa cells [8]) or chemically stressed (arsenite on KB cells [9]). The major physiological activator of p38 MAPK in chemically stressed or IL-1-stimulated KB cells and TNF stimulated monocytic cells, THP1, has been shown to be the MAPKK member MKK6 [10], previously identified by cDNA cloning [11, 12]. P38 MAPK can also be activated *in vitro* by two other MAPKKs, i.e., MKK3, which is specific for p38 MAPK, and MKK4, which can also phosphorylate and activate JNK *in vitro* [13]. The identity of the physiological upstream activator(s) of MKK3 and MKK6 has yet to be defined. A number of possible candidates have been suggested based on re-constitution experiments in transfected cells, for example, MAPKKK3 or MEKK3 [14] and TGFβ activated kinase 1, TAK1 [15]. How the p38 MAPK cascade is connected to events initiated at the cell surface still remains to be established. Most recently, TAK1 has been implicated in signalling proximal to the IL-1 receptor complex and thus may be a molecular link between p38 MAPK and this receptor [16].

As mentioned above, p38 MAPK phosphorylates MAPKAP kinase-2 in response to a number of stress stimuli. Two other related kinases, MAPKAP kinase-3 [17] and p38-regulated/activated kinase (PRAK) [18], are also substrates of the enzyme and like MAPKAP kinase-2, can phosphorylate Hsp27 in response to stress stimuli. P38 MAPK appears to have kinase substrates in common with ERK1/ 2, namely, mitogen- and stress-activated kinase 1 (MSK1) [19], and MAPK-interacting kinase 1 (Mnk1) [20]. These kinases are activated in response to both stress stimuli and growth factors by p38 MAPK and ERK1/2, respectively, though their downstream

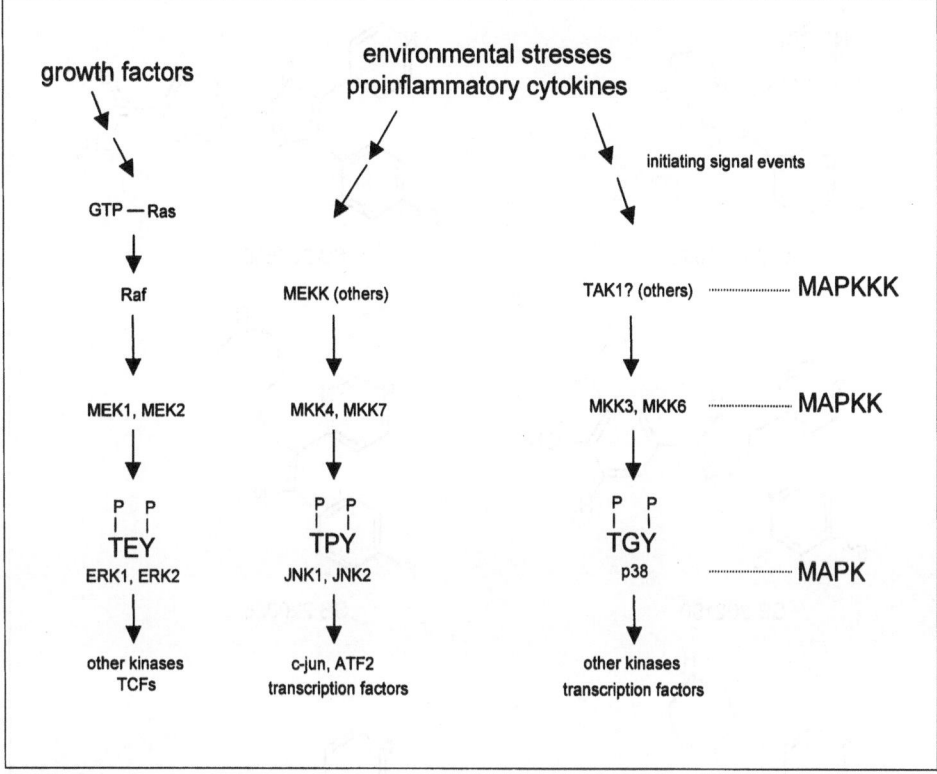

Figure 1

Schematic diagram illustrating the sequential activation of ERK, JNK, p38 MAPK enzymes by growth factors, cytokines and environmental stresses.

substrates are not known. In addition to kinases, there is evidence from *in vitro* kinase and reporter gene assays that p38 MAPK phosphorylates the transcription factors activated transcription factor-2 (ATF-2) [12], myocyte enhancer factor 2C (MEF2C) [21] and Sap1a [22]. However, the physiological significance of these observations remains to be established, since these transcription factors can also be activated independently of p38 MAPK.

p38 MAPK inhibitors

Chemistry: structure-activity relations

The prototype p38 MAPK inhibitor, SK&F 86002 (Fig. 2), was initially identified from a series of dual 5-lipoxygenase/cycloxygenase inhibitors which also blocked

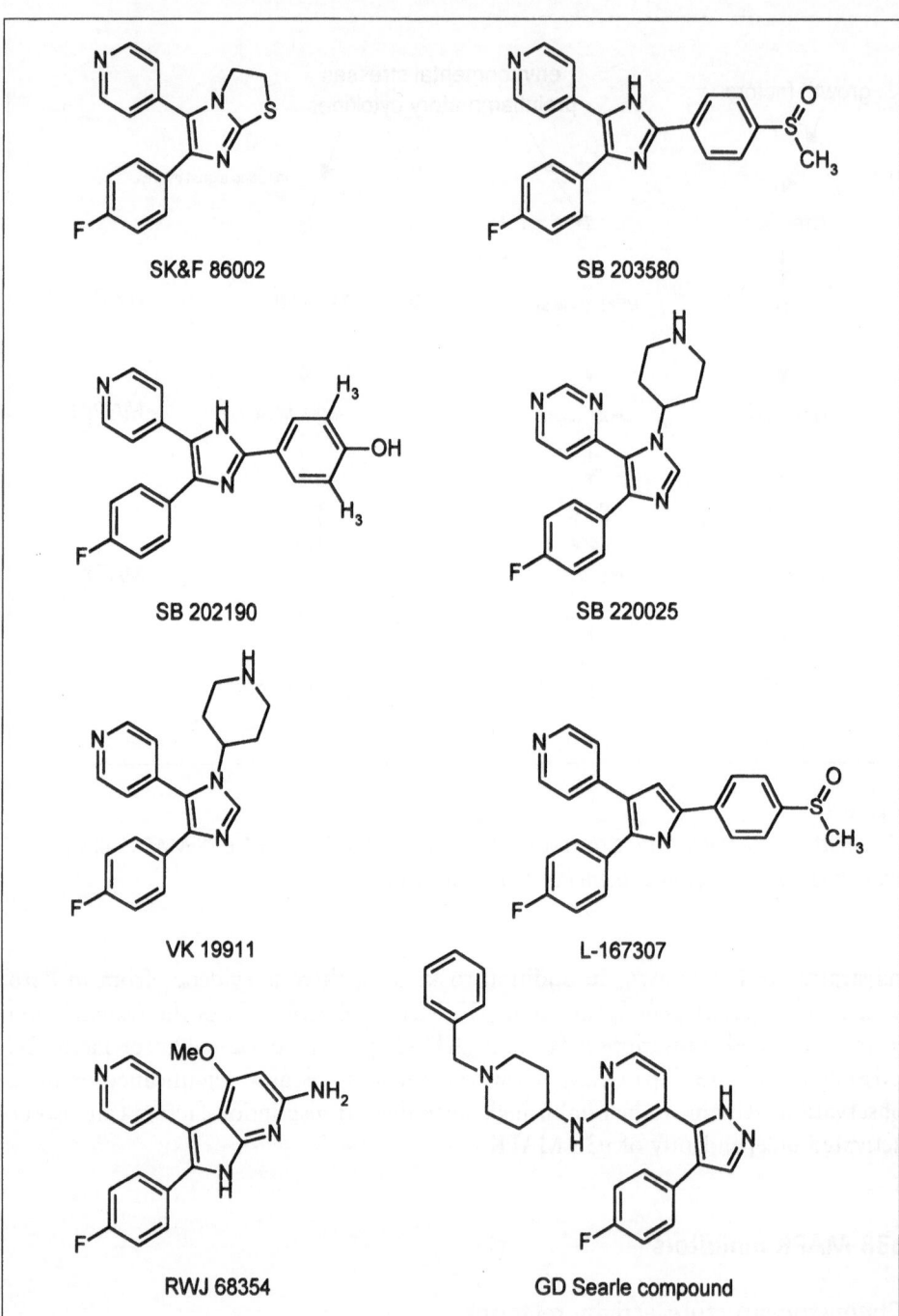

Figure 2
Structures of p38 MAPK inhibitors

the production of IL-1 and TNF both *in vitro* and *in vivo* [23, 24] This series of compounds was derived by combining the structurally related pharmacophores of two known anti-inflammatory agents, Flumizole and Levamisole. Optimisation of the resulting hybrid compounds led to the synthesis of SK&F 86002, a bicyclic imidazole, which demonstrated efficacy in a range of inflammatory disease models including collagen-induced arthritis (CIA) [25]. Recognition that the anti-cytokine properties of SK&F 86002 could not be accounted for by its inhibition of arachidonate metabolism led to the search for other molecular targets for this compound. A radiolabelled derivative (SB 202190, Fig. 2) with greater potency in the cytokine synthesis bioassay was used to isolate the binding protein from lysates of the human monocytic cell line, THP1. Subsequently, molecular cloning led to the identification of the binding protein as the kinase p38 MAPK [4]. The availability of both enzyme activity and binding assays has enabled the development of the p38 MAPK pharmacophore and hence identification of potent and selective inhibitors of the enzyme, devoid of 5-lipoxygenase/cycloxygenase inhibitory activity. The most potent analogue generated from this work was the compound SB 203580 (Fig. 2) [26]. This compound has been used in subsequent work on characterising the structure and function of p38 MAPK.

In general, p38 MAPK inhibitors possess a common structural feature, namely a core 5-membered ring nitrogen heterocycle (an azole) substituted with a 4-pyridinyl ring and a phenyl ring attached at adjacent positions. This type of structure is exemplified by SB 203580 in which the central ring is an imidazole. A variety of other heterocycles (Fig. 2) have been used in place of imidazole, for example, pyrroles (L-167307) [27], pyrrolopyridines (RWJ 68354) [28] and pyrazoles [29], to produce pharmacologically active p38 MAPK inhibitors (IC_{50}s of 5–10 nM against the isolated enzyme). A key feature of the published inhibitors is the presence of the 4-fluorophenyl substituent which appears to be critical for binding. Similarly, most structures contain a 4-pyridyl group, though this can be been replaced by a pyrimidinyl group without loss of activity (e.g., SB 220025) [30]. Where the central ring is an imidazole, it is important that the 6-membered ring nitrogen heterocycle is positioned beta to an unalkylated nitrogen of the central azole ring [24]. In the case of the pyrazole derivatives disclosed by Searle [29], a relatively simple template retaining the 4-pyridyl ring is active and further substitution of the pyridine ring leads to increased potency *in vitro* (Fig. 2).

Mechanism of p38 MAPK inhibition

Following activation of the enzyme by dual phosphorylation of threonine (Thr 180) and tyrosine (Tyr 182), p38 MAPK demonstrates a K_m (substrate binding constant) for ATP of 25 µM, which is relatively high for a serine/threonine kinase [31]. In addition it has been shown that for at least one substrate of the enzyme, ATF2,

ATP will only bind to the activated enzyme *in vitro* once it has associated with the substrate [31]. SB 203580 and related pyridinylimidazole inhibitors compete with ATP in binding to the active enzyme but also appear to bind to the inactive enzyme with equal affinity [32, 33]. This suggests that this class of inhibitor may block the biological activity of p38 MAPK by binding to the inactive form of the enzyme. Under these conditions ATP would not effectively compete with the inhibitors *in situ*. A combination of the high K_m for ATP and the mechanism of inhibition may account for the maintenance of inhibitor potencies in cell systems where the ATP concentration is considered to be very high. For example, the K_i (inhibitor binding constant) for SB 203580 *in vitro* is 21 nM [32] which correlates with an inhibition of LPS-induced TNF from human monocytes of 25 nM (Tab. 1) [28].

An early finding for the SB 203580 inhibitor was the relatively high degree of selectivity for p38 MAPK amongst a large panel of other kinases, including the related MAPKs, ERK1/2 and JNKs [34]. Subsequently it has been shown that certain isotypes of JNK (JNK2, IC_{50} = 280–1920 nM) and raf kinase (IC_{50} = 370 nM) are sensitive to the inhibitor but with a tenfold selectivity compared to p38 MAPK [27, 35]. Selectivity ratios of up to 1000-fold have been observed for the inhibition of representatives of other kinase classes, for example EGF receptor, *src* and *lck* tyrosine kinases, protein kinase A and protein kinase C [28]. The high degree of selectivity of the SB 203580 series of inhibitors has prompted studies of the molecular basis of inhibitor specificity. Comparison of the primary sequence of p38 MAPKα with the two isozymes that are not inhibited by SB 203580, p38 MAPKγ,δ [7] shows that there are three amino acid differences close to the ATP binding pocket, threonine 106 (Thr 106), histidine 107 (His 107) and leucine 108 (Leu 108). Two of these residues (Thr 106 and His 107) also differ between p38 MAPKα and the less closely related ERK and JNK enzymes. Site-directed mutagenesis has been used to substitute the p38 MAPKα amino acids at 106,107 and 108 with the corresponding residues in p38 MAPKγ,δ, with a consequent loss of sensitivity to SB 203580 [36]. Conversely, p38 MAPKγ,δ became SB 203580-sensitive when the reciprocal changes were made [36]. In fact, substitution of residue 106 with a threonine alone confers a degree of susceptibility to the pyridinylimadazole inhibitors not only on p38 MAPKγ but also on JNK1 [37,38]. Conversely, changing this amino acid in p38 MAPKα leads to a loss of inhibition, confirming that the molecular basis for the specificity of the p38 MAPK inhibitors can be largely attributed to a single residue in the enzyme.

X-ray crystallography of p38 MAPK/inhibitor complexes

P38 MAPK has proved highly amenable to study by X-ray crystallography and the structure of the inactive enzyme at atomic resolution has been reported by a number of groups [39,40]. More importantly, several structures of enzyme/inhibitor

Table 1 - Activity of p38 MAPK inhibitors (data from [28])

Inhibitor	p38 MAPK inhibition *in vitro* (IC_{50} nM)	Inhibition of LPS-induced TNF from human monocytes *in vitro* (IC_{50} nM)	Inhibition of TNF release from mice treated with LPS *in vivo* (% inhibition)	
			10 mg/kg p.o.	25 mg/kg p.o.
SB 203580	28	25	33	83
RWJ 68354	4	6.3	85	99

complexes have also been derived [41–43]. Again SB 203580 or close analogues have been used and examination of the resulting inhibitor-bound structures has confirmed the structure activity relations deduced from medicinal chemistry approaches and are consistent with the biochemical studies discussed above. Thus the pyridinylimidazole compounds (SB 203580 and VK 19911) (Fig. 2) all bind into the ATP pocket of the inactive enzyme. The binding mode of the inhibitors is similar to that of ATP, although there are significant differences. The pyridyl and imidazole rings of the inhibitors occupy positions in the ATP pocket that correspond to the adenine base. The nitrogen atom in the pyridine ring accepts a hydrogen bond from the backbone amide of Met 109. The importance of this residue in inhibitor binding has been confirmed by site-directed mutagenesis. Changing this amino acid to alanine results in a six-fold drop in SB 203580 inhibition of p38 MAPK [37]. The fluorophenyl group common to all the p38 MAPK inhibitors (Fig. 2) binds into a hydrophobic pocket forming favourable Van der Waal contacts with a number of hydrophobic residues, including Thr 106. Comparison of the structure of the unbound enzyme with the complex between p38 MAPK and SB 203580 shows that inhibitor binding induces a local conformational change in the enzyme which is associated with close contact between the fluorophenyl substituent and Thr 106 [43]. This suggests that the fluorinated ring makes a significant contribution to the potency of p38 MAPK inhibitors and reinforces the role for Thr 106 in the specificity of the compounds. Collectively the analysis of the interaction between enzyme and inhibitors shows that it is possible to achieve selective inhibition of kinase activity by targeting the ATP binding site.

Pharmacological activity of p38 MAPK inhibitors

As first shown for SB 203580, potent inhibitors of p38 MAPK catalytic activity potently inhibit the production of TNF and IL-1 by human monocytes treated with LPS (Tab. 1). This in turn translates into an inhibitory effect in an acute model of TNF production *in vivo* (Tab. 1). Mice are typically pre-treated with inhibitors, challenged with endotoxin and then levels of TNF in the blood serum are assayed 1–2 h later. In disease models of inflammation, SB 203580 has been shown to have a therapeutic effect; for example, in CIA in DBA/LACj mice, treatment with the inhibitor (50 mg/kg p.o., b.i.d.) resulted in a significant inhibition of paw inflammation and serum amyloid protein levels [44]. In this model of arthritis, it is well established that TNF and IL-1 contribute to the joint inflammation and erosions as shown by the therapeutic effect of both anti-TNF and anti-IL-1 antibodies [45, 46]. Therefore the efficacy of p38 MAPK inhibitors in this disease model is entirely consistent with the regulation of TNF and IL-1 signalling by p38 MAPK. The pharmacological activity of another p38 MAPK inhibitor, the pyrimidinylimidazole SB 220025 (Fig. 2), has also been reported by SmithKline Beecham. This compound was assessed in an inflammatory angiogenesis model (murine air pouch granuloma) in which elevated levels of IL-1 and TNF are observed during the chronic inflammatory phase when intense angiogenesis occurs. SB 220025 (30mg/kg p.o. b.i.d.) was shown to significantly reduce the levels of these cytokines in the affected tissue and inhibit angiogenesis by approximately 40% compared to vehicle-treated controls [30].

The efficacy of p38 MAPK inhibitors in animal models of inflammatory disease has prompted an investigation of the underlying mechanism(s) which could account for the effects of these inhibitors. The role of p38 MAPK in the response of cells to IL-1 and TNF has been investigated in a number of cell systems relevant to the inflammatory response using the selective inhibitor SB 203580. In human fibroblasts and vascular endothelial and fibroblast-like synoviocytes, it has been shown that inhibition of p38 MAPK activity is associated with an inhibition of IL-1-induced IL-6, a secondary cytokine implicated in the inflammatory response [47, 48]. Similarly, SB 203580 has been reported to block the TNF-induced production of IL6 from L929 and HeLa cells [49]. Modulation of prostaglandin synthesis by p38 MAPK inhibition has also been noted. Induction of cyclooxygenase-2 (COX-2) in human monocytes following treatment with a variety of stimuli, including zymosan, IL-1, TNF and LPS, is blocked by treatment of cells with p38 MAPK inhibitors [50,51]. This effect is correlated with a decrease in prostanoid production. The fact that COX-2 expression initiated by IL-1 and TNF is also blocked by p38 MAPK inhibition shows that the there is a direct suppressive effect on COX-2 and that it is not only a result of blocking cytokine production *per se*. Similarly, the inhibition of IL-1 induction of COX-2 by both fibroblasts and HeLa cells by SB 203580 indicates a direct involvement of p38 MAPK in the regulation of COX-2 [47,52].

An immunomodulatory role for p38 MAPK has also been investigated, since this would also contribute to the disease-modifying activity of selective inhibitors. It is well established that the manifestation of CIA is dependent upon the generation of a Th1 cellular immune response and production of anti-CII antibodies [53]. However, to date no clear picture has emerged regarding a pivotal role of p38

MAPK in immune cell function, though the enzyme appears to be modestly activated in response to mitogenic stimulation of both B [54] and T lymphocytes [55]. Badger et al. [44] compared the p38 MAPK inhibitor, SB 203580, with the immunosuppressive drug, rapamycin, in immune response assays. Mice (BALBc) immunised with ovalbumin were treated with either SB 203580 (60 mg/kg i.p.) or rapamycin (50 mg/ kg i.p.) for two weeks. The serum anti-(ovalbumin) antibody response was completely suppressed by rapamycin treatment and there was a significant reduction in specific antibody titre in the SB 203580-treated mice. However, no inhibition of lymph node cell proliferation *ex vivo* in response to either ovalbumin or mitogen ConA was observed for the SB 203580-treated mice, in contrast to the rapamycin-dosed animal, in which lymphocyte responses were suppressed. [44]. Subsequently, a number of *in vitro* studies using SB 203580 have indicated an involvement of p38 MAPK in T-cell function. Proliferation of primary human T cells in response to either IL-2 and IL-7 has been shown to be sensitive to SB 203580 with IC_{50}s of approximately 3 μM [56].This contrasts with the much lower potency for inhibition of cytokine production by monocytes and may not be entirely due to p38 MAPK inhibition. Certainly in another study where the effect of SB 203580 on the production of various cytokines from CD3/CD28 stimulated peripheral blood T lymphocytes was investigated, little effect on IL-2 was noted [57]. This is consistent with a lack of effect of p38 MAPK inhibitors on T-lymphocyte proliferation following T-cell receptor stimulation [4, 57]. However a variable effect on other T-cell cytokines has been observed, with the most marked being on IL-10, whose production was completely blocked by SB 203580 with an IC_{50} of 150 nM [57].

An interesting and perhaps significant connection has been established between p38 MAPK and CD23. CD23 is a type-II membrane glycoprotein receptor which is specifically cleaved to produce soluble receptor fragments which have immunostimulatory activity, for example, promoting the growth and differentiation of germinal centre B cells, enhancing IgE production and activating monocyte TNF production [58] . Originally identified as a low-affinity receptor for IgE, CD23 also binds to the complement receptor CD21 and to the β2 integrins CD11b-CD18 and CD11c-CD18 [58]. It has been shown that SB 203580 reduces the level of soluble CD23 formed by IL-4 stimulated human monocyte with an IC_{50} of 67 nM, indicating a direct effect *via* p38 MAPK [59]. The reduction in soluble CD23 levels reflects a reduction in surface expression of the protein, not an effect on the cleavage mechanism. Recent reports indicate that CD23 may play a significant role in the inflammatory process. Antibodies to CD23 have been shown to be disease-modifying in a CIA model [60] and CIA induced in transgenic mice homozygous negative for CD23, shows a reduced severity and delayed onset compared to normal mice [61]. An effect of p38 MAPK on CD23 therefore provides an additional rationale for the potent anti-inflammatory action of the p38 MAPK inhibitors.

Molecular mechanisms

Activation of p38 MAPK causes a selective increase in the expression of pro-inflammatory and immunomodulatory mediators. What are the molecular mechanisms involved and are they common to the different events? The p38 MAPK pathway may increase gene expression by a general effect on transcription, by activation of specific transcription factors or by post-transcriptional regulation (mRNA processing, nuclear export, mRNA stability and translation). The selective effect of p38 MAPK inhibitors argues against a non-specific action of the enzyme on gene transcription. As already referred to above, transcription factors feature amongst the potential substrates of p38 MAPK. However, it is not yet clear whether p38 MAPK activity is rate-limiting for the activation of these transcription factors *in vivo*. No studies of the effect of selective inhibitors of p38 MAPK on the activation of endogenous transcription factors by proinflammatory stimuli have been reported. By contrast, there is accumulating evidence that p38 MAPK may have a post-transcriptional role in the control of cytokine expression. It has been reported that in THP1 cells, SK&F 86002 inhibits the production of LPS-induced TNF but does not affect the level of TNF mRNA. This was interpreted to indicate an inhibition of TNF translation and/or recruitment of mRNA by the translational machinery [62]. However, in LPS-stimulated human monocytes, SB 203580 has been shown to dose-dependently inhibit expression of both TNF and its mRNA [52]. SB 203580 also blocks the induction of both COX-2 protein and COX-2 mRNA by IL-1 in HeLa cells and LPS in monocytes. Evidence has been obtained that this involves an effect on both COX-2 transcription and the stability of COX-2 mRNA [51, 52]. Thus in both cell types, treatment with SB 203580 was associated with a rapid disappearance of COX-2 mRNA. Similarly, the control of IL-6 expression by p38 MAPK also appears to involve an effect on mRNA turnover [48]. Collectively these results suggest a novel function for p38 MAPK in the regulation of mRNA stability.

The question arises as to which of the known substrates of p38 MAPK are involved in the post-transcriptional events linked to the action of p38 MAPK. As discussed above, there is good evidence that phosphorylation of the kinase, MAP-KAP kinase-2 by p38 MAPK is a consequence of both stress and cytokine treatment of a variety of cells. However, apart from the phosphorylation of Hsp27, the relationship of MAPKAP kinase-2 activation to the functional effects of p38 MAPK as revealed by selective inhibitors is not resolved. Recently mice homozygous negative for the expression of MAPKAP kinase-2 have been generated and most interestingly these animals showed increased stress resistance and survive LPS-induced endotoxic shock [63]. This is correlated with an approximately 90% reduction in the production of TNF from LPS-challenged spleen cells *ex vivo*. However, no changes in TNF mRNA expression or stability were observed in these cells compared to wild-type controls. These results suggest that there is another, as yet undefined post-transcriptional mechanism whereby p38 MAPK may regulate TNF production and directly implicate MAPKAP kinase-2 in this process.

Summary and future directions

Molecular biology approaches have significantly advanced an understanding of the inflammatory process over the last ten years. A network of signalling pathways links cell surface receptors to the nucleus and elicits the expression of new gene products.

Over the last few years p38 MAPK has emerged as a key component of this intracellular "wiring" and inhibition of the activity of p38 MAPK can maintain an non-activated state in the face of potent pro-inflammatory stimuli. This appears to involve effects on the synthesis and response of cells to two key inflammatory mediators, TNF and IL-1, at both transcriptional and post-transcriptional levels. Therefore selective inhibitors of p38 MAPK have been proposed as potential disease-modifying drugs with a novel mechansim of action. The progression of these compounds into clinical studies over the next few years will enable this proposal to be tested and may lead to a new class of anti-inflammatory drugs targeted against cytokine signal transduction pathways.

References

1   Han J, Lee J-D, Bibbs R, Ulevitch RJ (1994) A MAP Kinase targeted by endotoxin and hyperosmolarity in mammalian cells. *Science* 265: 808–811

2   Freshney NW, Rawlinson L, Guesdon F, Jones E, Cowley S, Hsuan J, Saklatvala J (1994) Interleukin-1 activates a novel protein kinase cascade that results in the phosphorylation of Hsp27. *Cell* 78: 1039–1049

3   Rouse J, Cohen P, Trigon S, Morange M, Alonso-Llamazares A, Zamanillo D, Hunt T, Nebreda AR (1994) A novel kinase cascade triggered by stress and heat shock that stimulates MAPKAP kinase-2 and phosphorylation of the small heat shock proteins. *Cell* 78: 1027–1037

4   Lee J, Laydon JT, McDonnell PC, Gallagher TF, Kumar S, Green D, McNulty D, Blumenthal MJ, Heys JR, Landvatter SW et al (1994) A protein kinase involved in the regulation of inflammatory cytokine biosynthesis. *Nature* 372: 739–746

5   Jiang Y, Chen C, Zhuangjie L, Guo W, Gegner JA, Lin S, Han J (1996) Characterisation of the structure and function of a new mitogen-activated protein kinase (p38B). *J Biol Chem* 271: 17920–17926

6   Li Z, Jiang Y, Ulevitch RJ, Han J (1996) the primary structure of p38γ: a new member of p38 group of MAP kinases. *Biochem Biophy Res Commun* 228: 334–340

7   Kumar S, McDonnell PC, Gum RJ, Hand AT, Lee JC, Young PR (1997) Novel homo-

logues of CSBP/p38 MAP kinase: activation, substrate specificity and sensitivity to inhibiton by pyridinyl imidazoles. *Biochem Biophys Res Commun* 235: 533–538

8   Raingeaud J, Gupta S, Rogers JS, Dickens M, Han J, Ulevitch RJ, Davis RJ (1995) Pro-inflammatory cytokines and environmental stress cause p38 mitogen-activated protein kinase activation by dual phosphorylation on tyrosine and threonine. *J Biol Chem* 270: 7420–7426

9   Doza YN, Cuenda A, Thomas GM, Cohen P, Nebreda AR (1995) Activation of the MAP kinase homologue RK requires the phosphorylation of Thr-180 and Tyr-182 and both residues are phosphorylated in chemically stressed KB cells. *FEBS Lett* 364: 223–228

10   Cuenda A, Alonso G, Morrice N, Jones M, Meier R, Cohen P, Nebreda AR (1996) Purification and cDNA cloning of SAPKK3, the major activator of RK/p38 stress- and cytokine-stimulated monocytes and epithelial cells. *EMBO J* 15: 4156–4164

11   Han J, Lee J-D, Jiang Y, Li Z, Feng L, Ulevitch R (1996) Characterisation of the structure and function of a novel MAP kinase kinase (MKK6). *J Biol Chem* 271: 2886–2891

12   Raingeaud J, Whitmarsh AJ, Barrett T, Derijard B, Davis RJ (1996) MKK3 and MKK6-regulated gene expression is mediated by p38 mitogen-activated protein kinase signal transduction pathway. *Mol Cell Biol* 16: 1247–1255

13   Derijard B, Raingeaud J, Barrett T, Wu I-H, Han J, Ulevitch RJ, Davis RJ (1995) Independent human MAP kinase signal transduction pathways defined by MEK and MKK isoforms. *Science* 267: 682–685

14   Deacon K, Blank JL (1997) Characterisation of the mitogen-activated protein kinase kinase 4 (MKK4)/c-Jun NH2-terminal kinase 1 and MKK3/p38 pathways regulated by MEK kinases 2 and 3. *J Biol Chem* 272: 14489–14496

15   Moriguchi M, Kuroyanagi N, Yamaguchi K, Gotoh Y, Irie K, Kano T, Shirakabe K, Muro Y, Shibuya H, Matsumoto K et al (1996) A novel kinase cascade mediated by mitogen activated protein kinase kinase 6 and MKK3. *J Biol Chem* 271: 13675–13679

16   Ninomiya-Tsuji J, Kishimoto K, Hiyama A, Inoue J-I, Cao Z, Matsumoto K (1999) The kinase TAK1 can activate the NIK-IκB as well as the MAP kinase cascade in the IL-1 signalling pathway. *Nature* 398: 252–256

17   McLaughlin M, Kumar S, McDonnell PC, Van Horn S, Lee JC, Livi GP, Young PR (1996) Identification of mitogen activated protein (MAP) kinase-activated protein kinase-3, a novel substrate of CSBP p38 MAP kinase. *J Biol Chem* 271: 8488–8492

18   New L, Jiang Y, Zhao M, Liu K, Zhu W, Flood LJ, Kato Y, Parry GC, Han J (1998) PRAK, a novel protein kinase regulated by p38 MAP kinase. *EMBO J* 17: 3372–3384

19   Deak M, Clifton AD, Lucocq JM, Alessi DR (1998) Mitogen- and stress-activated protein kinase-1 (MSK-1) is directly activated by MAPK and SAPK/p38 and may mediate the activation of CREB. *EMBO J* 17: 4426–4441

20   Waskiewicz AJ, Flynn A, Proud CG, Cooper JA (1997) Mitogen-activated protein kinases activate the serine/threonine kinases Mnk1 and Mnk2. *EMBO J* 16: 1909–1920

21   Han J, Jiang Y, Li Z, Kravchenko V, Ulevitch RJ (1997) Activation of the transcription factor MEF2C by the MAP kinase p38 in inflammation. *Nature* 386: 296–299

22 Janknecht R and Hunter T (1997) Convergence of MAP kinase pathways on a ternary complex factor Sap-1a. EMBO ϑ 16: 1620–1627

23 Lee JC, Griswold D, Votta B, Hanna N (1988) Inhibition of monocyte Il-1 production by the anti-inflammatory compound, SK&F 86002. *Int J Immunopharmacol* 10: 835– 843

24 Lee JC, Badger AM, Griswold DE, Dunnington D, Truneh A, Votta B, White JR, Young PR, Bender PE (1993) Bicyclic imidazoles as a class of cytokine biosynthesis inhibitors. *Ann NY Acad Sci* 696: 149–170

25 Griswold DE, Hillegrass LM, Meunier PC, DiMartino MJ, Hanna N (1988) The effect of inhibitors of eicosanoid metabolism in murine collagen-induced arthritis. *Arthritis Rheum* 31: 1406–1412

26 Gallagher TF, Fier-Thompson SM, Garigipati RS, Sorenson ME, Smietana JM, Lee D, Bender PE, Lee JC, Laydon JT, Griswold DE et al (1995) 2,4,5-triarylimidazole inhibitors of IL-1 biosynthesis. *Bioorg Med Chem Lett* 5: 1171–1176

27 de Laszlo SE, Visco D, Agarwal L, Cahng L, Chin J, Croft G, Forsyth A, Flectcher D, Frantz B, Hacker C et al (1998) Pyrroles and other heterocycles as inhibitors of p38 kinase. *Bioorg Med Chem Lett* 8: 2689–2694

28 Henry JR, Rupert KC, Dodd JH, Turchi IJ, Wadsworth SA, Cavender DE, Fahmy B, Olini GC, Davis JE, Pellegrino-Gensey JL et al (1998) 6-amino-2-(4-fluorophenyl)-4-methoxy-3-(4-pyridyl)-1*H*-pyrrolo2,3-*b*]pyridine (RWJ 68354): a potent and selective p38 kinase inhibitor. *J Med Chem* 41: 4196–4198

29 Aantanarayan A, Clare M, Geng L, Hanson G, Partis RA, Stealey MA, Weier RM (1998) 3(5)-heteroaryl substituted pyrazoles as p38 kinase inhibitors. GD Searle & Co patent application WO9852937

30 Jackson JR, Bolognese B, Hillegass L, Kassis S, Adams J, Griswold DE, Winkler JD (1998) Pharmacological effects of SB 220025, a selective inhibitor of p38 mitogen activated protein kinase, in angiogenesis and chronic inflammatory disease models. *J Pharmacol Exp Ther* 284: 687–692

31 LoGrasso PV, Frantz B, Rolando AM, O'Keefe SJ, Hermes JD, O'Neill EA (1997) Kinetic mechanism for p38 MAP kinase. *Biochemistry* 36: 10422–10427

32 Young PR, Mclaughlin M, Kumar S, Kassis S, Doyle ML, McNulty D, Gallagher TF, Fisher S, McDonnell PC, Carr SA et al (1997) Pyridinylimidazole inhibitors of p38 mitogen activated protein kinase bind in the ATP site. *J Biol Chem* 272:12116–12121

33 Frantz B, Klatt MP, Parsons J, Rolando A, Williams H, Tocci MJ, O'Keefe SJ O'Neill EA (1998) The activation of p38 mitogen activated protein kinase determines the efficiency of ATP competition for the pyridinylimidazole inhibitor binding. *Biochemistry* 37: 13846–13853

34 Cuenda A, Rouse J, Doza YN, Meier R, Cohen P, Gallagher TF, Young PR, Lee JC (1995) SB 203580 is as a specific inhibitor of a MAP kinase homologue which is stimulated by cellular stresses and interleukin-1. *FEBS Lett* 364: 229–233

35 Clerk A, Sugden PH (1998) The p38 MAPK inhibitor, SB 203580, inhibits cardiac stress-activated protein kinases/c-jun N-terminal kinases (SAPKs/JNKs). *FEBS Lett* 426: 93–96

36 Gum RJ, McLaughlin M, Kumar S, Wang Z, Bower MJ, Lee JC, Adams JL, Livi GP,

Goldsmith EJ, Young PR (1998) Acquisition of sensitivity of stress-activated protein kinases to the p38 inhibitor, SB 203580, by alteration of one or more amino acids within the ATP binding pocket. *J Biol Chem* 273; 15605–15610

37 Lisnock J, Tebben A, Frantz B, O'Neill EA, Croft G, O'Keefe SJ, Li B, Hacker C, de Laszlo S, Smith A et al (1998) Molecular basis for p38 protein kinase inhibitor specificity. *Biochemistry* 37: 16573–16581

38 Eyers PA, Craxton M, Morrice N, Cohne P, Goedert M (1998) Conversion of SB 203580-sensitive MAP kinase family members to drug-sensitive forms by a single amino acid substitution. *Chem Biol* 5: 321–328

39 Wilson KP, Fitzgibbon MJ, Caron PR, Griffith JP, Chen W, McCaffrey PG, Chambers SP, Su M (1996) Crystal structure of p38 mitogen activated protein kinase. *J Biol Chem* 271: 27696–27700

40 Wang Z, Harkins PC, Ulevitch RJ, Han J, Cobb MH, Goldsmith EJ (1997) The structure of mitogen activated protein kinase p38 at 2.1-A resolution. *Proc Natl Acad Sci USA* 94: 2327–2332

41 Tong L, Pav S, White DM, Rogers S, Crane KM, Cywin CL, Brown ML, Pargellis CA (1997) A highly specific inhibitor of human p38 MAP kinase binds in the ATP pocket. *Nature Struct Biol* 4: 311–316

42 Wislon KP, McCaffrey PG, Hsiao K, Pazhanisamy S Galullo V, Bemis GW, Fitzgibbon MJ, Caron PR, Murcko MA, Su M (1997) The structural basis for the specificity of pyridinylimidazole inhibitors of p38 MAP kinase. *Chem Biol* 4: 423–431

43 Wang Z, Canagarajah BJ, Boehm JC, Kassisa S, Cobb MH, Young PR, Abdel-Meguid S, Adams JL, Goldsmith EJ (1998) Structural basis of inhibitor selectivity in MAP kinases. *Structure* 6: 1117–1128

44 Badger AM, Bradbeer JM, Votta B, Lee JC, Adams JL, Griswold DE (1996) Pharmacological profile of SB 203580, a selective inhibitor of cytokine suppressive binding protein/p38 kinase, in animal models of arthritis, bone resorption, endotoxin shock and immune function. *J Pharmacol Exp Ther* 279: 1453–1461

45 Williams RO, Feldmann M, Maini RN (1992) Anti-tumour necrosis factor ameliorates joint disease in murine collagen-induced arthritis. *Proc Natl Acad Sci USA* 89: 9784–9788

46 Van den Berg WB, Joosten LA, Helsen M, Van de Loo FA (1994) Amelioration of established murine collagen-induced arthritis with anti-IL-1 treatment. *Clin Exp Immunol* 95: 237–243

47 Ridley SH, Sarsfield SJ, Lee JC, Bigg HF, Cawston TE, Taylor DJ, DeWitt DL, Saklatvala J (1997) Actions of IL-1 are selectively controlled by p38 mitogen activated protein kinase. *J Immunol* 158: 3165–3173

48 Miyazawa K, Mori A, Miyata H, Akahane M, Ajisawa Y, Okudaira H (1998) Regulation of interleukin-1β-induced interleukin-6 gene expression in human fibroblast-like synoviocytes by p38 mitogen activated protein kinase. *J Biol Chem* 273: 24832–24838

49 Beyaert R, Cuenda A, Vanden Berghe W, Plaisance S, Lee JC, Haegeman JC, Cohen P, Fiers W (1996) The p38/RK mitogen activated protein kinase pathway regulates interleukin-6 synthesis response to tumour necrosis factor. *EMBO J* 15: 1914–1923

50 Pouliot M, Baillargeon J, Lee JC, Cleland LG, James MJ (1997) Inhibition of prostaglandin endoperoxide synthase-2 expression in stimulated human monocytes by inhibitors of p38 mitogen activated protein kinase. *J Immunol* 158: 4930–4937

51 Dean JL, Brook M, Clark AR, Saklatvala J (1999) p38 mitogen activated protein kinase regulates cyclooxygenase-2 mRNA stability and transcription in lipopolysaccharide-treated human monocytes. *J Biol Chem* 274: 264–269

52 Ridley S, Dean JL, Sarsfield SJ, Brook M, Clark AR, Saklatvala J (1998) A p38 map kinase inhibitor regulates stability of interleukin-1-induced cyclooxygenase-2 mRNA. *FEBS Lett* 439: 75–80

53 Seki N, Sudo Y, Yoshioka T, Sugihara S, Fujitsu T, Sakuma S, Ogawa T, Hamaoka T, Senoh H, Fujiwara H (1988) Type II collagen-induced murine arthritis. I. Induction and perpetuation of arthritis require synergy between humoral and cell-mediated immunity. *J Immunol* 140: 1477–1484

54 Craxton A, Shu G, Graves JD, Saklatvala J, Krebs EG, Clark EA (1998) p38 MAPK is required for CD40-induced gene expression and proliferation in B lymphocytes. *J Immunol* 161: 3225–3236

55 Schafer PH, Wang L, Wadsworth SA, Davis JE, Siekierka JJ (1999) T cell activation signals up-regulate p38 mitogen activated protein kinase activity and induce TNFα production in a manner distinct from LPS activation of monocytes. *J Immunol* 162: 659–668

56 Crawley JB, Rawlinson L, Lai FV, Page TH, Saklatvala J, Foxwell BM (1997) T cell proliferation in response to interleukins 2 and 7 requires p38 MAP kinase activation. *J Biol Chem* 272: 15023–15027

57 Koprak S, Staruch MJ, Dumont FJ (1999) A specific inhibitor of p38 mitogen activated protein kinase affects differentially the production of various cytokines by activated human T cells: dependence on CD28 signalling and preferential inhibition of IL-10 production. *Cell Immunol* 192: 87–95

58 Bonnefoy JY, Lecoanet-Henchoz S, Gauchat JF, Graber P, Aubry JP, Jeannin P, Plater-Zyberk C (1997) Structure and functions of CD23. *Int Rev Immunol* 16: 113–128

59 Marshall LA, Hansbury MJ, Bolognese BJ, Gum RJ, Young PR, Mayer RJ (1998) Inhibitors of the p38 mitogen activated kinase modulate IL-4 induction of low affinity IgE receptor (CD23) in human monocytes. *J Immunol* 161: 6005–6013

60 Plater-Zyberk C, Bonnefoy JF (1995) Marked amelioration of established collagen-induced arthritis by treatment with antibodies to CD23 *in vivo*. *Nat Med* 1: 781–785

61 Kleinu S, Martinsson P, Gustavsson S, Heyman B (1999) Importance of CD23 for collagen-induced arthritis: delayed onset and reduced severity in CD23-deficient mice. *J Immunol* 162: 4266–4270

62 Pritchett W, Hand A, Shields J, Dunnington D (1995) Mechanism of action of bicylic imidazoles defines a translational regulatory pathway for tumour necrosis factor alpha. *J Inflamm* 45: 97–105

63 Kotlyarov A, Neininger A, Schubert C, Eckert R, Birchmeier C, Volk H-D, Gaestel M (1999) MAPKAP kinase-2 is essential for LPS-induced TNFα biosynthesis. *Nat Cell Biol* 1: 94–97

TGFβ and IL-10: inhibitory cytokines regulating immunity and the response to infection

Christian Bogdan[1], Yoram Vodovotz[2] and John Letterio[3]

[1]Institute for Clinical Microbiology, Immunology, and Hygiene, University of Erlangen-Nuremberg, Wasserturmstrasse 3, D-91054 Erlangen, Germany; [2]Department of Surgery, University of Pittsburgh, 200 Lothrop St., Room E1558, Pittsburgh, PA 15213, USA; [3]Lab of Cell Regulation and Carcinogenesis, Building 41, Room C629, National Cancer Institute, National Institutes of Health, Bethesda, MD 20892-5055, USA

Introduction

A complex interplay between cytokines provides plasticity to our systems of host defense and immune surveillance, but also presents a tremendous challenge to our understanding of the host immune response. These cytokines are typically pleiotropic. They may amplify their own expression, influence the production of other cytokines and effector molecules, and exert an end-effect that is often defined by the array of other signals acting in their presence. Determining those participants which negatively regulate these effector responses may enable us to manipulate the exaggerated response, whether to pathogens or to self, and to enhance an ineffective response by interfering with their activity.

Among the few cytokines known to suppress immune responses, transforming growth factor-β (TGFβ) and interleukin-10 (IL-10) have been studied in greatest detail, and have clearly been implicated in mechanisms of disease pathogenesis. While the spectrum of functional activities for TGFβ is much broader than that known for IL-10, their effects on the immune system are often overlapping and sometimes interdependent [1]. Both cytokines are important in the definition of Th responses, control of antigen presentation, and the regulation of macrophage function. This chapter is focused on the most notable and best-described activities of these inhibitory cytokines, and we aim to summarize our current understanding of their role in disorders of immune function and host response.

Novel Cytokine Inhibitors, edited by Gerry A. Higgs and Brian Henderson
© 2000 Birkhäuser Verlag Basel/Switzerland

Effects of TGFβ1 on leukocyte activation and function

Effects of TGFβ1 on macrophages and antigen-presenting cells

Studies of mechanisms of immune suppression in malignancy have documented the ability of tumor cells or their secreted products to suppress the function of tumor-infiltrating macrophages [2, 3]. Similar effects have been ascribed to various intracellular pathogens, such as *Leishmania* or *Schistosoma*, which infect macrophages and whose growth may be reduced by these effector cells [2, 4]. TGFβ1 is clearly the most commonly identified host factor implicated as facilitating tumor progression and metastasis, as well as aiding in the survival of such intracellular pathogens. The effects of TGFβ1 on macrophages range from influence over chemotaxis and antigen presentation to regulation of activation and function. It is essentially impossible to truly separate these effects since they are often mutually dependent. Functionally, the modulation of these processes by TGFβ1 may confer a benefit on the macroorganism insofar as the often toxic products of the immune system can cause great damage if left unchecked. However, it is possible that throughout evolution numerous pathogens whose survival would be impaired by vigorously activated macrophages have adapted to usurp these pathways for their own advantage, and that tumors may also gain a survival advantage through modulation of macrophage effector functions.

Modulation of macrophage cytokine production by TGFβ
It is often difficult to discern the direct effects of TGFβ from those which are secondary to the modulation of either pro- or anti-inflammatory cytokines. While the impact of TGFβ on antigen presentation or free radical production may well be a consequence of the latter, *in vivo* studies in transgenic or gene knockout animals suggest that each is important, as will be discussed below.

Effects on the production of proinflammatory cytokines
In vitro, TGFβ1 has been described as a potent suppressor of tumor necrosis factor-α (TNFα) production [5, 6]. In mouse thioglycollate-elicited macrophages, TGFβ1 suppresses TNFα production by post-transcriptional mechanisms [6]. Interestingly, McCartney-Frances et al. had suggested that TGFβ1 might actually enhance TNFα production under some circumstances in peripheral blood monocytes, which suggests a complex interplay between temporal and spatial factors [7]. In mice which over-express TGFβ1 (Alb/TGFβ1), it was found that lipopolysaccharide (LPS) stimulation actually enhanced the production of TNFα [8]. TNFα levels present in sera of endotoxaemic Alb/TGFβ1 mice were several-fold higher than those of endotoxaemic wild-type mice, and remained elevated up to 12 h following challenge with LPS, whereas TNFα levels in wild-type mice peaked at 1–3 h [8]. These findings raise the

possibility that chronic exposure to elevated levels of active endogenous TGFβ1 may alter the systemic response to endotoxin. TNFα is known to be regulated both post-transcriptionally and post-translationally, and either mechanism may operate in the presence of sustained systemic exposure to activated TGFβ1.

Since TNFα and IL-1 are often regulated concomitantly, it is not surprising that *in vitro* TGFβ1 can also suppress the expression of IL-1 [5, 6]. Interestingly, TGFβ1 may also reduce the expression of the IL-1 receptor [9], and has a similar effect on a number of other cytokine receptors, including those for the colony stimulating factor G-CSF and for the stem cell factor (c-kit), the IL-2 receptor, and the interferon-γ (IFNγ) receptor [10]. Many of the actions of IFNγ are inhibited by TGFβ1, such as induction of antigens of major histocompatibility complex (MHC) class I and class II [11, 12]. Studies in the TGFβ1 knockout mouse demonstrated an increased expression of both class I and class II MHC antigens, though this was not correlated with an overall difference in IFNγ mRNA levels prior to the onset of inflammation [13]. As already mentioned above, TGFβ can also stimulate production of pro-inflammatory cytokines, a process which may depend on the state of activation of the cell. For example, resting human peripheral blood monocytes are stimulated by picomolar concentrations of TGFβ to produce increased levels of mRNAs for both IL-1α, -1β, and TNFα, as well as PDGF-BB, and bFGF (reviewed in [14]).

Anti-inflammatory cytokines induced by TGFβ

One of the potentially most important effects of TGFβ1 is its capacity to induce its own production through the presence of AP-1 and EGR-1 binding sites in its promoter [15]. Thus, TGFβ1 elaborated during lymphocyte activation [16] may act in an autocrine fashion to suppress the proliferation and differentiation of lymphocytes [17] and lymphokine-activated killer cells [18]. In cardiac myocytes, autocrine TGFβ1 protects against the negative inotropic effects of IL-1 by suppression of the expression of NOS2 [19]. Production of TGFβ1 by tumor cells may be pro-metastatic through the induction of proteases [20] and by the direct suppression of cytotoxic T lymphocytes (CTL) [21], but it also enhances the production of IL-10 by macrophages, thereby contributing to a shift toward Th2-type respones and the general immunosuppression associated with the tumor-bearing state [1].

Effects of TGFβ on antigen presentation

TGFβ1 suppresses the expression of the MHC class I and class II genes, both *in vivo* and *in vitro* [11, 12]. Altered expression of these antigens in the TGFβ1 knockout mouse suggests a role for this cytokine in the developmental regulation of these molecules, including their expression on nonprofessional antigen-presenting cells (APCs) [13]. These potentially suppressive effects of TGFβ on antigen presentation are balanced by positive effects, which appear to be exerted in a context-dependent fashion. For example, the phagocytic activity of monocytes/macrophages (and perhaps other APCs) is activated by TGFβ. The ability of TGFβ to upregulate FcγRIII expression and macrophage recognition of phosphatidyl serine enhances both immunophagocytosis and recognition of apoptotic cells, facilitating antigen uptake and presentation. Moreover, the development of highly specialized populations of APCs, such as epidermal dendritic cells, is dependent on TGFβ. Their expansion *in vitro* is greatly enhanced by the ability of TGFβ to inhibit apoptosis of differentiating progenitors. The absence of Langerhans cells in the epidermis of TGFβ1 knockout mice implicates this cytokine as a key participant in dendritic cell ontogeny (reviewed in [14]).

In addition to direct effects on macrophages and APCs, local production of active TGFβ by macrophages, natural killer (NK) cells, and suppressor T cells can clearly modify the response of lymphocytes in contact with these APCs. A dramatic example is provided by a recent investigation into the protective effect of donor-derived, activated natural killer (NK) cells in a murine model of graft-versus-host disease (GvHD). The severity of GvHD following allogeneic bone marrow transplantation was greatly reduced in mice that received donor NK cells, an effect that was completely reversed if the mice were concomitantly treated with systemic administration of anti-TGFβ-blocking antibodies, or if the NK cell administration was delayed [22]. Whether TGFβ is produced directly by the NK cells remains to be shown, but the results suggest that the presence of TGFβ at the time of APC interaction with allogeneic lymphocytes can induce a state of tolerance, a concept which may be useful in crossing the HLA barrier to transplantation, and in establishing tolerance to self antigens in the setting of autoimmune disorders [14, 23].

Suppression of free radical generation by TGFβ
Two weapons in the anti-pathogen and anti-tumor arsenal of activated macrophages are the free radicals superoxide (O_2^-) and nitric oxide (NO) [2, 4, 24, 25], derived enzymatically from NADPH:ubiquinone oxidoreductase (NADPH oxidase) and the inducible nitric oxide synthase (NOS2; iNOS). Among the first immunosuppressive effects of TGFβ1 to be reported was the suppression of O_2^- production by activated macrophages [26]. Subsequently, it was demonstrated that TGFβ1 also suppressed NO production by activated macrophages [27–30]. In fact, TGFβ1 as well as TGFβ2 and TGFβ3 were found to suppress NO production in numerous cell types *in vitro* [31, 32]. *In vivo*, TGFβ1 appears to be a primary negative regulator of NO production, as determined from the spontaneously elevated expression of NOS2 in TGFβ1 null mice [33] and from the blunted NO production exhibited by endotoxaemic Alb/TGFβ1 mice [8]. In macrophages, the actions of TGFβ1 appear to be predominantly post-transcriptional and post-translational and thus can effect a reduction of fully-formed NOS2 [30]. A role has also been described for endogenous TGFβ1 in reducing the NO-producing capacity of macrophages in response to

elevated systemic NO production [34]. Nonetheless, there are several reports documenting the ability of TGFβ to enhance NO production by activated macrophages, most notably in cows [35, 36].

Autocrine regulatory loops control TGFβ activity in macrophages

A unique feature of the TGFβ system is the fact that cytokines whose expression is suppressed by TGFβ1 also serve to induce or activate this molecule. IL-2 induces transcription of TGFβ1 in macrophages, and IL-1α induces murine splenic T cells to secrete active TGFβ without an effect on levels of TGFβ1 message. Activation of macrophages or lymphocytes with either LPS or concanavalin A (conA) leads to increased secretion of active TGFβ by these cells, without significant changes in TGFβ1 mRNA (reviewed in [37]). This would appear to serve as an efficient regulatory loop, acting to prevent the excessive proliferation and pathological consequences of unfettered functional activities of these cell populations. Recent studies now indicate that the free radical effector mechanisms of macrophages can also bring about increased expression or activation of latent TGFβ1, either directly or indirectly. Activated macrophages express predominantly active TGFβ1 despite a constitutive level of total TGFβ1 mRNA, through what appear to be post-translational mechanisms [38, 39]. A role for free radicals in this process is suggested by the observation that O_2^- can activate latent TGFβ1 directly, raising the possibility of a rapid negative feedback response to oxidative stress [39]. This form of autocrine feedback may be specific to the activating signal. The treatment of thioglycolate-elicited macrophages with IFNγ leads to a prominent reduction of the release of latent TGFβ without an increase in activation or secretion, in a process which requires endogenous NO [40, 41].

Suppressive effects of TGFβ in lymphocyte function

TGFβ influences normal B and T lymphopoiesis, beginning with inhibitory effects on stem cells and their committed progenitors, and extending to control over their activation and their phenotype in the terminally differentiated state. The mechanisms which underlie these effects include regulation of cell cycle progression, control of accessory molecule and receptor expression, and both the induction and prevention of apoptosis. Studies detailing these events have recently been reviewed [14, 37], thus we will highlight the suppressive effects of a specialized group of TGFβ-producing lymphocytes.

Impact on Th subset determination and cytokine production

The factors involved in the generation of TGFβ-producing cells represent an active area of research. The importance of these suppressor cells in controlling inflamma-tory responses is supported by the demonstration that several different lymphocyte subsets can fulfill this role via production of active TGFβ; these include both CD4+ and CD8+ T cells and NK cells, each of which are critical determinants of an effec-tive immune response, suppressed autoreactivity, or clearance of an infectious process. As mentioned above, TGFβ is presumed to mediate the suppression of murine GvHD by NK cells, and recent studies of human peripheral blood mononu-clear cells (PBLs) indicate that NK cells can serve as the source of this cytokine [22]. Mitogenic combinations of anti-CD2 antibodies fail to stimulate immunoglobulin production in these PBL cultures, unless NK cells are removed, or if an anti-TGFβ blocking antibody is added at the time of stimulation [42]. Short overnight exposure of CD8+ T cells to TGFβ will prime them to suppress Ig production in these cultures (in the absence of NK cells), revealing a complex regulatory circuit, in which TGFβ from one source can be primed for its production by another, and affect the func-tion of a distinct population.

The natural occurrence of TGFβ-producing, CD8+-suppressor T cells has been demonstrated in studies of experimental allergic encephalomyelitis (EAE) [43]. Oral tolerization to myelin basic protein induces peripheral tolerance through mecha-nisms that include the generation of CD8+ T cells, as well as CD4+ TH2-like cells producing IL-4, IL-10, and secreting TGFβ in an antigen-specific response (reviewed in [14]). More recent studies have focused on factors priming CD4+ T cells for pro-duction of TGFβ in response to antigen. *In vitro* studies show that APC stimulation of CD4+ T cells in the presence of blocking antibodies to IL-12 and IFNγ results in significant production of active TGFβ. *In vivo*, the administration of anti-IL12 dur-ing oral tolerization resulted in a several-fold increase in TGFβ production, clearly polarizing the Th1-determining IL-12 effects from the suppressive, Th2-specific effects of TGFβ [23]. While the addition of IL-4 to priming cultures of CD4+ T cells can also augment their production of TGFβ on restimulation, exposure to IL-4 (or other cytokines) has not been shown to be critical for the generation of TGFβ-pro-ducing suppressor T cells (reviewed in [14]). Moreover, TGFβ itself can regulate its own production [44], and may be a key factor in the differentiation of naive CD4+ T cells into TGFβ-producing cells. The unique nature and potentially special func-tion of this T cell subset has prompted the designation "Th3" for cells predomi-nantly secreting TGFβ, distinguishing them from the IL-4- and IL-10-producing Th2 subsets that may also produce TGFβ.

TGFβ in disease pathogenesis and treatment

The profound immune modulatory effects of TGFβ have obvious functional conse-quences for processes ranging from autoimmunity to tumor surveillance and the host response to infectious diseases. *In vivo* studies of these processes in murine models and human disease have established important concepts regarding the function of this cytokine. The role of TGFβ in autoimmune disease has recently been reviewed [14] and thus will not be considered here.

Duality of TGFβ function in response to infectious diseases.

Among the many similarities between TGFβ and IL-10 is their sharing the dual capacity to enhance the establishment of opportunistic and intracellular infections while protecting the host from the potentially harmful effects of an unchecked, chronic inflammatory response. TGFβ is now recognized as an important parasite escape mechanism in all forms of human and murine leishmaniosis (for which macrophages serve as an important cellular target) and for *Trypanosoma cruzi* (which infect all nucleated cells).

Studies in *T. cruzi* suggest that autocrine TGFβ signaling is required for parasite entry into epithelial cells [45]. In *Leishmania*, there is a correlation between macrophage production of TGFβ upon infection with promastigotes and both parasite proliferation as well as strain virulence (reviewed in [14]). In both rodents and humans, the suppression of NO production by TGFβ1 has been implicated in increased susceptibility to infection by intracellular parasites which infect macrophages [28, 29, 46–53]. Other mechansims include the inhibition of cell-mediated immunity by suppression of the effects of IL-4 and IFNγ on the expression of class II MHC antigen on APCs, and by direct suppression of CD8+ cytotoxic T cells which are important in eliminating infected cells. The ability of TGFβ1 to downregulate the expression of the IFNγ receptor [10] may account for the decreased responsiveness to IFNγ in macrophages treated with TGFβ1 [54]. This may also contribute to the shift toward Th2 responses in susceptible strains, with increased production of IL-4 and IL-10. This is in contrast to resistant strains in which a Th1 response predominates and includes production of IFNγ and IL-2. While the sustained production of active TGFβ may serve to limit the destructive sequelae associated with a chronic inflammatory state, it has also been implicated in the fibrotic complications often associated with these conditions, such as the granulomatous reaction seen in *Schistosomiasis* [55].

Tumor progression and TGFβ-mediated immune suppression

This dichotomy in the activity of TGFβ is perhaps best exemplified in the setting of malignancy. TGFβ maintains a negative effect on the growth of most cell types, and members of the entire TGFβ signaling pathway (including ligands, receptors, and signaling intermediates) each function as true tumor suppressors [56]. In a paradigm that is now well established, cells which escape the cell cycle controls exerted by

TGFβ typically acquire the capacity to elaborate an actived form of TGFβ. This clearly contributes to invasion and progression through increased angiogenesis and matrix production, but also provides for the escape of tumor cells from cell-mediated immune responses.

TGFβ-mediated mechanisms for tumor-induced immunosuppression are also quite complex. Early studies demonstrated that TGFβ1 suppressed macrophage tumor cytocidal activity *in vitro* [57], and later *in vivo* studies confirmed these findings [58–60]. Studies of xenotransplants in athymic mice have shown that production of TGFβ by both colonic and breast carcinomas suppresses cytotoxicity of activated monocytes and NK cells [61]. This effect is reversed by systemic administration of blocking antibodies to TGFβ, which inhibits development of primary tumors and distant metastases [61, 62]. T-cell-mediated immunity is also generally impaired in tumor-bearing hosts. The suppression of CD4+ T-cell function associated with the growth of the MH134 hepatoma *in vivo* can be reversed by systemic administration of anti-TGFβ antibodies [21]. In transfection experiments, stable expression of TGFβ1 in a highly immunogenic fibrosarcoma completely impairs the ability of the tumor to elicit CTL responses, and results in a tumorigenic phenotype [63]. Less is known about effects of TGFβ on presentation of tumor antigens by APCs, but this is also an area of great interest. Studies of EL-4 tumor-bearing mice reveal a link between TGFβ production and increased IL-10 expression, producing a bias toward Th2 responses, possibly through inhibitory effects on APCs [1]. Recent studies of modified tumor vaccines document the ability of IL-10 to prevent APC from obtaining access to tumor antigens [64].

Effects of interleukin-10

IL-10 was originally described as a product of mouse type 2 CD4+ T-helper lymphocytes (Th2), which potently inhibits the secretion of IFNγ and IL-2 by Th1 cells [65, 66]. In the meantime, IL-10 has been shown to be generated by many other cell types including monocytes and macrophages, CD5+ and CD5- B lymphocytes, CD8+ T cells, human Th1 and Th2 cells, NK cells, mast cells, eosinophils and keratinocytes. An equally broad spectrum of cells (monocytes/macrophages, dendritic cells, neutrophils, mast cells, γδ-T cells, CD8+ T cells, B cells, NK cells, endothelial cells, mesangial cells) have been identified as targets of IL-10 [65–72]. IL-10 inhibits both innate and T-cell-mediated specific immune inflammation chiefly through mechanisms that involve downmodulation of cytokine production by macrophages and the inhibition of accessory functions of APCs in T-cell activation. There are also direct effects of IL-10 on T cells, which may mediate the aberrant T-helper-cell activation and impaired T-cell function seen in autoimmunity and infection [73, 74]. In the following sections, we review the major activities of IL-10 in the regulation of macrophages and other accessory cells involved in T-cell activation.

Effects of IL-10 on monocytes/macrophages and other antigen-presenting cells

The key functions of monocytes/macrophages are the phagocytosis and killing of microbial pathogens, the production of cytokines, and the processing and MHC-dependent presentation of antigens. Each of these functions can be modulated by IL-10 (Tabs. 1 and 2). Although direct and inhibitory effects of IL-10 appear to predominate, additional aspects of the function of IL-10 need to be considered. First, part of the immunosuppressive effects of IL-10 on macrophages can be indirect, e.g., *via* inhibition of secretion of IFNγ or macrophage migration inhibitory factor (MIF) by Th1 cells [75, 76] , *via* reduction of the release of T-cell-activating cytokines such as IL-1 [6, 77, 78] or IL-12 [79-81], or by inhibiting the release of T cell-attracting chemokines (e.g., macrophage inflammatory protein (MIP)-1α and MIP-1β [82, 83]. Second, some macrophage functions can be positively or negatively regulated by IL-10 depending on the type of macrophage, the source of IL-10 (expressed in *E. coli* vs eukaryotic cells) and the stimulation conditions applied *in vitro*. The production of NO, for example, can be inhibited, enhanced, or remain unaltered by IL-10 (reviewed in [25]). Third, IL-10 can also have immunostimulatory properties. These have been best demonstrated *in vivo*, where transgenic expression of IL-10 promoted the destruction of β cells in the pancreatic islets and increased the prevalence of diabetes in nonobese diabetic (NOD) mice and treatment with IL-10 inhibited the metastasis of tumor cells [84, 85].

Modulation of macrophage antimicrobial activity
In most *in vitro* models, treatment of mouse or human macrophages with IL-10 not only impeded the production of reactive oxygen intermediates (O_2^-, H_2O_2) and the generation of NO by iNOS (NOS2) (reviewed in [25]), but in parallel also interfered with the killing of extra- and intracellular pathogens. The list of microbes which show enhanced survival in the presence of IL-10 now includes *Aspergillus fumigatus*, *Candida albicans*, *Listeria monocytogenes*, *Mycobacterium bovis*, *Leishmania major*, schistosomula of *Schistosoma mansoni*, *Toxoplasma gondii* and *Trypanosoma cruzi* [68, 86–89]. Whereas the mechanism of suppression of iNOS by IL-10 has not yet been analysed, there is one study which shows that the reduced production of superoxid anion (O_2^-) by human monocytes after treatment with IL-10 is accompanied by a downregulation of the mRNA expression of two components of the NADPH oxidase (p47 phox and gp91 phox) [90]. More recently, IL-10 was reported to also reduce the oxidative burst and bactericidal activity of human neutrophils [91].

Modulation of macrophage cytokine production by IL-10
The spectrum of cytokines which are regulated by IL-10 is broad and comprises

Table 1 - Modulation of macrophage cytokine production by IL-10

Cytokine	Mouse macrophages	Human monocytes/ macrophages	References
TNFα	↓	↓	[77, 78, 116]
IL-1α, IL-1β	↓	↓	[77, 78]
IL-1RA	n.d.	↑	[96, 117]
IL-6	↓	↓	[77,78]
IL-10	n.d.	↓	[78]
IL-12 (p40, p70)	↓	↓	[79–81]
IL-15	↑	n.d.	[118]
GM-CSF, G-CSF	n.d.	↓	[78]
M-CSF	n.d.	↓	[119]
MIP-1α	↓	↓	[82, 83]
MIP-1β	↓	n.d.	[83]
s-TNF-R75	n.d.	↑	[120]

both pro- and anti-inflammatory mediators (Tab. 1). In all cases, the inhibitory effect of IL-10 was strongest when it was added before or simultaneously with the cytokine-inducing stimulus. In contrast, fully activated macrophages were only weakly responsive to IL-10 [6]. Apart from monocytes and macrophages, IL-10 was also reported to affect the cytokine production by dendritic cells or neutrophils. For example, IL-10 antagonized the anti-CD40-induced IL-12 production by mouse splenic dendritic cells [92, 93] and suppressed the production of several chemokines (MIP-1α, MIP-1β, IL-8) by human neutrophils [94].

Several studies have addressed the mechanism(s) by which IL-10 affects the expression of the various cytokines. In virtually all cases where IL-10 reduced the cytokine protein levels, it also inhibited the expression of cytokine mRNA. In different systems (human vs mouse, macrophages vs neutrophils) and for different cytokines, IL-10 was either shown to inhibit cytokine gene transcription [95] or to accelerate the decay of cytokine mRNA [6, 94, 96, 97]. Various signalling pathways are modulated by IL-10 and might therefore mediate its effects. In human monocytes IL-10 inhibited the p56[lyn] tyrosine kinase activation and the subsequent activation of Ras [98], induced tyrosine phosphorylation of tyk2, Jak1, STAT1α and STAT3 [99], and suppressed the nuclear translocation of NF-κB [100]. However, it is important to point out that the suppression of cytokine production and of any of

Table 2 - Effect of IL-10 on the expression of MHC and co-stimulatory molecules

APC surface molecule	monocytes/macrophages	dendritic cells	References
MHC class I	M: ↓	H: (↑)	[102,104]
MHC class II	H: ↓	H: ↓	[121–123]
	M: ±	M: ±	
CD40	n.d.	H: ((↓))	[124]
CD54 (ICAM-1)	H: ↓	H: ±	[123–125]
	M: ±	M: (↓)	
CD80 (B7-1)	H: ↓ or (↑)	H: ±	[123, 124, 126]
	M: ↓	M: ↓	
CD80 (B7-2)	H: ↓	H: ↓ or ±*	[122–124, 126–128]
	M: ↓	M: ±	

M, mouse; H, human
*LPS-induced CD86 expression was suppressed, whereas CD40 ligand-triggered expression of CD86 remains unaltered by IL-10

these pathways can be parallel but independent events, as recently demonstrated for the IL-10-mediated inhibition of TNFα release and of NF-κB activity [101].

Effects of IL-10 on antigen presentation

Efficient antigen-presentation by macrophages or dendritic cells and the subsequent activation of T cells requires antigen capture, processing and loading, surface expression of MHC class I or II and costimulatory molecules, and the production of T-cell-activating cytokines. A large number of studies has shown that IL-10 inhibits the cytokine (IFNγ) production and/or proliferation of mouse or human T lymphocytes (CD4+ or CD8+, Th1) in response to syngeneic or allogeneic epidermal Langerhans cells, blood or splenic dendritic cells, or monocytes/macrophages. Although IL-10 can exert direct effects on T lymphocytes, it is believed that in co-cultures its main target is the antigen-presenting cell (APC) (reviewed in [65, 68]). Depending on the type of antigen-presenting cell, IL-10 interferes with various steps of accessory cell function. Known alterations include the decrease of IL-1 and IL-12 production (see above), the inhibition of TAP-dependent translocation of processed peptides to the endoplasmic reticulum (due to reduced expression of the TAP1 and TAP2 peptide transporters) [102], the downregulation of MHC class II on human monocytes (caused by an impaired exocytosis and recycling of the MHC molecules) [103], and the suppression of B7-1 (CD80), B7-2 (CD86) and/or ICAM-1 (CD54) on the surface of different APCs ([104] and references therein). Table 2 summarizes our current knowledge on the influence of IL-10 on the expression of MHC and costimulatory molecules. It is likely that any of these modulatory effects contributes to the induction of tolerance, anergy, and immune privilege by IL-10.

The role of IL-10 in infectious diseases *in vivo*

Results from experimental mouse studies
From the *in vitro* effects of IL-10 discussed above, it can be predicted that IL-10 exerts a dual function in infectious diseases. On the one hand, the suppression of the antimicrobial activity of macrophages by IL-10 is likely to promote infections with intracellular pathogens which are otherwise controlled by the oxidative and non-oxidative killing mechanisms of macrophages. On the other hand, IL-10 is a potent downregulator of the production of proinflammatory cytokines, which contribute to tissue damage during acute infections and are also implicated in the progression of chronic infectious diseases. Therefore, the expression of IL-10 might also be beneficial to the host by protecting against an overzealous inflammatory response. This dichotomy of the function of IL-10 has been documented in a large array of experimental mouse studies (Tab. 3) in which recombinant IL-10, neutralizing anti-IL-10 antibodies, or IL-10 transgenic or knockout (k.o.) mice were used. In several instances, protective (i.e., reduced microbial burden in the tissue) and counter-protective effects (i.e., hyperinflammation and reduced survival) were simultaneously observed after IL-10 depletion of infected mice, which illustrates the difficulty to foresee the net outcome of IL-10-based therapeutic strategies in animals and men.

Results from clinical studies of IL-10 expression in disease
The expression of IL-10 has been investigated in humans with various infectious diseases. In several instances (e.g., malaria, HIV), elevated levels of IL-10 were found, but these analyses did not conclusively determine whether IL-10 played a protective role (by reducing the pathogen-induced inflammatory response) or a detrimental one (by impeding the antimicrobial activity of granulocytes, monocytes and macrophages). In the case of HIV-infected individuals, it is still not clear whether IL-10 blocks the replication of the virus in macrophages via inhibition of the autocrine loop of TNFα induction (which is a potent activator of HIV transcription) or whether it functions to activate HIV-1 from infected cells [105, 106]. However, in

some infectious diseases, the expression of IL-10 was either correlated with the course and the outcome of the infection or was clearly shown to suppress the protective cellular immune response. For example, in human visceral leishmaniosis, IFNγ production and proliferation of peripheral blood mononuclear cells (PBMC) from untreated patients were restored by the addition of anti-IL-10 to the cultures *in vitro* [107, 108]. A similar hyperexpression and immunosuppressive role of IL-10 was also found in patients with the lepromatous, multibacillary form (but not with the tuberculous form) of leprosy [109].

Perhaps the most intriguing findings on IL-10 and human infections were obtained with patients suffering from meningococcal meningitis. Several independent groups reported that patients with fulminant and ultimately fatal meningococcal septic shock had drastically elevated plasma levels of IL-10 as compared to patients with non-fatal meningitis without shock [110–112]. The idea that the degree of IL-10 production might be causally linked to the fatal outcome of the disease is now further supported by the recent analysis of the LPS-triggered IL-10 production of whole blood samples from 190 first-degree relatives of 61 patients with meningococcal disease. Families characterized by high IL-10 production had a 20-fold increased risk for a fatal outcome of the meningococcal infection [113]. Although these results might argue for an anti-IL-10 treatment rationale, it is important to bear in mind that the current regimen for bacterial meningitis in childhood includes the use of an immunosuppressive agent (dexamethasone) together with the antibiotics. This recommendation is based on the observation that partial inhibition of the inflammatory response improves the overall clinical outcome for the children. Therefore, the systemic use of anti-IL-10 for the treatment for bacterial meningitis is not yet on the horizon, especially as in animal models of meningitis IL-10 (rather than anti-IL-10) turned out to be beneficial [114, 115].

Conclusions

In this overview of the functions of TGFβ and IL-10, we have considered the major suppressive effects that these cytokines exert on the many participants of an immune response. The great degree of similarity in their activity is consistent with the observation that they are often coexpressed by a single cell, whether this be a tumor cell effecting suppression of CTL or a lymphocyte acting in response to orally administered antigen. The efficacy of these cytokines as biological response modifiers in the clinical setting is complicated by their pleiotropic behavior, particularly their ability to exert completely opposing effects at various times within the course of an immune response. By gaining a greater understanding of their behavior in the setting of disease, and of their downstream effectors, we may eventually tap into their pathways with agents capable of either amplifying or inhibiting their suppressive actions.

Table 3 - The function of IL-10 in experimental infectious disease models

Model	Effect of IL-10 treatment	Effect of IL-10 depletion	References
Toxins			
Endotoxin (LPS shock)	protective (death rate ↓)[3]	counter-protective (death rate ↑)[2]	[129, 130]
Staphylococcal enterotoxin B	protective (death rate ↓)[3]	counter-protective (superantigenic response ↑)[1]	[131, 132]
Viruses			
Herpes simplex virus type 1	protective (stromal keratitis ↓; viral clearance ±)3)		[133]
Bacteria			
Brucella abortus		protective (bacterial load ↓)[2]	[134]
Chlamydia pneumoniae		protective (bacterial load ↓)[2]	[135]
Klebsiella pneumoniae		protective (bacterial load ↓)[2]	[136]
Listeria monocytogenes	counter-protective (bacterial load ↑)[3]	protective (acute) or counter-protective (late stage)[2]	[137, 138]
Mycobacterium avium		protective (bacterial load ↓)[2]	[139, 140]
Mycobacterium bovis BCG	counter-protective (bacterial load ↑)[4]	none[1]	[141]
Pseudomonas aeruginosa		counter-protective (tissue damage ↑ in chronic infection)[1]	[142]
Salmonella choleraesuis		protective (bacterial load ↓)[2]	[143]
Salmonella typhimurium		none (unaltered infection)[2]	[144]
Streptococcus pneumoniae	counter-protective (bacterial load ↑ in pneumonitis) or protective (patho-physiological changes in meningitis ↓)[3]	protective (bacterial load ↓, survival ↑ in pneumonitis)[2]	[115, 145]
Streptococci group B	protective (enhanced survival)[3]	none (unaltered lethality)[2]	[146]
Protozoa			
Leishmania major	none[4]	none[2]	[147–148]
Toxoplasma gondii		counter-protective (acute phase)[1]; protective (chronic phase)[2]	[149–151]

Table 3 - continued

Model	Effect of IL-10 treatment	Effect of IL-10 depletion	References
Trypanosoma cruzi (Y strain)		protective (parasite load ↓)[1]	[152]
Trypanosoma cruzi (Tulahuen strain)		protective (parasite load ↓) and counter-protective (increased mortality)[1,2]	[153, 154]
Helminths			
Schistosoma mansoni		protective (acute liver pathology ↓; Th2 polarization)[1]	[155]
Fungi			
Candida albicans	counter-protective (fungal load ↑, survival ↓)[3]	protective (fungal load ↓, survival ↑)[2]	[156, 157]

[1] *genetic deletion of IL-10 (IL-10 k.o. mice)*
[2] *application of neutralizing anti-IL-10 antibodies*
[3] *application of recombinant IL-10*
[4] *transgenic expression of IL-10 from T cells*

References

1   Maeda H, Shiraishi A (1996) TGF-β contributes to the shift toward Th2-type respons-es through direct and IL-10-mediated pathways in tumor-bearing mice. *J Immunol* 156 (1): 73–78

2   Nathan C (1991) Mechanisms and modulation of macrophage activation. *Behring Inst Res Commun* 88: 200–207

3   Mantovani A, Bottazzi B, Colotta F, Sozzani S, Ruco L (1992) The origin and function of tumor-associated macrophages. *Immunology Today* 13: 265–270

4   Nathan CF (1986) Macrophage activation: some questions. *Ann Inst Pasteur/Immunol* 137: 345–351

5   Chantry D, Turner M, Abney E, Feldmann M (1989) Modulation of cytokine produc-tion by transforming growth factor-β. *J Immunol* 142: 4295–4300

6   Bogdan C, Paik J, Vodovotz Y, Nathan C (1992) Contrasting mechanisms for suppres-

sion of macrophage cytokine release by transforming growth factor-β and interleukin-10. *J Biol Chem* 267: 23301–23308

7   McCartney-Francis N, Mizel D, Wong H, Wahl L, Wahl S (1990) TGF-β regulates production of growth factors and TGF-β by human peripheral blood monocytes. *Growth Factors* 4: 27–35

8   Vodovotz Y, Kopp JB, Takeguchi H, Shrivastav S, Coffin D, Lucia MS, Mitchell JB, Webber R, Letterio J, Wink D et al (1998) Increased mortality, blunted production of nitric oxide, and increased production of tumor necrosis factor-α in endotoxemic transforming growth factor-β1 transgenic mice. *J Leukoc Biol* 63: 31–39

9   Dubois CM, Ruscetti FW, Palaszynski EW, Falk LA, Oppenheim JJ, Keller JR (1990) Transforming growth factor β is a potent inhibitor of interleukin 1 (IL-1) receptor expression: Proposed mechanism of inhibition of IL-1 action. *J Exp Med* 172: 737–744

10 Pinson DM, LeClaire RD, Lorsbach RB, Parmely MJ, Russell SW (1992) Regulation by transforming growth factor-β1 of expression and function of the receptor for IFNγ on mouse macrophages. *J Immunol* 149: 2028–2034

11 Schluesener HJ (1990) Transforming growth factors type $β_1$ and $β_2$ suppress rat astrocyte autoantigen presentation and antagonize hyperinduction of class II major histocompatibility complex antigen expression by interferon-γ and tumor necrosis factor-α. *J Neuroimmunol* 27: 41–47

12 Czarniecki CW, Chiu HH, Wong GHW, McBabe SM, Palladino MA (1988) Transforming growth factor-$β_1$ modulates the expression of class II histocompatibility antigens on human cells. *J Immunol* 140: 4217–4223

13 Geiser AG, Letterio JJ, Kulkarni AB, Karlsson S, Roberts AB, Sporn MB (1993) Transforming growth factor $β_1$ (TGF-$β_1$) controls expression of major histocompatibility genes in the postnatal mouse: Aberrant histocompatibility antigen expression in the pathogenesis of the TGF-β1 null mouse phenotype. *Proc Natl Acad Sci USA* 90: 9944–9948

14 Letterio J, Roberts A (1998) Regulation of Immune Responses by TGF-β. *Annu Rev Immunol* 16: 137–161

15 Kim SJ, Angel P, Lafyatis R, Hattori K, Kim KY, Sporn MB, Karin M, Roberts AB (1990) Autoinduction of transforming growth factor Beta1 is mediated by the AP-1 complex. *Mol Cell Biol* 10: 1492–1497

16 Lucas C, Bald LN, Fendly BM, Mora-Worms M, Figari IS, Patzer EJ, Palladino MA (1990) The autocrine production of transforming growth factor-Beta1 during lymphocyte activation: A study with a monoclonal antibody-based ELISA. *J Immunol* 145: 1415–1422

17 Kehrl JH, Taylor A, Kim SJ, Fauci AS (1991) Transforming growth factor-β is a potent negative regulator of human lymphocytes. *Ann NY Acad Sci* 628: 345–353

18 Kasid A, Bell GI, Director EP (1988) Effects of transforming growth factor-β on human lymphokine-activated killer cell precursors: Autocrine inhibition of cellular proliferation and differentiation to immune killer cells. *J Immunol* 141: 690–698

19 Roberts AB, Vodovotz Y, Roche NS, Sporn MB, Nathan CF (1992) Role of nitric oxide

in antagonistic effects of transforming growth factor-β and interleukin-1β on the beating rate of cultured cardiac myocytes. *Mol Endocrinol* 6: 1921–1930

20 Samuel SK, Hurta RAR, Kondaiah P, Khalil N, Turley EA, Wright JA, Greenberg AH (1992) Autocrine induction of tumor protease production and invasion by a metallothionein-regulated TGF-β$_1$ (Ser223, 225). *EMBO J* 11: 1599–1605

21 Tada T, Ohzeki S, Utsumi K, Takiuchi H, Muramatsu M, Li X, Fujiwara H, Hamaoka T (1991) Transfroming growth factor-β-induced inhibition of T cell function. *J Immunol* 146: 1077–1082

22 Asai O, Longo D, Tian Z, Hornung R, Taub D, Ruscetti F, Murphy W (1998) Suppression of graft-versus-host disease and amplification of graft-versus-tumor effect by activated natural killer cells after allogeneic bone marrow transplantation. *J Clin Invest* 101 (9): 1835–1842

23 Marth T, Strober W, Seder R, Kelsall B (1997) Regulation of transforming growth factor-β production by interleukin-12. *Eur J Immunol* 27 (5): 1213–1220

24 Nathan C (1992) Nitric oxide as a secretory product of mammalian cells. FASEB J 6: 3051–3064

25 Bogdan C, Vodovotz Y, Xie Q, Nathan C, Röllinghoff M (1994) Regulation of inducible nitric oxide synthase in macrophages by cytokines and microbial products. In: N Masihi (ed): *Immunotherapy of infections*. Marcel Dekker, New York, 37–54

26 Tsunawaki S, Sporn M, Ding A, Nathan C (1988) Deactivation of macrophages by transforming growth factor-β. *Nature* 334: 260–262

27 Ding A, Nathan CF, Graycar J, Derynck R, Stuehr DJ, Srimal S (1990) Macrophage deactivating factor and transforming growth factors-β1, -β2, and -β3 inhibit induction of macrophage nitrogen oxide synthesis by IFN-γ. *J Immunol* 145: 940–944

28 Nelson BJ, Ralph P, Green SJ, Nacy CA (1991) Differential susceptibility of activated macrophage cytotoxic effector reactions to the suppressive effects of transforming growth factor-β1. *J Immunol* 146: 1849–1857

29 Gazzinelli RT, Oswald IP, Hieny S, James SL, Sher A (1992) The microbicidal activity of interferon-τ-treated macrophages against *Trypanosoma cruzi* involves an L-arginine-dependent, nitrogen-oxide-mediated mechanism inhibitable by interleukin 10 and transforming growth factor-β. *Eur J Immunol* 22: 2501–2506

30 Vodovotz Y, Bogdan C, Paik J, Xie Q, Nathan C (1993) Mechanisms of suppression of macrophage nitric oxide release by transforming growth factor-β. *J Exp Med* 178: 605–613

31 Vodovotz Y, Bogdan C (1994) Control of nitric oxide synthase expression by transforming growth factor-β: Implications for homeostasis. *Prog Growth Factor Res* 5: 341–351

32 Vodovotz Y (1997) Control of nitric oxide production by transforming growth factor-β1: Mechanistic insights and potential relevance to human disease. *Nitric Oxide: Biol and Chem* 1: 3–17

33 Vodovotz Y, Geiser AG, Chesler L, Letterio JJ, Campbell A, Lucia MS, Sporn MB, Roberts AB (1996) Spontaneously increased production of nitric oxide and aberrant

expression of the inducible nitric oxide synthase *in vivo* in the transforming growth factor-β1 null mouse. *J Exp Med* 183: 2337–2342

34 Vodovotz Y, Letterio JJ, Geiser AG, Chesler L, Roberts AB, Sparrow J (1996) Control of nitric oxide production by endogenous transforming growth factor-β1 and systemic nitric oxide in retinal pigment epithelial cells and peritoneal macrophages. *J Leukoc Biol* 60: 261–270

35 Gilbert RS, Herschmann HR (1993) Transforming growth factor beta differentially modulates the inducible nitric oxide synthase gene in distinct cell types. *Biochem Biophys Res Commun* 195: 380–384

36 Adler H, Frech B, Thöny M, Pfister H, Peterhans E, Jungi TW (1995) Inducible nitric oxide synthase in cattle: Differential cytokine regulation of nitric oxide synthase in bovine and murine macrophages. *J Immunol* 154: 4710–4718

37 Stavnezer J (1996) Transforming growth factor-β. In: C Snapper (ed): *Cytokine regulation of humoral immunity: Basic and clinical aspects.* Wiley, New York, 289–324

38 Nunes I, Shapiro RL, Rifkin DB (1995) Characterization of latent TGF-β activation by murine peritoneal macrophages. *J Immunol* 155: 1450–1459

39 Barcellos-Hoff MH, Dix TA (1996) Redox mediated activation of latent transforming growth factor-β1. *Mol Endocrinol* 10: 1077–1083

40 Diefenbach A, Schindler H, Donhauser N, Lorenz E, Laskay T, MacMicking J, Röllinghoff M, Gresser I, Bogdan C (1998) Type 1 interferon (IFNα/β) and type 2 nitric oxide synthase regulate the innate immune response to a protozoan parasite. *Immunity* 8: 77–87

41 Schindler H, Diefenbach A, Röllinghoff M, Bogdan C (1998) IFN-γ inhibits the production of latent transforming growth factor-β1 by mouse inflammatory macrophages. *Eur J Immunol* 28: 1181–1188

42 Gray J, Hirokawa J, Ohtsuka K, Horwitz D (1998) Generation of an inhibitory circuit involving CD8+ T cells, IL-2, and NK cell-derived TGF-β: contrasting effects of anti-CD2 and anti-CD3. *J Immunol* 160 (5): 2248–2254

43 Chen Y, Inobe J, Weiner H (1995) Iduction of oral tolerance to myelin basic protein in CD8-depleted mice: both CD4+ and CD8+ cells mediate active suppression. *J Immunol* 155: 910–916

44 Seder RS, Marth T, Sieve MC, Strober W, Letterio J, Roberts AB, Kelsall B (1998) Factors involved in the differentiation of TGF-β-producing cells from naive CD4+ T cells: IL-4 and IFN-γ have opposing effects, while TGF-β positively regulates its own production. *J Immunol* 160 (12): 5707–5718

45 Ming M, Ewen M, Pereira M (1995) Trypanosome invasion of mammalian cells requires activation of the TGF-β signaling pathway. *Cell* 82: 287–296

46 Silva JS, Twardzik DR, Reed SG (1991) Regulation of *Trypanosoma cruzi* infections *in vitro* and *in vivo* by transforming growth factor β. *J Exp Med* 174: 539

47 Barral-Netto M, Barral A, Brownell CE, Skeiky YAW, Ellingsworth LR, Twardzik DR, Reed SG (1992) Transforming growth factor-β in leishmanial infection: A parasite escape mechanism. *Science* 257: 545–548

48   Oswald IP, Gazzinelli RT, Sher A, James SL (1992) IL-10 synergizes with IL-4 and trans-forming growth factor-b to inhibit macrophage cytotoxic activity. *J Immunol* 148: 3578–3582

49   Barral A, Barral-Netto M, Yong EC, Brownell CE, Twardzik DR, Reed SG (1993) Transforming growth factor β as a virulence mechanism for *Leishmania braziliensis*. *Proc Natl Acad Sci USA* 90: 3442–3446

50   Bermudez LE (1993) Production of transforming growth factor-β by *Mycobacterium avium*-infected human macrophages is associated with unresponsiveness to IFN-γ. *J Immunol* 150: 1838–1845

51   Stenger S, Thüring H, Röllinghoff M, Bogdan C (1994) Tissue expression of inducible nitric oxide synthase is closely associated with resistance to *Leishmania major*. *J Exp Med* 180: 783–793

52   Bläuer F, Groscurth P, Schneeman M, Schoedon G, Schaffner A (1995) Modulation of the antilisterial activity of human blood-derived macrophages by activating and deacti-vating cytokines. *J Interferon Cytokine Res* 15: 105–114

53   Tomioka H, Sato K, Maw WW, Saito H (1995) The role of tumor necrosis factor, inter-feron-γ, transforming growth factor-β, and nitric oxide in the expression of immuno-suppressive functions of splenic macrophages induced by *Mycobacterium avium* com-plex infection. *J Leukoc Biol* 58: 704–712

54   Hausmann EHS, Hao S, Pace J, Parmely MJ (1994) Transforming growth factor β1 and gamma interferon provide opposing signals to lipopolysaccharide-activated mouse macrophages. *Infect Immun* 62: 3625–3632

55   Wahl S, Frazier-Jessen M, Jin W, Kopp J, Sher A, Cheever A (1997) Cytokine regulation of schistosome-induced granuloma and fibrosis. *Kidney International* 51: 1370–1375

56   Markowitz S, Roberts A (1996) Tumor supressor activity of the TGF-beta pathway in human cancers. *Cytokine Growth Factor Rev* 7: 93–102

57   Haak-Frendscho M, Wynn TA, Czuprynski CJ, Paulnock D (1990) Transforming growth factor-beta1 inhibits activation of macrophage cell line RAW 2647 for cell killing. *Clin Exp Immunol* 82: 404–410

58   Alleva DG, Burger CJ, Elgert KD (1994) Tumor-induced regulation of suppressor macrophage nitric oxide and TNF-α production: Role of tumor derived IL-10, TGF-β, and prostaglandin E_2. *J Immunol* 153: 1674–1686

59   Maeda H, Tsuru S, Shiraishi A (1994) Improvement of macrophage dysfunction by administration of anti-transforming growth factor-β antibody in EL4-bearing hosts. *Jpn J Cancer Res* 85: 1137–1143

60   Alleva DG, Walker TM, Elgert KD (1995) Induction of macrophage suppressor activity by fibrosarcoma-derived transforming growth factor-β1: contrasting effects on resting and activated macrophages. *J Leukoc Biol* 57: 919–928

61   Hoefer M, Anderer F (1995) Anti-transforming growth factor-β antibodies with prede-fined specificity inhibit metastasis of highly tumorigenic human xenotransplants in nu/nu mice. *Cancer Immunol Immunother* 41:302–308

62   Arteaga C, Hurd S, Winner A, Johnson M, Fendly B, Forbes J (1993) Anti-transforming

growth factor-β antibodies inhibit breast cancer cell tumorigenicity and increase mouse spleen natural killer cell activity. *J Clin Invest* 92: 2569–2576

63 Torre-Amione G, Beuchamp R, Koeppen H, Park B, Schreiber H, Moses H, Rowley D (1990) A highly immunogenic tumor transfected with a murine transforming growth factor type β1 cDNA escapes immune surveillance. *Proc Natl Acad Sci USA* 87: 1486–1490

64 Quin Z, Noffz G, Mohaupt M, Blankenstein T (1997) Interleukin-10 prevents dendritic cell accumulation and vaccination with granulocyte-macrophage colon-stimulating factor gene-modified tumor cells. *J Immunol* 159 (2): 770–776

65 Moore KW, O'Garra A, de Waal Malefyt R, Vieira P, Mosmann TR, Mosmann TR (1993) Interleukin-10. *Ann Rev Immunol* 11: 165–190

66 Mosmann TR (1994) Properties and functions of interleukin-10 Adv Immunol 56: 1-26

67 Banchereau J (1995) Converging and diverging properties of human interleukin-4 and interleukin-10. *Behring Inst Res Commun* 96: 58–77

68 Bogdan C, Nathan C (1993) Modulation of macrophage function by transforming growth factor-β, interleukin 4 and interleukin 10. *Ann NY Acad Sci* 685: 713–739

69 Nakajima H, Gleich GJ, Kita H (1996) Constitutive production of IL-4 and IL-10 and stimulated production of IL-8 by normal peripheral blood eosinophils. *J Immunol* 156: 4859–4866

70 Mehrotra PT, Donnelly RP, Wong S, Kanegane H, Geremew A, Mostowski HS, Furuke K, Siegel JP, Bloom ET (1998) Production of IL-10 by human natural killer cells stimulated with IL-2 and/or IL-12. *J Immunol* 160: 2637–2644

71 Pechhold K, Wesch D, Schondelmaier S, Kabelitz D (1994) Primary activation of Vγ9-expressing γδ T cells by *Mycobacterium tuberculosis*. Requirement for Th1-type CD4 T cell help and inhibition by IL-10. *J Immunol* 152: 4984–4991

72 Sironi M, Munoz C, Pollicino T, Siboni A, Sciacca FL, Bernasconi S, Vecchi A, Colotta F, Mantovani A (1993) Divergent effects of interleukin-10 on cytokine production by mononuclear phagocytes and endothelial cells. *Eur J Immunol* 23: 2692–2695

73 Georgescu L, Vakkalanka RK, Elkon KB, Crow MK (1997) Interleukin-10 promotes activation-induced cell death of SLE lymphocytes mediated by Fas ligand. *J Clin Invest* 100 (10): 2622–2633

74 Estaquier J, Marguerite M, Sahuc F, Bessis N, Auriault C, Ameisen JC (1997) Interleukin-10-mediated T cell apoptosis during the T helper type 2 cytokine response in murine *Schistosoma mansoni* parasite infection. *Eur Cytokine Netw* 8 (2) : 153–160

75 Fiorentino DF, Bond MW, Mosmann TR (1989) Two types of mouse T helper cell IV Th2 clones secrete a factor that inhibits cytokine production by Th1 clones. *J Exp Med* 170: 2081–2095

76 Wu J, Cunha FQ, Liew FY, Weiser WY (1993) IL-10 inhibits the synthesis of migration inhibitory factor and migration inhibitory factor-mediated macrophage activation. *J Immunol* 151: 4325–4332

77 Fiorentino DF, Zlotnik A, Mosmann TR, Howard M, O'Garra A (1991) IL-10 inhibits cytokine production by activated macrophages. *J Immunol* 147: 3815–3822

78   de Waal Malefyt R, Abrams J, Bennett B, Figdor CG, de Vries JE (1991) Interleukin 10
      (IL-10) inhibits cytokine synthesis by human monocytes: an autoregulatory role of IL-
      10 produced by monocytes. *J Exp Med* 174: 1209–1220

79   D'Andrea A, Aste-Amezaga M, Valiante NM, Ma X, Kubin M, Trinchieri G (1993) IL-
      10 inhibits human lymphocyte interferon-g production by suppressing natural killer cell
      stimulatory factor/IL-12 synthesis in accessory cells. *J Exp Med* 178: 1041–1048

80   Skeen MJ, Miller MA, Shinnick TM, Ziegler HK (1996) Regulation of murine
      macrophage IL-12 production. Activation of macrophages *in vivo*, restimulation *in
      vitro*, and modulation by other cytokines. *J Immunol* 156: 1196–1206

81   Snijders A, Hilkens CMU, van der Pouw Kraan TCTM, Engel M, Aarden LA, Kapsen-
      berg ML (1996) Regulation of bioactive IL-12 production in lipopolysaccharide-stimu-
      lated human monocytes is determined by the expression of the p35 subunit. *J Immunol*
      156: 1207–1212

82   Berkman N, John M, Roesems G, Jose PJ, Barnes PJ, Chung KF (1995) Inhibition of
      macrophage inflammatory protein-1α expression by IL-10. *J Immunol* 155: 4412–4418

83   Horton MR, Burdick MD, Strieter RM, Bao C, Noble PW (1998) Regulation of
      Hyaluron-induced chemokine gene expression by IL-10 and IFN-γ in mouse macro-
      phages. *J Immunol* 160: 3023–3030

84   Zheng LM, Ojcius DM, Garaud F, Roth C, Maxwell E, Li Z, Rong H, Chen J, Wang
      XY, Catino JJ, King I (1996) Interleukin-10 inhibits tumor metastasis through an NK
      cell-dependent mechanism. *J Exp Med* 184: 579–584

85   Wogensen L, Lee M-S, Sarvetnick N (1994) Production of IL-10 by islet cells accelerates
      immune-mediated destruction of β cells in nonobese diabetic mice. *J Exp Med* 179:
      1379–1384

86   Flesch IEA, Hess JH, Oswald IP, Kaufmann SHE (1994) Growth inhibition of *Mycobac-
      terium bovis* by IFN-γ stimulated macrophages: regulation by endogenous tumor necro-
      sis factor-α and by IL-10. *Int Immunol* 6: 693–700

87   Vieth M, Will A, Schröppel K, Röllinghoff M, Gessner A (1994) Interleukin-10 inhibits
      antimicrobial activity against *Leishmania major* in murine macrophages. *Scand J
      Immunol* 40: 403–409

88   Meier-Osusky I, Schoedon G, Bläuer F, Schneemann M, Schaffner A (1996) Compari-
      son of the antimicrobial activity of deactivated human macrophages challenged with
      Aspergillus fumigatus and *Listeria monocytogenes*. *J Infect Dis* 174: 651–654

89   Roilides E, Dimitriadou A, Kadiltsoglou I, Sein T, Karpouzas J, Pizzo PA, Walsh TJ
      (1997) IL-10 exerts suppressive and enhancing effects on antifungal activity of mononu-
      clear phagocytes against *Aspergillus fumigatus*. *J Immunol* 158: 322–329

90   Kuga S, Otsuka T, Niiro H, Nunoi H, Nemoto Y, Nakano T, Ogo T, Umei T, Niho Y
      (1996) Suppression of superoxide anion production by interleukin-10 is accompanied
      by a downregulation of the genes for subunit proteins of NADPH oxidase. *Exp Hema-
      tol* 24: 151–157

91   Laichalk LL, Danforth JM, Standiford TJ (1996) Interleukin-10 inhibits neutrophil
      phagocytic and bactericidal activity. *FEMS Immunol Med Microbiol* 15: 181–187

92   Koch F, Stanzl U, Jennewein P, Janke K, Heufler C, Kämpgen E, Romani N, Schuler G (1996) High level IL-12 production by murine dendritic cells: upregulation via MHC class II and CD40 molecules and downregulation by IL-4 and IL-10. *J Exp Med* 184: 741–746

93   Kelsall BL, Stuber E, Neurath M, Strober W (1996) Interleukin-12 production by dendritic cells The role of CD40-CD40L interactions in Th1 T-cell responses. *Ann NY Acad Sci* 795: 116–126

94   Kasama T, Strieter RM, Lukacs NW, Burdick MD, Kunkel SL (1994) Regulation of neutrophil-derived chemokine expression by IL-10. *J Immunol* 152: 3559–3667

95   Wang P, Wu P, Siegel MI, Egan RW, Billah MM (1994) IL-10 inhibits transcription of cytokine genes in human peripheral blood mononulcear cells. *J Immunol* 153: 811–816

96   Cassatella MA, Meda L, Gasperini S, Calzetti F, Bonora S (1994) IL-10 upregulates IL-1 receptor antagonist production from lipopolysaccharide-stimulated human polymorphonuclear leukocytes by delaying mRNA degradation. *J Exp Med* 179: 1695–1699

97   Brown CY, Lagnado CA, Vadas MA, Goodall GJ (1996) Differential regulation of the stability of cytokine mRNAs in lipopolysaccharide-activated blood monocytes in response to IL-10. *J Biol Chem* 271: 20108–20112

98   Geng Y, Gulbins E, Altman A, Lotz M (1994) Monocyte deactivation by interleukin-10 via inhibition of tyrosine kinase activity and the Ras signaling pathway. *Proc Natl Acad Sci USA* 91: 8602–8606

99   Finbloom DS, Winestock KD (1995) IL-10 induces the tyrosine phosphorylation of tyk2 and Jak1 and the differential assembly of STAT1α and STAT3 complexes in human T cells and monocytes. *J Immunol* 155: 1079–1090

100 Lentsch AB, Shanley TP, Sarma V, Ward PA (1997) *In vivo* suppression of NF-κB and preservation of IκBα by interleukin-10 and interleukin-13. *J Clin Invest* 100: 2443–2448

101 Clarke CJP, Hales A, Hunt A, Foxwell BMJ (1998) IL-10-mediated suppression of TNF-α production is independent of its ability to inhibit NF-κB activity. *Eur J Immunol* 28: 1719–1726

102 Salazar-Onfray F, Charo J, Petersson M, Freland S, Noffz G, Qin Z, Blankenstein T, Ljunggren H-G, Kiessling R (1997) Down-regulation of the expression and function of the transporter associated with antigen processing in murine tumor cell lines expressing IL-10. *J Immunol* 159: 3195–3202

103 Koppelman B, Neefjes JJ, de Vries JE, de Waal Malefyt R (1997) Interleukin-10 downregulates MHC class II αβ peptide complexes at the plasma membrane of monocytes by affecting arrival and recycling. *Immunity* 7: 861–871

104 Morel A-S, Quaratino S, Douek DC, Londei M (1997) Split activity of interleukin 10 on antigen capture and antigen presentation by human dendritic cells: definition of a maturative step. *Eur J Immunol* 27: 26–34

105 Weissman D, Poli G, Fauci AS (1994) Interleukin-10 blocks HIV replication in macrophages by inhibiting the autocrine loop of tumor necrosis factor-α and IL-6 induction. *AIDS Res Human Retroviruses* 10: 1199–1206

106 Finnegan A, Roebuck KA, Nakai BE, Gu DS, Rabbi MF, Song S, Landay AL (1996) IL-10 cooperates with TNF-α to activate HIV-1 from latently and acutely infected cells of monocyte/macrophage lineage. *J Immunol* 156: 841–851

107 Holaday BJ, de Lima Pompeu MM, Jeronimo S, Texeira MJ, de Queiroz Sousa A, Vasconcelos AW, Pearson RD, Abrams JS, Locksley RM (1993) Potential role for interleukin-10 in the immunosuppression associated with kala-azar. *J Clin Invest* 92: 2626–2632

108 Ghalib HW, Piuvezam MR, Skeiky YAW, Siddig M, Hashim FA, El-Hassan AM, Russo DM, Reed SG (1993) Interleukin-10 production correlates with pathology in human *Leishmania donovani* infections. *J Clin Invest* 92: 324–329

109 Sieling PA, Abrams JS, Yamamura M, Salgame P, Bloom BR, Rea TH, Modlin RL (1993) Immunosuppressive roles for IL-10 and IL-4 in human infection. *In vitro* modulation of T cell responses in leprosy. *J Immunol* 150: 5501–5510

110 Lehmann AK, Halstensen A, Sornes S, Rokke O, Waage A (1995) High levels of IL-10 in serum are associated with fatality in infectious diseases. *Infect Immun* 63: 2109–2112

111 Derkx B, Marchant A, Goldman M, Bijlmer R, van Deventer S (1995) High levels of IL-10 during the initial phase of fulminant meningococcal septic shock. *J Infect Dis* 171: 229–232

112 Brandtzaeg P, Osnes L, Ovstebo R, Joo GB, Westvik A-B, Kierulf P (1996) Net inflammatory capacity of human septic shock plasma evaluated by a monocyte-based target cell assay: identification of IL-10 as a major functional deactivator of human monocytes. *J Exp Med* 184: 51–60

113 Westendorp RGJ, Langermans JAM, Huizinga TWJ, Elouali AH, Verweij CL, Boomsma DI, Vandenbrouke JP (1997) Genetic influence on cytokine production and fatal meningococcal disease. *Lancet* 349: 170–173

114 Paris MM, Hickey SM, Trujillo M, Ahmed A, Olsen K, McCracken GH (1997) The effect of IL-10 on meningeal inflammation in experimental bacterial meningitis. *J Infect Dis* 176: 1239–1246

115 Koedel U, Bernatowicz A, Frei K, Fontana A, Pfister H-W (1996) Systemically (but not intrathecally) administered IL-10 attenuates pathophysiologic alterations in experimental pneumococcal meningitis. *J Immunol* 157: 5185–5191

116 Bogdan C, Vodovotz Y, Nathan C (1991) Macrophage deactivation by IL-10. *J Exp Med* 174: 1549–1555

117 Chomarat P, Vannier E, Dechanet J, Rissoan MC, Banchereau J, Dinarello CA, Miossec P (1995) Balance of IL-1 receptor antagonist/IL-1β in rheumatoid synovium and its regulation by IL-4 and IL-10. *J Immunol* 154: 1432–1439

118 Doherty TM, Seder RA, Sher A (1996) Induction and regulation of IL-15 expression in murine macrophages. *J Immunol* 156: 735–741

119 Gruber MF, Williams CC, Gerrard TL (1994) Macrophage-colony-stimulating factor expression by anti-CD45 stimulated human monocytes is transcriptionally up-regulated by IL-1β and inhibited by IL-4 and IL-10. *J Immunol* 152: 1354–1361

120 Joyce DA, Steer JH (1996) IL-4, IL-10 and IFN-γ have distinct, but interacting, effects

on differentiation-induced changes in TNF-α and TNF-receptor release by cultured human monocytes. *Cytokine* 8: 49–57

121 Fiorentino DF, Zlotnik A, Vieira P, Mosmann TR, Howard M, Moore KW, O'Garra A (1991) IL-10 acts on the antigen-presenting cell to inhibit cytokine production by Th1 cells. *J Immunol* 146: 3444–3451

122 Chang C-H, Furue M, Tamaki K (1994) B7-1 expression of Langerhans cells is up-regulated by proinflammatory cytokines, and is down-regulated by interferon-γ or by interleukin-10 Eur. *J Immunol* 25: 394–398

123 Buelens C, Willems F, Delvaux A, Pierard G, Delville J-P, Velu T, Goldman M (1995) Interleukin-10 differentially regulates B7-1 (CD80) and B7-2 (CD86) expression on human peripheral blood dendritic cells. *Eur J Immunol* 25: 2668–2672

124 Ozawa H, Aiba S, Nakagawa S, Tagami H (1996) Interferon-γ and interleukin-10 inhibit antigen-presentation by Langerhans cells for T helper type 1 cells by suppressing their CD80 (B7-1) expression. *Eur J Immunol* 26: 648–652

125 Ding L, Linsley PS, Huang L-Y, Germain RN, Shevach EM (1993) IL-10 inhibits macrophage costimulatory activity by selectively inhibiting the up-regulation of B7 expression. *J Immunol* 151: 1224–1234

126 Creery WD, Diaz-Mitoma F, Filion L, Kumar A (1996) Differential modulation of B7-1 and B7-2 isoform expression on human monocytes by cytokines which influence the development of T helper cell phenotype. *Eur J Immunol* 26: 1273–1277

127 Buelens C, Verhasselt V, de Groote D, Thielemans K, Goldmann M, Willems F (1997) Human dendritic cell responses to lipopolysaccharide and CD40 ligation are differentially regulated by interleukin-10. *Eur J Immunol* 27: 1848–1852

128 Flores Villanueva PO, Reiser H, Stadecker MJ (1994) Regulation of T helper cell responses in experimental murine schistosomiasis by IL-10: effect on expression of B7 and B7-2 costimulatory molecules by macrophages. *J Immunol* 153: 5190–5199

129 Howard M, Muchamuel T, Andrade S, Menon S (1993) Interleukin-10 protects mice from lethal endotoxemia. *J Exp Med* 177: 1205–1208

130 Gerard C, Bruyns C, Marchant A, Abramowicz D, Vandenabeele P, Delvaux A, Fiers W, Goldman M, Velu T (1993) Interleukin-10 reduces the release of tumor necrosis factor and prevents lethality in experimental endotoxemia. *J Exp Med* 177: 547–550

131 Bean AGD, Freiberg RA, Andrade S, Menon S, Zlotnik A (1993) Interleukin-10 protects mice against staphylococcal enterotoxin B-induced lethal shock. *Infect Immun* 61: 4937–4939

132 Hasko G, Virag L, Egnaczyk G, Salzman AL, Szabo C (1998) The crucial role of IL-10 in the suppression of the immunological response in mice exposed to staphylococcal enterotoxin B. *Eur J Immunol* 28: 1417–1425

133 Tumpey TM, Elner VM, Chen S-H, Oakes JE, Lausch RN (1994) Interleukin-10 treatment can suppress stromal keratitis induced by Herpes simplex virus type 1. *J Immunol* 153: 2258–2263

134 Fernandes DM, Baldwin CL (1995) Interleukin-10 downregulates protective immunity to *Brucella abortus*. *Infect Immun* 63: 1130–1133

135 Yang X, HayGlass KT, Brunham RC (1996) Genetically determined differences in IL-10 and IFN-γ responses correlate with clearance of *Chlamydia trachomatis* mouse pneumonitis infection. *J Immunol* 156: 4338–4344

136 Greenberger MJ, RM Strieter, SL Kunkel, JM Danforth, RE Goodman, TJ Standiford (1995) Neutralization of IL-10 increases survival in a murine model of *Klebsiella pneumonia. J Immunol* 155: 722–729

137 Wagner RD, Maroushek NM, Brown JF, Czuprynski CJ (1994) Treatment with anti-interleukin-10 monoclonal antibody enhances early resistance to but impairs complete clearance of *Listeria monocytogenes* infection in mice. *Infect Immun* 62: 2345–2353

138 Kelly JP, Bancroft GJ (1996) Administration of interleukin-10 abolishes innate resistance to *Listeria monocytogenes. Eur J Immunol* 26: 356–364

139 Denis M, Ghadirian E (1993) IL-10 neutralization augments mouse resistance to systemic *Mycobacterium avium* infections. *J Immunol* 5425–5430

140 Bermudez LE, Champsi J (1993) Infection with *Mycobacterium avium* induces production of IL-10, and administration of anti-IL-10 antibody is associated with enhanced resistance to infection in mice. *Infect Immun* 61: 3093–3097

141 Erb KJ, Kirman J, Delahunt B, Chen W, Le Gros G (1998) IL-4, IL-5 and IL-10 are not required for the control of *M. bovis* BCG infection in mice. *Immunol Cell Biol* 76: 41–46

142 Yu H, Hanes M, Chrisp CE, Boucher JC, Deretic V (1998) Microbial pathogenesis in cystic fibrosis: pulmonary clearance of mucoid *Pseudomonas aeruginosa* and inflammation in a mouse model of repeated respiratory challenge. *Infect Immun* 66: 280–288

143 Arai T, Hiromatsu K, Nishimura H, Kimura Y, Kobayashi N, Ishida H, Nimura Y, Yoshikai Y (1995) Effects of *in vivo* administration of anti-IL-10 monoclonal antibody on the host defence mechanism against murine *Salmonella* infection. *Immunology* 85: 381–388

144 Pie S, Matsiota-Bernard P, Truffa-Bachi P, Nauciel C (1996) Gamma interferon and interleukin-10 gene expression in innately susceptible and resistant mice during the early phase of *Salmonella typhimurium* infection. *Infect Immun* 64: 849–854

145 van der Poll T, Marchant A, Keogh CV, Goldman M, Lowry SF (1996) Interleukin-10 impairs host defense in murine pneumococcal pneumonia. *J Infect Dis* 174: 994–1000

146 Cusumano V, Genovese F, Mancuso G, Carbone M, Fera MT, Teti G (1996) Interleukin-10 protects neonatal mice from lethal group B streptococcal infection. *Infect Immun* 64: 2850–2852

147 Coffman RL, Varkila K, Scott P, Chatelain R (1991) Role of cytokines in the differentiation of CD4+ T-cell subsets *in vivo. Immunol Reviews* 123: 189

148 Hagenbaugh A, Sharma S, Dubinett SM, Wei SH-Y, Aranda R, Cheroutre H, Fowell DJ, Binder S, Tsao B, Locksley RM et al (1997) Altered immune responses in interleukin-10 transgenic mice. *J Exp Med* 185: 2101–2110

149 Neyer LE, Grunig G, Fort M, Remington JS, Rennick D, Hunter CA (1997) Role of IL-10 in regulation of T-cell-dependent and T-cell-independent mechanisms of resistance to *Toxoplasma gondii. Infect Immun* 65: 1675–1682

150 Gazzinelli RT, Wysocka M, Hieny S, Scharton-Kersten T, Cheever A, Kuhn R, Müller W, TrinchieriG, Sher A (1996) In the absence of endogenous IL-10, mice acutely infected with *Toxoplasma gondii* succumb to a lethal response dependent on CD4⁺ T cells and accompanied by overproduction of IL-12, IFN-γ, and TNF-α. *J Immunol* 157: 798–805

151 Deckert-Schlüter M, Buck C, Weiner D, Kaefer N, Rang A, Hof H, Wiestler OD, Schlüter D (1997) Interleukin-10 downregulates the intracerebral immune response in chronic *Toxoplasma encephalitis*. *J Neuroimmunol* 76: 167–176

152 Abrahamsohn IA, Coffman RL (1996) *Trypanosoma cruzi*: IL-10, TNF, IFN-gamma, and IL-12 regulate innate and acquired immunity to infection. *Exp Parasitol* 84: 231–244

153 Reed SG, Brownell CE, Russo DM, Silva JS, Grabstein KH, Morrissey PJ (1994) IL-10 mediates susceptibility to *Trypanosoma cruzi* infection. *J Immunol* 153: 3135–3140

154 Hunter CA, Ellis-Neyes LA, Slifer T, Kanaly S, Grunig G, Fort M, Rennick D, Araujo FG (1997) IL-10 is required to prevent immune hyperactivity during infection with *Trypanosoma cruzi*. *J Immunol* 158: 3311–3316

155 Whynn TA, Cheever AW, Williams ME, Hieny S, Caspar P, Kühn R, Müller W, Sher A (1998) IL-10 regulates liver pathology in acute murine *Schistosomiasis mansoni* but is not required for immune down-modulation of chronic disease. *J Immunol* 160: 4473–4480

156 Romani L, Puccetti P, Mencacci A, Cenci E, Spaccapelo R, Tonnetti L, Grohmann U, Bistoni F (1994) Neutralization of IL-10 upregulates nitric oxide production and protects susceptible mice from challenge with *Candida albicans*. *J Immunol* 152: 3514–3521

157 Tonnetti L, Spaccapelo R, Cenci E, Mencacci A, Puccetti P, Coffman RL, Bistoni F, Romani L (1995) Interleukin-4 and -10 exacerbate candidiasis in mice. *Eur J Immunol* 25: 1559–1565

Therapeutic control of cytokines: lessons from microorganisms

Brian Henderson

Cellular Microbiology Research Group, Eastman Dental Institute, University College London, 256 Gray's Inn Road, London WC1X 8LD, UK

Introduction

The systems of innate and acquired immunity that protect us from the vast numbers of potential parasites that live in our environment (prions, viruses, bacteria, protozoa, fungi and parasitic worms) have been shaped by these parasites as part of the evolutionary struggle for survival. Thus the subdivision of the CD4 lymphocyte subpopulation into Th1 and Th2 cells is an evolutionary recognition that parasitic microorganisms (viruses, bacteria and protozoa) can live either inside cells or in intercellular environments. The action of CD8 and natural killer cells is further evolutionary recognition that microorganisms can enter and reside within the cells of the host.

It has been recognised for many years that microorganisms can evade host immune responses. Probably the best known examples of such evasion are: (i) the capsules of bacteria, complex surface coatings that prevent antibodies or complementary components from reaching the bacterial cell surface and (ii) antigenic and phase variation, in which surface antigens on viruses, bacteria and protozoa can be altered to avoid antibody-mediated attack [1]. During the past decade many more examples of immune evasion mechanisms have been discovered, including those which involve the evasion of cytokine-controlled processes.

It is now recognised that our protective systems of innate and acquired immunity are controlled by a large number of local hormone-like molecules that are grouped under the catch-all title of cytokines. As has been discussed in earlier chapters, these cytokines have been subdivided, on the basis of supposed biological function, into six main groupings [2, 3]. The interleukins are, for example, largely composed of proteins that act as lymphoid growth factors. Obvious exceptions are members of the interleukin-1 family (IL-1α, IL-1β, IL-1ra, and IL-18, also known as IL-1γ) and the chemokine, IL-8. These cytokines are intimately involved in inflammatory responses. There are a growing number of cytokines with homology

Novel Cytokine Inhibitors, edited by Gerry A. Higgs and Brian Henderson to tumour necrosis factor (TNF) and TNF receptors. The prototypic member of this family, TNFα, is a key pro-inflammatory cytokine involved in host defence responses to exogenous pathogens. One of the first cytokines to be discovered was interferon. We now recognise that there are three 'families' of interferon molecules (α,β,γ) which play major roles in antiviral defences and in immune control. The most recent group of cytokines to be discovered is the chemokines - peptides involved in the cell-selective trafficking of leukocytes to sites of inflammation. The colony-stimulating factors, and a large group of cytokines known as growth factors complete the panoply of cytokines produced by mammals.

While it had been recognised for decades that microorganisms could evade immune and inflammatory responses, it was only in the 1990s that evidence for immune evasion by the dysregulation of cytokine networks was discovered. There is now abundant evidence that viruses can manipulate host cytokine networks by a variety of gene products that have been variously named virokines and viroceptors [4]. Protozoa have also been found to produce certain components that have the ability to interfere with host protective cytokines [3, 5]. The term trypanokine has been coined for one such protein from a trypanosome [6]. Bacteria are only now being recognised to produce molecules that can affect cytokine network control [3]. The various products of microorganisms, which can interact with the cytokine networks of the host and interefere with protective immune and inflammatory responses, are themselves cytokine-like in nature and the generic term – microkines – has been suggested as a catch-all title for these molecules [3, 7, 8].

Drug discovery owes much to natural products and has often been helped by studying how Nature "tackles the problem". The finding that microorganisms have evolved mechanisms to subvert pro-inflammatory cytokine networks may provide clues as to the best targets for the blockade of cytokine-induced pathology. In this chapter the methods that microorganisms use to evade cytokine-driven inflammation will be described and the therapeutic potential of these natural mechanisms discussed.

Viral evasion of immunity

Viruses are obligately intracellular microorganisms that are dependent on the living host cell to provide the mechanism for their replication [9]. Although the largest viruses contain only a few hundred genes (compared to the estimated 100,000, plus, genes in the human genome) it is established that viruses can take control of many of the functions of the eukaryotic cell in order to survive. This includes gene transcription, protein synthesis, cytoskeletal function, and so on [9]. Mounted against the virus is a powerful array of cellular and humoral defence mechanisms. Cellular defences include the CD8, so-called cytolytic, T lymphocyte and the natural killer (NK) cells. These leukocytes attempt to kill virally-infected cells by inducing them

Table 1 - Viral immune evasion mechanisms

Inhibition of cell apoptosis
Inhibition of complement activation and complement receptors
Suppression of MHC class I and antigen presentation (multiple mechanisms)
Mimicking Fc-binding proteins
Inhibition of B cell function
Synthesis of steroid hormones via a virally-encoded 3β-hydroxysteroid dehydrogenase
Expression of homologues of adhesion protein, CD2
Expression of superantigen
Interference with cytokine, including interferon, networks

to become apoptotic. Other leukocytes are also involved in anti-viral defences, although more indirectly. These include macrophages, dendritic cells, B lymphocytes and CD4 T lymphocytes. The generation of antibodies and the activation of complement components play their roles in anti-viral defences, as does the induction of the synthesis of so-called anti-viral cytokines such as the interferons. It is now established that viruses have genes that encode proteins that can aid in the evasion of virtually the complete armamentarium of the innate and acquired immune response (Tab. 1) [10–12].

Viral genes encoding cytokine-evading proteins

The 1990s have witnessed the discovery that viruses can evade cytokine-controlled defence mechanisms, and a number of distinct mechanisms for such evasion are now known (Tab. 2). These include cytokine (virokine) and cytokine receptor (viroceptor) mimicry, inhibition of cytokine processing or transcription and interference with cytokine-induced intracellular signalling.

Although earlier reports had appeared of cytokine-like genes in viruses, the trigger which led to the active study of such genes was the three reports in 1992 in the journal, *Cell*, on the ability of cowpox and vaccinia viruses to inhibit the action of the key pro-inflammatory cytokine, IL-1β. Two of these papers involved workers from the biotechnology company, Immunex, which had been heavily involved in developing means of blocking IL-1 action [13–15]. This work followed on from the discovery by Geoff Smith in Oxford (UK) that the vaccinia virus genome contained two open reading frames (ORFs – called *B15R* and *B18R*) that had homology to the type-I IL-1 receptor (IL-1R) [16]. Further analysis revealed that these two ORFs had greater homology to the recently discovered type-II IL-1R [17]. *B15R* encodes a

Table 2 - Viral mechanisms for evading cytokine-controlled defences

Mechanism	Virus	Protein
Virokines	Epstein-Barr virus	vIL-10
	Poxviruses	vEGF, MGF
	HHV-8	vIL-6
	Cowpox	crmA (ICE inhibitor)
	Molluscum contagiosum	chemokine homologue (MC148R)
Viroceptors	Vaccinia	B15R (IL-1β-R)
Interferon inhibitors	Vaccinia	E3L and K3L
Inhibition of JAK-STAT signalling	Human T-cell leukaemia virus	tax protein
Inhibition of cytokine gene expression	Adenovirus	E1A

33 kDa secreted glycoprotein that binds IL-1β, but not IL-1α, with high affinity (Kd 200–300 pM), giving insight into how evolutionary processes view the IL-1 family. Binding is associated with loss of the biological activity of IL-1β. Individual virally infected cells are calculated to produce 10^5 viral soluble IL-1β receptors per day, which would swamp the 10^2 to 10^3 cell surface IL-1 receptors expressed by most cells [13]. Deletion of B15R created a virus that was reported to be less virulent when administered intracranially [14] but was more virulent when administered by the intranasal route [13]. Perhaps the most surprising facet of B15R is that it inhibits the pyrogenic response to vaccinia virus infection in mice [18], suggesting that it has the appropriate pharmacokinetic and pharmacodynamic properties to be considered as a therapeutic agent.

The other viral gene product reported in *Cell* in 1992 was *crm*A (cytokine response modifier A) which encodes a serpin that specifically inhibits the activity of caspase 1, a protease also known as pro-IL-1β converting enzyme (ICE) [15]. This enzyme is involved in the activation of the IL-1β and IL-18 precursor proteins [19]. Caspase 1 also plays a role in the process of apoptosis. The reader should refer to the chapter by Croucher et al., this volume, for more details about cytokine processing. Baculovirus also produces an inhibitor of caspase 1 termed p35 [20]. Interestingly, the *B13* gene in vaccinia virus strain Western Reserve, which encodes a protein almost identical to crmA, does not prevent fever in infected mice [21]. Since these pioneering discoveries, a large number of reports have appeared delineating the different mechanisms that viruses utilise to evade cytokine-driven host defences. This chapter will only briefly review the available literature.

Virokines

A number of viral gene products have cytokine-like action and in consequence have been termed virokines. The first such virokine to be discovered was an epidermal growth factor (EGF)/transforming growth factor α (TGFα)-like gene in myxoma virus and the closely related members of the leporipoxvirus genus [22]. Myxoma growth factor (MGF) was shown to have EGF-like activity in bioassays and to bind to the EGF receptor [23, 24]. Insertional inactivation of MGF resulted in viruses which grew normally in culture but were significantly less virulent in susceptible rabbits [25]. Human herpesvirus 8 (HHV-8) produces a homologue of IL-6 that promotes the growth of transformed B cells [26]. Epstein-Barr virus expresses a protein, BCRF1, which has IL-10-like activity [27, 28]. IL-10 is now recognised to be a potent macrophage downregulator (see the chapter by Bogdan et al., this volume) and this virokine may have the potential to downregulate macrophage-driven antiviral mechanisms. Indeed, the viral IL-10 gene has recently proved to be beneficial in the treatment of animals with collagen-induced arthritis [29]. The human cutaneous poxvirus *Molluscum contagiosum* contains a gene resembling a human chemokine, but the protein acts as an antagonist of MIP-1α [30]. Thus this virokine is an anti-cytokine. Understanding the action of this protein could be useful in the design of chemokine inhibitors.

Viroceptors

The soluble IL-1β receptor produced by poxvirus has already been described. This has turned out to be simply one of a number of cytokine receptors, both soluble and cell bound, produced by a variety of viruses (Tab. 3). Vaccinia virus and poxvirus have been most intensively studied (reviewed in [11]). These viruses contains genes which have homology to the TNF receptor, although not all strains contain such genes [11]. A number of these receptors have been expressed and are secreted from cells as the soluble form of the receptor. In this form they bind and inactivate TNFα [31–33]. Vaccinia virus encodes a soluble receptor for interferon γ (IFNγ) which binds and inactivates the biological activity of this cytokine. Of interest is the finding that this viroceptor has a broad specificity range, being able to inhibit the actions of IFNγ of the rat, cow and human, but not the mouse [34, 35]. Deletion of this virokine had no effect on the pathogenicity of vaccinia virus [35] but resulted in myxoma virus attenuation [36]. However, as it has been found that the myxoma virus IFN-γR also binds to chemokines [37], it is not clear what the attenuation is due to. The αβ IFNs are also targeted by vaccinia and pox viruses with these viruses producing soluble proteins able to bind and inactivate both of these anti-viral IFNs [38].

Chemokines are recognised as very important proteins in inflammation allowing the correct trafficking of leukocytes to sites of infection. Four structurally distinct families of chemokines, based on the position and number of conserved cysteines in

Table 3 - Virally-encoded cytokine receptors

Virus	Protein	Function
Poxviruses	B15R	soluble IL-1β-R
	T2, crmB, G2R, crmC	soluble TNF-R
	T7, BBR, C6L, B8R	soluble IFNγ-R
	B17R, B18R	soluble IFNα/β-R
	T7	soluble chemokine binding protein
	B29R, T1	soluble chemokine binding protein
	Tanapox glycopeptide	soluble binding protein for IL-2/IL-5/IFNγ
Human cytomegalovirus	ORF78	soluble IL-8-R
Herpesvirus Saimiri	US28	soluble IL-8-R
EBV		CSF-1-R
Adenovirus		Four genes encoding TNF-neutralising proteins

the peptide, have been identified: C chemokines (e.g., lymphotactin), CC chemokines (e.g., macrophage inflammatory protein (MIP)-1α); CXC chemokines (e.g., IL-8) and CX3C chemokines (e.g., fractalkine). Poxviruses have genes encoding proteins that can interfere with the activity of various chemokines. Myxoma virus produces a protein which, as well as binding IFNγ, also binds to C, CC and CXC chemokines [37] and can modulate leukocyte infiltration [36]. Poxviruses also produce a second gene that encodes a chemokine-binding protein. In myxoma virus the protein, termed M-T1, binds RANTES (regulated on expression, normal T cell expressed and secreted) and IL-8 [39]. The HIV envelope glycoprotein, gp120, has recently been suggested to be an antagonist of the chemokine receptors CXCR4 and CCR5 which act as coreceptors with CD4 to enable HIV to enter cells [40]. Blockade of these chemokine receptors could have therapeutic benefit in the treament of HIV infection [41].

In addition to these viroceptors, which recognise and inactivate the anti-viral IFNs or the pro-inflammatory cytokines, IL-1β, TNF and IFNγ, a number of other viral genes with anti-cytokine effects have been discovered. Epstein-Barr virus (EBV) produces a protein that acts as a soluble receptor for colony-stimulating factor 1 (CSF-1), a member of the CSF group of cytokines involved in the generation and activation of myeloid cells [42]. Tanapox virus-infected cells secrete a glycopeptide with the capacity to inactivate the functions of a variety of lymphokines (IL-2, IL-5 and IFNγ) [43].

The potential for viroceptors to be therapeutic agents is obvious and the report that the lack of a pyretic response, in mice infected with vaccinia virus, is due to the viral soluble IL-1βR supports this contention. The vaccinia virus 35kDa chemokine-binding protein (vCKBP) has been shown to block eotaxin-induced eosinophil migration in guinea pigs [44], again supporting the hypothesis that viroceptors could have therapeutic potential.

Viral inhibition of cytokine gene expression and cytokine signalling

The importance of IFNs as antiviral agents is demonstrated by the fact that viruses can inhibit the synthesis of IFNs or their receptor-mediated signalling. Hepatitis B virus has two genes that encode proteins able to inhibit the cell expression of IFNβ [45]. Vaccinia virus uses two proteins (products of the genes *E3L* and *K3L*) to block the IFN-induced intracellular anti-viral mechanisms. These gene products interfere with PKR, a double stranded RNA-dependent protein kinase which is involved in the inhibition of viral protein synthesis, and 2'5' oligo A synthetase, which stimulates the degradation of viral RNA [46, 47]. The IFNs act through receptors that are linked to the Janus kinase-signal transducer and activator of transcription (JAK-STAT) intracellular signalling pathways. Murine polyoma virus can inhibit the function of JAK1 *via* the large T antigen [48]. Cytomegalovirus can inhibit JAK-STAT signalling by inducing the proteolysis of JAK1 [49]. Adenoviruses produce the protein E1A which blocks IFN signalling by downregulating p48 and STAT1 [50]. The former forms a complex with IFN-activated STATs inside the nucleus to allow binding to the IFN-stimulated response element (ISRE) and ensure appropriate gene transcription. E1A also represses the transcription of IL-6 and IL-6-dependent genes [51, 52].

Bacterial evasion of immunity

Surprisingly, since bacteria were discovered before viruses, we understand less about the mechanisms evolved by bacteria to evade innate and acquired immune responses. The development of the new interdisciplinary science of cellular microbiology – a combination of cell biology, microbiology and molecular biology [1] is rapidly shedding light on the complex mechanisms by which bacteria and host eukaryotic cells interact. This work is highlighting the fact that bacteria, although generally physically smaller, and with genomes containing significantly less DNA than eukaryotic cells, are complex organisms in their own right, able to communicate both with themselves (e.g., by the process known as quorum sensing [53]) and with eukaryotic cells. Interactions with eukaryotic cells can lead to the bacteria taking over essential cell processes such as cytoskeletal organisation in order to infect their hosts [1]. As part of this process of discovery we are gaining much more information about the interactions between bacteria and cytokines.

Bacteria and host cytokine synthesis

Bacterial infections are invariably associated with inflammation and inflammation is driven and controlled by cytokines. For many years it was thought that the major cytokine-inducing components of bacteria were cell wall components such as lipopolysaccharide (LPS) peptidoglycan and lipoarabinomannan. However, over the past decade or so it has become clear that bacteria produce a very large number of molecules, of all chemical classes (carbohydrates, nucleic acids, proteins, lipids, peptides, glycopeptides, lipopeptides, etc.) that stimulate most, if not all, cells to produce cytokines. The author recently suggested that these variegated cytokine-inducing bacterial molecules constituted a novel form of virulence mechanism as they produced tissue pathology through their ability to induce pro-inflammatory cytokine synthesis. It has been proposed that bacterial virulence is due to the action of molecules known as bacterial virulence factors and these are currently divided into four classes: adhesins, invasins, aggressins and impedins. Cytokine-inducing bacterial molecules would modulate the activity of the cell producing the cytokines (and in doing so produce pathology) and so they have been termed modulins [7]. The modulins appear to fall into two classes. The first are non-proteinaceous molecules like LPS. These molecules need to bind the protein, CD14, found both in serum and on the surfaces of certains cells (e.g., monocytes). The role of CD14 in modulin action can be determined by the use of anti-CD14 monoclonals. However, CD14 is not the receptor for LPS and related molecules. The true receptor for LPS seems to be a molecule (or molecules) related to the *Drosophila* cell surface receptor, Toll, a protein involved in the control of dorso-ventral patterning. These proteins are known as Toll-like receptors [54]. The other modulins do not interact with cells *via* CD14 and the nature of the receptors for these many molecules remains to be defined. Included among these non CD14-dependent modulins are the essential intracellular protein-folding proteins known as molecular chaperones. These proteins are present in all cells and are upregulated during cell stress. For example, the oligomeric molecular chaperone, chaperonin 60, represents 1% of the protein in *E. coli*. However, when this bacterium is stressed, the intracellular content of chaperonin 60 increases to 10% of the cell protein content [55]. The author has demonstrated that highly purified *E. coli* chaperonin 60 (known as groEL) is a potent inducer of human monocyte pro-inflammatory cytokine synthesis and proteolysis of this oligomeric protein does not reduce its cytokine-inducing activity significantly [56].

One of the most striking findings to have arisen from the study of bacterial induction of cytokines has been that all bacterial toxins examined have the capacity to stimulate cytokine synthesis. Indeed, the most potent cytokine-inducing bacterial proteins are toxins, some of which are active at femto- to atto-molar concentrations. Moreover, some of these toxins are very much more potent as inducers of cytokine synthesis than they are as toxins. The biological significance of the ability of bacterial toxins to induce cytokine synthesis is not clear. In addition to stimulating cytokine synthesis, certain bacterial toxins can inhibit the synthesis of specific cytokine synthesis. The reader is referred to [57] for further information.

Having discovered that so many bacterial gene products can promote cytokine synthesis, it will be important to decipher the communication between bacteria and cytokine networks [58]. It is recognised that certain cytokines are involved in particular forms of cellular interactions. For example, the Th1 and Th2 CD4 T lymphocytes produce different patterns of cytokines and the maturation of these cells from the naïve Th0 precursor requires the production of particular cytokines (IL-12 for Th1 development and IL-4 for Th2 development). It is also likely that bacteria can manipulate this Th1:Th2 balance for their own advantage (reviewed in [58]) and examples will be given later in this review.

Bacterial cytokine inhibitors

As described earlier, viruses can block the action of host cytokines by producing: (i) viral homologues (virokines), (ii) soluble cytokine receptors (viroceptors), (iii) ICE inhibitors, (iv) inhibitors of cytokine gene expression, (v) inhibitors of cytokine signal transduction. Many, if not most, of the viral cytokine inhibitors appear to be eukaryotic genes which have been "pirated" by viruses by processes which are not yet fully defined [59]. Less is known about cytokine-inhibitory proteins in bacteria. However, it is expected that such proteins will have evolved independently and therefore may provide novel mechanisms for anti-inflammatory activity.

Many bacterial molecules can stimulate cytokine synthesis and therefore have properties associated with the pro-inflammatory cytokines IL-1 and TNF. The nature of the cellular receptors for these bacterial cytokine inducers is generally unknown. These receptors, for example the TLRs, represent important therapeutic targets for downregulating inflammation. Thus far, only one bacterial protein has been reported to have cytokine-like actions. This is the redox protein, thioredoxin, which was independently discovered as the cytokine, adult T-cell leukaemia-derived factor [60, 61], a protein that synergises with other lymphokines to induce lymphocyte growth. Thioredoxin has a CXXC motif, which is similar, but not identical, to that found in the chemokines. However, it has been a surprise to realise that thioredoxin has potent chemotactic activity [62]. HIV-infected cells produce large amounts of thioredoxin and it has been found that a proportion of AIDS patients have large amounts of thioredoxin in their blood. It is suggested that the presence of large amounts of a potent chemoattractant in the blood is likely to interfere with normal leukocyte trafficking to inflammatory foci [62]. If bacteria release thioredoxin during infection, it may also interfere with chemokine-mediated leukocyte trafficking and be a protective cytokine-evading process. It is assumed that other bacterial proteins with cytokine activity will be discovered and a bacterial growth factor, known as resuscitation promoting factor (Rpf), has recently been isolated from the bacterium, *Micrococcus luteus*, and termed a bacterial cytokine. The *M.*

tuberculosis genome contains five genes homologous to Rpf and suggests that bacterial growth factors are important for the survival of certain microorganisms [63].

No reports have yet appeared suggesting that bacteria produce soluble cytokine receptors. However, it has been reported that bacteria can bind various cytokines and that binding appears to be specific. Thus virulent, but not avirulent, strains of *E. coli* have been reported to specifically bind IL-1 and binding kinetic studies suggest that each bacterium contains 20–40,000 receptors [64]. Binding of IL-1 was associated with cell growth. A number of bacteria, including *Shigella flexneri*, *E. coli* and *Salmonella typhimurium*, have been reported to exhibit specific receptors for TNFα [65]. The nature of these receptors has not been established. One bacterial cytokine receptor that has been identified is the Mycobacterium tuberculosis cell surface receptor that binds human epidermal growth factor (EGF) and promotes growth of this bacterium. This high affinity receptor has been identified as the glycolytic enzyme – glyceraldehyde 3-phosphate dehydrogenase [66]. There is no evidence that these putative cytokine receptors can be shed from cells to bind and inactivate cytokines. However, it may be sufficient for them to bind cytokines to the bacterial cell surface in order to block cytokine activity.

CrmA, the viral serpin and inhibitor of ICE, was discussed earlier in this review. The function of viral ICE inhibitors are to prevent apoptosis and IL-1β/IL-18-induced inflammation. Bacterial inhibitors of ICE have not yet been described. However, a number of bacteria produce proteins which activate the pro-form of ICE and induce macrophage apoptosis. Thus *S. flexneri* when phagocytosed by macrophages releases the protein – invasion plasmid antigen B (IpaB) – into the cytosol, where it interacts with and activates ICE, resulting in apoptosis [67]. The direct effect of inducing macrophage apoptosis is the release of pro-inflammatory cytokines, increased inflammation, and the destruction of the epithelial barrier. The result of this network of interactions is dysentery. Mice in which ICE has been knocked out are insensitive to IpaB and intestinal inflammation and pathology is much reduced [68].

A number of bacterial proteins have been reported to block the synthesis of various cytokines by human cells (reviewed in [1, 3, 7, 69]) (see Tab. 4).

Bacterial toxins as cytokine inhibitors

While many bacterial toxins can potently induce pro-inflammatory cytokine synthesis, a number are able to block the cellular production of cytokines. These include anthrax oedema toxin and cholera toxin, which block LPS-induced TNFα synthesis, and pertussis toxin which blocks LPS-induced IL-1 synthesis [57]. Two bacterial toxins are of particular interest. Cholera toxin is an ADP-ribosylating toxin that is now recognised as a potent mucosal vaccine adjuvant able to induce Th2 responses. It is now established that cholera toxin is a potent inhibitor of monocyte and dendritic cell IL-12 p70 production, active at the level of gene tran-

Table 4 - Bacterial cytokine-inhibiting proteins

Bacterium	Protein	Cytokines inhibited
E. coli	Lymphostatin	inhibits lymphokine synthesis
Actinobacillus actinomycetemcomitans	14 kDa protein	inhibits lymphokine synthesis
Brucella suis	50 kDa protein	inhibits TNFα release by monocytes
Vibrio cholerae	Cholera toxin	inhibits TNFα and IL-12 p70 synthesis
Pseudomonas aeruginosa	ADP-ribosylating toxin	inhibits IL-1, TNFα and IFNγ synthesis
Salmonella typhimurium	protein	inhibits IL-2 synthesis
Yersinia enterocolitica	YopP/YopJ	inhibits TNFα synthesis

scription. *E. coli* heat-labile toxin (LT) is also an inhibitor of monocyte IL-12 p70 production [70]. Administration of *E. coli* LT to mice with collagen-induced arthritis was demonstrated to inhibit induction of this experimental autoimmune disease. Protection was associated with a shift in the Th1/Th2 balance and with a reduction in the extent of the anti-type II collagen immune response [71]. It is therefore clear that bacterial toxins have a major role in the manipulation of immune responses and that such activity could be of therapeutic value.

Other bacterial cytokine-inhibiting molecules

As shown in Table 4, a number of bacterial molecules have been reported to inhibit the production of various cytokines. The mechanism of many of these proteins is not known. A 14 kDa protein from the oral bacterium *Actinobacillus actinomycetemcomitans* has been reported to inhibit the release of a range of lymphokines (IL-2, IL-4, IL-5 and IFNγ) from concanavalin A-treated murine CD4 cells [72]. It is possible that this protein is thioredoxin, although the activity reported for it is at variance with reports of the activity of the human and *E. coli* thioredoxins [60, 61] which are stimulators of lymphocyte proliferation and cytokine synthesis. Another activity able to inhibit lymphokines has been reported in lysates of certain strains of enteropathogenic bacteria [73]. This activity has recently been cloned and expressed and named Lymphostatin. This toxin is of 366 kDa and inhibits synthesis of lymphocyte IL-2, IL-4 and IFNγ synthesis [73]. *Brucella suis* produces a 45–50 kDa protein that inhibits TNFα release by *E. coli*-infected macrophages [74] and *Salmonella typhimurium* contains a protein which inhibits IL-2 production by murine T lymphocytes [75].

The concept of quorum sensing, by which bacteria can recognise their cell density, was introduced earlier in this article [53]. Among the major quorum sensing mediators are the acyl homoserine lactones (AHLs). It is now known that at least one of these AHLs is able to inhibit TNFα and IL-12 release from LPS-stimulated human monocytes [76].

Unfortunately, the mechanism of action of these various cytokine inhibitory proteins (and AHL) is not understood. However, the inhibitory activity of the *Yersinia enterocolitica* outer proteins, the Yops, has recently been examined. The Yops are encoded in a 70 kb virulence plasmid and consist of 12 genes encoding proteins that are involved in the interaction of the bacterium with myeloid cells *via* a type-III secretion system. This is a system that bacteria have evolved for directly injecting proteins into eukaryotic cells. It was initially thought that one of these proteins, YopB, which forms the protein translocation system, was able to inhibit the synthesis of TNFα and IFNγ [77]. However, it is now known that it is YopP (also known as YopJ) that is the key protein inhibiting TNFα synthesis. It has been established that the ability of the Yops to inhibit TNFα is due to their ability to inhibit the breakdown of the NFκB inhibitory proteins IκBα and IκBβ [78]. It is the breakdown of these proteins that allows NFκB to enter the nucleus and switch on the synthesis of selected cytokines.

Protozoan parasites and immune evasion

Many common and life-threatening diseases are caused by protozoan parasites. These include: malaria, amoebiasis, giardiasis, leishmaniasis, sleeping sickness, trichomoniasis and Chagas' disease. This last condition is believed to have been contracted by Charles Darwin. Protozoa are nucleated single-celled organisms, which often reside inside cells such as erythrocytes or macrophages. Protozoan parasites have evolved effective mechanisms for evading many aspects of the defensive measures of the immune response of the human host. Most readers will be familiar with the story of the African trypanosome and its ability to evade immune recognition by changing the antigenicity of its major surface coat protein, known as the variable surface glycoprotein (see [79] for review of protozoan immune evasion). Only one example of protozoan immune evasion will be discussed. This is the mechanism of action of the surface lipophosphoglycan from *Leishmania*.

Leishmania lipophosphoglycan and IL-1β gene expression

Leishmania are obligately intracellular parasites of human macrophages and infection is associated with inhibition of macrophage function [80]. Lipophosphoglycan (LPG) is the most abundant glycolipid on the surface of this protozoan. It was shown that *Leishmania* strains deficient in surface LPG had a reduced capacity to deactivate macrophages and had a poor capacity for survival inside the macrophage. Passive transfer of LPG to deficient mutants overcame this inability to survive within macrophages [81]. It has now been shown that LPG has major effects on both macrophages and endothelial cells. In particular, LPG can inhibit the synthesis of pro-inflammatory cytokines including IL-1β, TNFα and monocyte chemoattractant protein (MCP)-1 [82–84]. The mechanism of action of LPG in inhibiting IL-1β has been studied in some detail [82]. Using human monocytes and THP-1 cells, it was shown that LPG could inhibit IL-1β production induced by LPS, heat-killed *Staphylococcus epidermidis* and TNFα. However, it did not inhibit IL-1β synthesis induced by phorbol 12-myristate 13-acetate (PMA). The LPG was shown to act at the level of gene transcription on a unique DNA sequence in the IL-1β promoter within the position –357 to –57. This sequence in the IL-1β promoter acts as a "gene silencer" in response to LPG . Structure-function studies of components of LPG have identified the *lyso*-alkyl-phosphatidylinositol and the repeating phosphodisaccharide as exhibiting much of the biological activity of the parent molecule [83]. Chemical manipulation of the minimal LPG structure may be one method of devising a low molecular mass selective IL-1 inhibitor.

Conclusions

Inflammation is a part of the pathology of all diseases and for many of the major diseases of humanity (arthritis, asthma, psoriasis) it is the inflammatory process that is the target. The inflammatory process evolved as a response to the large number of microbial infectious agents that share the environment with multicellular eukaryotes. Thus, it is sensible to look to these microbes for clues to inhibit inflammation. This can be seen as mimicking the early pharmacologists who started with the chemical manipulation of natural products – the best example being sodium salicylate - to produce acetylsalicylic acid (aspirin). The last few decades have seen the discovery of many examples of mechanisms used by viruses, bacteria and protozoa to inhibit or avoid both the innate and acquired immune responses by evading complement, NK cells, specific antibody production, CD8 cells, etc. Much of the inflammatory process is driven and integrated by cytokines and within the last decade it has become obvious that viruses are accomplished evaders of cytokine-driven processes. This appears to be due to the ability of certain viruses to copy and modify host genes [59]. Can therapeutic products be derived from the various viral genomes that encode virokines, viroceptors and other cytokine-modulatory proteins? The answer to this question is not obvious other than the fact that the cytokine or cytokines targeted by the virus should be looked at in more detail by the pharmacologist. There is obvious interest in using soluble cytokine receptors for the treatment of human diseases and this approach is being taken with rheumatoid arthritis and other inflammatory diseases. An obvious problem with soluble receptors is their short half-life in the circulation. This can be overcome by making fusion proteins with the Fc region of IgG. Do viral forms of soluble receptors offer any better pharmacokinetic or pharmacodynamic properties? The finding that mice infected with vaccinia virus fail to mount a pyrogenic response [18], due to the virus producing a soluble IL-1β receptor, suggests that this protein has appropriate pharmacological properties to act as a therapeutic agent. Unfortunately, the likely immune response to this protein would limit its usefulness. However, it may be possible to determine the structural basis of B15R and devise low molecular mass isosteres which only target IL-1β. Similar possibilities exist with the various gene products produced by viruses to defeat the activity of chemokines.

The last decade has seen the discovery of many viral genes that encode cytokine-inhibitory proteins. It is certain that many more genes exist and so the search is still on for novel mechanisms by which viruses can block the action of human cytokines. These could lead to novel therapeutic modalities.

The various viral cytokine inhibitors have a close evolutionary relationship to the homologous genes in the host. Bacteria and protozoa, as far as we can tell at present, appear to have independently evolved "inhibitors" of cytokines. It is still early days but it is possible that bacterial and protozoan cytokine inhibitors may have greater novelty than the virokines/viroceptors and be more likely to lead to the production of therapeutic agents. Bacterial toxins such as cholera toxin, the *E. coli* LT and the Yops could have therapeutic potential, as could the LPG of *Leishmania*. It is only at the present time that attention is focusing on how bacteria and protozoa address the problem of cytokine networks. A relatively small number of cytokine-inhibiting bacterial proteins have been found but it is likely that many more await discovery, any one of which could have therapeutic potential.

References

1   Henderson B, Wilson M, McNab R, Lax A (1999) *Cellular microbiology: Bacteria-host interactions in health and disease.* John Wiley & Sons, Chichester

2   Meager T (1998) *The molecular biology of cytokines.* John Wiley & Sons, Chichester

3   Henderson B, Poole S, Wilson M (1998) *Bacteria-cytokine interactions in health and disease.* Portland Press, London

4   Kalvakolanu DV (1999) Virus interception of cytokine-regulated pathways. *Trends Microbiol* 7: 166–171

5   Damian RT (1997) Parasite immune evasion and exploitation: reflections and projections. *Parasitology* 115: S169–175

6   Vaidya T, Bakhiet M, Hill KL, Olsson T, Kristensson K, Donelson JE (1997) The gene for a T lymphocyte triggering factor from African trypanosomes. *J Exp Med* 186: 433–438

7   Henderson B, Poole S, Wilson M (1996) Bacterial modulins: a novel class of virulence factors which cause host tissue pathology by inducing cytokine synthesis. *Microbiol Revs* 60: 316–341

8   Henderson B, Wilson M (1996) Homo bacteriens and a network of surprises. *J Med Microbiol* 45: 393–394

9   Cann AJ (1997) *Principles of molecular virology*, 2nd ed. Academic Press

10 Smith GL (1995) Virus strategies for evasion of the host response to infection. *Trends Microbiol* 2: 81–88

11 Smith GL, Symons JA, Khanna A, Vanderplasschen A, Alcami A (1997) Vaccinia virus immune evasion. *Immunol Revs* 159: 137–154

12 Ploegh HL (1998) Viral strategies of immune evasion. *Science* 280: 248–253

13 Alcami A, Smith GL (1992) A soluble receptor for interleukin-1β encoded by Vaccinia virus: a novel mechanism of virus modulation of the host response to infection. *Cell* 71: 153–167

14 Spriggs MK, Hruby DE, Maliszewski CR, Pickup DJ, Sims JE, Buller RML, VanSlyke J (1992) Vaccinia and cowpox viruses encode a novel secreted interleukin-1β binding protein. *Cell* 71: 145–152

15 Ray CA, Black RA, Kronheim SR, Greenstreet TA, Sleath PR, Salvesen GS, Pickup DJ (1992) Viral inhibition of inflammation: cowpox virus encodes an inhibitor of the interleukin-1β converting enzyme. *Cell* 69: 597–604

16 Smith GL, Chan YS (1991) Two vaccinia virus proteins structurally related to interleukin-1 receptor and the immunoglobulin superfamily. *J Gen Virol* 72: 511–518

17 McMahan CJ, Slack JL, Mosley B, et al (1991) A novel IL-1 receptor, cloned from B cells by mammalian expression is expressed in many cell types. *EMBO J* 10: 2821–2832

18 Alcami A, Smith GL (1996) A mechanism for the inhibition of fever by a virus. *Proc Natl Acad Sci USA* 93: 11029–11034

19 Ghayur T et al (1997) Caspase-1 processes IFN-γ-inducing factor and regulates LPS-induced IFN-γ production. *Nature* 386: 619–623

20 Bump NJ, Hackett M, Hugunin M et al (1995) Inhibition of ICE family proteases by baculovirus antiapoptotic protein p35. *Science* 269: 1885–1888

21 Kettle S, Alcami A, Khanna A, Ehret R, Jassoy C, Smith GL (1997) Vaccinia virus serpin B13R (SPI-2) inhibits interleukin-1β-converting enzyme and protects virus-infected cells from TNF- and Fas-mediated apoptosis but does not prevent IL-1β-induced fever. *J Gen Virol* 78: 677–685

22 Upton C, Macen JL, McFadden G (1987) Mapping and sequencing of a gene from myxoma virus that is related to those encoding epidermal growth factor and transforming growth factor alpha. *J Virol* 61: 1271–1275

23 Lin Y-Z, Le X-H, Tam JP (1991) Synthesis and structure/activity study of myxoma growth factor. *Biochemistry* 27: 5640–5645

24 McFadden G, Graham K, Ellison K et al (1995) Interruption of cytokine networks by poxviruses: lessons from myxoma virus. *J Leukoc Biol* 57: 731–738

25 Opgenorth A, Strayer D, Upton C, McFadden G (1992) Deletion of the growth factor gene related to EGF and TGFα reduces virulence of malignant rabbit fibroma virus. *Virology* 186: 175–191

26 Burger R, Neipel F, Fleckenstein B, Savino R, Ciliberto G, Kalden JR, Gramatzki M

(1998) Human herpesvirus type 8 interleukin-6 homologue is functionally active on human myeloma cells. *Blood* 91: 1858–1863

27   Moore KW, Vieira P, Fiorentino DF, Trounstine ML, Khan TA, Mossman TR (1990) Homology of cytokine synthesis inhibitory factor (IL-10) to the Epstein-Barr virus gene BCRF1. *Science* 248: 1230–1234

28   Hsu D-W, De Waal Malefyt R, Fiorentino DF et al (1990) Expression of interleukin-10 activity by Epstein-Barr virus protein BCRF1. *Science* 250: 830–832

29   Apparailly F, Verwaerde C, Jacquet C, Auriault C, Sany J, Jorgensen C (1998) Adenovirus-mediated transfer of viral IL-10 gene inhibits murine collagen-induced arthritis. *J Immunol* 160: 5213–5220

30   Krathwol MD, Hromas R, Brown DR, Broxmeyer HE, Fife KH (1997) Functional characterization of the C–C chemokine-like molecules encoded by *Molluscum contagiosum* virus types 1 and 2. *Proc Natl Acad Sci USA* 94: 9875–9880

31   Smith CA et al (1991) T2 open reading frame from Shope fibroma virus encodes a soluble form of the TNF receptor. *Biochem Biophys Res Commun* 176: 335–342

32   Upton C, Macen JL, Schreiber M, McFadden G (1991) Myxoma virus expresses a secreted protein with homology to the tumor necrosis factor gene family that contributes to bacterial virulence. *Virology* 184: 370–382

33   Schreiber M, McFadden G (1994) The myxoma virus TNF-receptor homologue (T2) inhibits tumor necrosis factor-alpha in a species-specific fashion. *Virology* 204: 692–705

34   Mossman K, Upton C, Buller RM, McFadden G (1995) Species specificity of ectromelia virus and vaccinia virus interferon-gamma binding proteins. *Virology* 208: 762–769

35   Alcami A, Smith GL (1995) Vaccinia, cowpox and camelpox viruses encode soluble interferon-γ receptors with novel broad species specificity. *J Virol* 69: 4633–4639

36   Mossman K, Nation P, Macen J, Garbutt M, Lucas A, McFadden G (1996) Myxoma virus M-T7, a secreted homolog of the interferon-gamma receptor, is a critical virulence factor for the development of myxomatosis in European rabbits. *Virology* 215: 17–30

37   Lalani AS et al (1997) The purified myxoma virus gamma interferon receptor homolog M-T7 interacts with the heparin-binding domains of chemokines. *J Virol* 71: 4356–4363

38   Symons JA, Alcami A, Smith GL (1995) Vaccinia virus encodes a soluble type I interferon receptor of novel structure and broad species specificity. *Cell* 81: 551–560

39   Graham KA et al (1997) The T1/35kDa family of poxvirus-secreted proteins bind chemokines and modulate leukocyte influx into virus-infected tissues. *Virology* 229: 12–24

40   Madani N, Kozak SL, Kavanaugh MP, Kabat D (1998) gp120 envelope glycoproteins of human immunodeficiency viruses competitively antagonize signalling by coreceptors CXCR4 and CCR5. *Proc Natl Acad Sci USA* 95: 8005–8010

41   Proudfoot AE, Wells TN, Clapham PR (1999) Chemokine receptors – future therapeutic targets for HIV? *Biochem Pharmacol* 57: 451–463

42   Strockbine LD, Cohen JI, Farrah T, Lyman SD, Wagener F, DuBose RF, Armitage RJ, Spriggs MK (1998) The Epstein-Barr virus BARF1 gene encodes a novel, soluble colony-stimulating factor-1 receptor. *J Virol* 72: 4015–4021

43 Essani K, Chalasani S, Eversole R, Beuving L, Birmingham L (1994) Multiple anti-cytokine activities secreted from tanapox virus-infected cells. *Microb Pathog* 17: 347–353

44 Alcami A, Symons JA, Collins PD, Williams TJ, Smith GL (1998) Blockade of chemokine activity by a soluble chemokine binding protein from vaccinia virus. *J Immunol* 160: 624–633

45 Whitten TM, Quets AT, Schloemer RH (1991) Identification of the hepatitis B virus factor that inhibits expression of the beta inteferon gene. *J Virol* 65: 4699–4704

46 Chang H-W, Watson JC, Jacobs BL (1992) The E3L gene of vaccinia virus encodes an inhibitor of the interferon-induced, double stranded RNA-dependent protein kinase. *Proc Natl Acad Sci USA* 89: 4825–4829

47 Beattie E, Tartaglia J, Paoletti E (1991) Vaccinia virus encoded elF-2α homolog abrogates the antiviral effect of interferon. *Virology* 183: 419–422

48 Weihua X, Ramanujam S, Lindner DJ, Kudaravalli RD, Freund R, Kalvakolanu DV (1998) The polyoma virus T antigen interferes with interferon-inducible gene expression. *Proc Natl Acad Sci USA* 95: 1085–1090

49 Miller DM, Rahill BM, Boss JM, Lairmore MD, Durbin JE, Waldman JW, Sedmak DD (1998) Human cytomegalovirus inhibits major histocompatibility complex class II expression by disruption of the Jak/Stat pathway. *J Exp Med* 187: 675–683

50 Leonard GT, Sen GC (1997) Restoration of interferon responses of adenovirus E1A-expressing HT1080 cell lines by overexpression of p48 protein. *J Virol* 71: 5095–5101

51 Janaswami PM, Kalvakolanu DV, Zhang Y, Sen GC (1992) Transcriptional repression of interleukin-6 gene by adenoviral E1A proteins. *J Biol Chem* 267: 24886–24891

52 Takeda T, Nakajima K, Kojima H, Hirano T (1994) E1A repression of IL-6 induced gene activation by blocking the assembly of IL-6 response element binding complexes. *J Immunol* 153: 4573–4582

53 Fucqua C, Greenberg EP (1998) Self perception in bacteria: quorum sensing with homoserine lactones. *Curr Opin Microbiol* 1: 183–189

54 Wright SD (1999) Toll, a new piece in the puzzle of innate immunity. *J Exp Med* 189: 605–609

55 Mayhew M, Hartl FU (1997) The CPN60 and CPN10 families – an overview. In: M-J Gething (ed): *Guidebook to molecular chaperones and protein-folding catalysts*. Oxford University Press, Oxford, 167–172

56 Tabona P, Reddi K, Khan S, Nair SP, Crean SV, Meghji S, Wilson M, Preuss M, Miller AD, Poole S et al (1998) Homogeneous *Escherichia coli* chaperonin 60 induces IL-1β and IL-6 gene expression in human monocytes by a mechanism independent of protein conformation. *J Immunol* 161: 1414–1421

57 Henderson B, Wilson M, Wren B (1997) Are bacterial exotoxins cytokine network regulators. *Trends Microbiol* 5: 454–458

58 Henderson B, Seymour R, Wilson M (1998) The cytokine network in infectious diseases. *J Immunol Immunopharmacol* 18: 7–14

59 Murphy PM (1994) Molecular piracy of chemokine receptors by Herpesviruses. *Infect Agents and Dis* 3: 137–154

60 Tagaya Y, Maeda Y, Mitsui A et al (1989) ATL-derived factor (ADF), an IL-2 receptor/Tac inducer homologous to thioredoxin: possible involvement of dithiol-reduction in the IL-2 receptor induction. *EMBO J* 8: 757–764

61 Wakasugi N, Tagaya H, Wakasugi H, Mitsui A, Maeda M, Yodoi J, Turse T (1990) Adult T cell leukemia-derived factor/thioredoxin, predicted by both human T-lymphotrophic virus-type-1 and Epstein-Barr virus-transformed lymphocytes, acts as an autocrine growth factor and synergizes with interleukin-1 and interleukin-2. *Proc Natl Acad Sci USA* 87: 8282–8286

62 Bertini R, Zack Howard OM, Dong H-F et al (1999) Thioredoxin, a redox enzyme released in infection and inflammation, is a unique chemoattractant for neutrophils, monocytes and T cells. *J Exp Med* 189: 1783–1789

63 Mukamolova G, Kaprelyants AS, Young DI, Kell DB (1998) A bacterial cytokine. *Proc Natl Acad Sci USA* 95: 8916–8921

64 Porat R, Clark BD, Wolff SM, Dinarello CA (1991) Enhancement of the growth of virulent strains of Escherichia coli by interleukin-1. *Science* 254: 430–432

65 Luo G, Niesel DW, Shahan RA, Grimm EA, Klimpel GR (1993) Tumour necrosis factor alpha binding to bacteria: evidence for high affinity receptor and alteration of bacterial virulence properties. *Infect Immun* 61: 830–835

66 Bermudez LE, Petrofsky M, Shelton K (1996) Epidermal growth factor-binding protein in *Mycobacterium avium* and *Mycobacterium tuberculosis*: a possible role in the mechanism of infection. *Infect Immun* 64: 2917–2922

67 Chen Y, Smith MR, Thirumalai K, Zylinsky A (1996) A bacterial invasin induces macrophage apoptosis by binding directly to ICE. *EMBO J* 15: 3853–3860

68 Hilbi H, Moss JE, Hersh D, Chen Y et al (1998) Shigella-induced apoptosis is dependent on caspase-1 which binds to IpaB. *J Biol Chem* 273: 32895–32900

69 Wilson M, Seymour R, Henderson B (1998) Bacterial perturbation of cytokine networks. *Infect Immun* 66: 2401–2409

70 Braun MC, He J, Wu C-Y, Kelsall BL (1999) Cholera toxin suppresses interleukin (IL)-12 production and IL-12 receptor β1 and β2 chain expression. *J Exp Med* 189: 541–552

71 Williams NA, Stasiuk LM, Nashar TO, Richards CM, Lang AK, Day MJ, Hirst TR (1997) Prevention of autoimmune disease due to lymphocyte modulation by the B-subunit of *Escherichia coli* heat-labile enterotoxin. *Proc Natl Acad Sci USA* 94: 5290–5295

72 Kurita-Ochiai T, Ochiai K (1996) Immunosuppressive factor from *Actinobacillus actinomycetemcomitans* down regulates cytokine production. *Infect Immun* 63: 2248–2254

73 Klapproth JM, Scaletsky IC, McNamara BP, Lai LC, Malstrom C, James SP, Donnenberg MS (2000) A large toxin from pathogenic *Escherichia coli* strains that inhibits lymphocyte activation. *Infect Immun* 68: 2148–2155

74 Caron EA, Gross J-P, Dornand J (1996) Brucella species release a specific protease-sensitive inhibitor of TNFα expression, active on human macrophage-like cells. *J Immunol* 156: 2885–2893

75 Matsui K (1996) A purified protein from Salmonella typhimurium inhibits proliferation of murine splenic anti-CD3 antibody-activated T lymphocytes. *FEMS Immunol Med Microbiol* 14: 121–127

76 Telford G, Wheeler D, Williams P, Tomkins PT, Appleby P, Sewell H, Stewart GSAB, Bycroft BW, Pritchard DI (1998) The *Pseudomonas aeruginosa* quorum sensing signal molecule, N-(3-oxodecanoyl)-L-homoserine lactone has immunomodulatory activity. *Infect Immun* 66: 36–42

77 Beuscher HU, Rodel A, Fosberg A, Rollinghoff M (1995) Bacterial evasion of host immune defence. *Yersinia enterocolitica* encodes a suppressor for tumor necrosis factor alpha expression. *Infect Immun* 63: 1270–1277

78 Schesser K, Splik A-K, Dukuzumuremyi J-M, Neurath MF, Pettersson S, Wolf-Watz H (1998) The yopJ locus is required for Yersinia-mediated inhibition of NF-κB activation and cytokine expression: YopJ contains a eukaryotic SH2-like domain that is essential for its repressive activity. *Molec Microbiol* 28: 1067–1079

79 Taverne J, Bradley JE (1998) Immunity to protozoa and worms. In: I Roitt, J Brostoff, D Male (eds): *Immunology*, 5th ed. Mosby, London, 243–261

80 Russel DG, Talamas-Rohana P (1989) Leishmania and the macrophage: a marriage of inconvenience. *Immunol Today* 10: 328–333

81 McNeely TB, Turco S (1990) Requirement of lipophosphoglycan for intracellular survival of *Leishmania donovani* within human monocytes. *J Immunol* 144: 2745–50

82 Hatzigeorgiou DE, Geng J, Zhu B, Zhang Y, Liu K, Rom WM, Fenton MJ, Turco SJ, Ho JL (1996) Lipophosphoglycan from Leishmania suppresses agonist-induced interleukin-1β gene expression in human monocytes *via* a unique promoter sequence. *Proc Natl Acad Sci USA* 93: 14708–14713.

83 Ho JL, Kim H-K, Sass PM, He S, Geng J, Xu H, Zhu B, Turco SJ, Lo SK (1996) Structure-function analysis of Leishmania lipophosphoglycan. *J Immunol* 157: 3013–3020

84 Lo SK, Bovis L, Matura R, Zhu B, He S, Lum H, Turco SJ, Ho JL (1998) Leishmania lipophosphoglycan reduces monocyte transendothelial migration: modulation of cell adhesion molecules, intercellular junctional proteins, and chemoattractants. *J Immunol* 160: 1857–1865

Index